ARDUINO PLAYGROUND

Geeky Projects for the Experienced Maker

by Warren Andrews

San Francisco

Printed in USA

First printing

21 20 19 18 17   1 2 3 4 5 6 7 8 9

ISBN-10: 1-59327-744-X
ISBN-13: 978-1-59327-744-4

Publisher: William Pollock
Production Editor: Laurel Chun
Cover Illustration: Josh Ellingson
Interior Design: Octopod Studios
Developmental Editor: Jennifer Griffith-Delgado
Technical Reviewer: Scott Collier
Copyeditor: Julianne Jigour
Compositor: Laurel Chun
Proofreader: James Fraleigh
Indexer: BIM Creatives, LLC

For information on distribution, translations, or bulk sales, please contact No Starch Press, Inc. directly:
No Starch Press, Inc.
245 8th Street, San Francisco, CA 94103
phone: 1.415.863.9900; info@nostarch.com
www.nostarch.com

Library of Congress Cataloging-in-Publication Data

```
Names: Andrews, Warren, author.
Title: Arduino playground : geeky projects for the experienced maker / by
   Warren Andrews.
Description: San Francisco : No Starch Press, [2017] | Includes index.
Identifiers: LCCN 2016013631 (print) | LCCN 2016019891 (ebook) | ISBN
   9781593277444 (pbk.) | ISBN 159327744X (pbk.) | ISBN 9781593277857 (epub)
   | ISBN 9781593277864 ( mobi) | ISBN 1593277865 (mobi)
Subjects: LCSH: Arduino (Programmable controller) | Electric
   circuits--Computer-aided design.
Classification: LCC TJ223.P76 A54 2017 (print) | LCC TJ223.P76 (ebook) | DDC
   621.381--dc23
LC record available at https://lccn.loc.gov/2016013631
```

To all the electronic hobbyists
and amateur radio operators who,
throughout the years, have taken their
hobbies to new levels

About the Author

Warren Andrews received his first amateur radio license at age 12 and has been an inveterate electronics hobbyist ever since. He has been writing about electronics for more than 30 years, and his work has been featured in publications like *EE Times*, *Electronic Design*, and *Computer Design*. He managed several publications in the RTC group and contributed extensively to all of their publications. Warren has also done extensive design and engineering development on a variety of commercial projects and was a technical consultant for several major corporations, including Motorola and GE. He holds one US Patent.

About the Technical Reviewer

Scott Collier is a self-taught Arduino Enthusiast who has created and published over 50 free Arduino tutorials on his blog: *http://arduinobasics.blogspot.com*. With more than 3.5 million views, it is very likely you have come across some of his work. Scott is an active online member of the Arduino community and fully supports the maker movement ideology.

BRIEF CONTENTS

CONTENTS IN DETAIL

ACKNOWLEDGMENTS

I would like to acknowledge all the friends, colleagues, and fellow editors that offered help and encouragement during the production of my book. In particular, I'd like to thank my friend Pete Yeatman, who consistently offered technical advice and counsel as I put together the various projects. In addition, I'd like to thank the editors and production staff at No Starch Press for doing such an excellent job bringing the book to publication. And of course, what acknowledgment would be complete without thanking my wife and family who have put up with me during the difficult as well as the easy times.

INTRODUCTION

Welcome to *Arduino Playground*! This book provides a broad spectrum of projects demonstrating the flexibility and versatility of the Arduino family of microcontroller boards. Each project contains everything you need to know to build it, including a schematic, a component list, and any *sketches* (that's what Arduino folks call programs). I also endeavored to provide all information about the mechanical parts of each project, including a list of supplies, so you can complete any enclosures, moving parts, skeletons, and so on. Any special tools required are also described in the projects.

I have tried to make the projects more than just recipes for assembling the parts by including some background explanations of how I came up with the projects and how the technology works. I hope that the projects can be useful end products by themselves and, with some ingenuity, perhaps even serve as a launching pad for you to create projects of your own.

Who This Book Is For

Building the projects in this book does not require an engineering degree, advanced mechanical aptitude, or programming expertise. That said, there are some basic requirements you should have to get the most out of the book:

- An understanding of basic electronics, including the ability to read a schematic diagram and recognize elements such as resistors and capacitors
- An understanding of how to use computers and write software (Von Neumann architecture if you want to be snobby); knowledge of Arduino or other microcontroller architectures helps
- Experience soldering connections and wires
- Limited mechanical skills, such as how to operate an electric drill, various saws, and so on

My hope is that both beginner and experienced Arduino users will learn something new about electronics in these projects.

How This Book Is Organized

Each chapter focuses on one project and describes how to prototype it on a breadboard for testing, briefly discusses how the sketch works, and finally shows how to construct the final product.

- **Chapter 0: Setting Up and Useful Skills** provides you with some basic knowledge that you'll use throughout the book, including how to prepare Arduino boards, how to program them, and how to use PCB software and make your own PCBs.
- In **Chapter 1: The Reaction-Time Machine**, you take advantage of the real-time capabilities of the Arduino microcontroller by measuring a user's reaction to a stimulus. This project is quick and easy to build, and you'll learn some of the fundamentals of using a controller for timing—with lots of opportunity to experiment with the sketch. The finished unit will provide hours of fun and entertainment for you and your friends and family.
- **Chapter 2: An Automated Agitator for PCB Etching** shows how you can use a change in current drain to make things happen in a circuit. In this case, the change reverses the direction of a motor so that it can be used to agitate printed-circuit boards in an etchant solution. Etching PCBs is but one application for the technology, as explained in the chapter.
- The project in **Chapter 3: The Regulated Power Supply** may well turn out to be one of the most frequently used products on your workbench. It's a regulated variable-voltage power supply with a digital readout for voltage and current. The design is simple yet effective, and it's fun to build.

- The project in **Chapter 4: A Watch Winder** is one of my favorites. It serves the utilitarian function of keeping automatic (self-winding) watches wound when not being worn, but the cool design also makes it a great kinetic sculpture. The Watch Winder uses an Arduino Nano to handle all the timing functions for keeping the watches wound and includes a multi-colored LED light display. Some of the assembly techniques may challenge a beginner, but the effort is more than worth it.
- The project in **Chapter 5: The Garage Sentry Parking Assistant** is a high-tech device designed to help you park your vehicle in your garage. It's the electronic version of a tennis-ball-on-a-string contraption that's designed to measure the distance you want to pull into your garage. It introduces ultrasonic transmitters and receivers and illustrates how they can be integrated with the Arduino controller. While this is a very practical application of the technology, other applications (such as liquid measurement) are limited only by your imagination.
- In **Chapter 6: The Battery Saver**, you make a device to help keep lead-acid storage batteries from being ruined by accidental discharge. The design is basically a high-current switch in series with a battery that disconnects when the battery reaches a dangerous level. While most automobiles today incorporate such circuitry, I have found this project particularly useful on boats and utility vehicles (tractors, mowers, and so on), and it can save you from having to replace these expensive batteries needlessly.
- In **Chapter 7: A Custom pH Meter**, you build a precision instrument for measuring pH in a variety of liquids. While the Custom pH Meter uses a professional probe, the electronics and readout are based on the Arduino processor. If you're into home brewing, winemaking, hydroponics, or aquariums, or if you're just managing the chemistry in your pool, the Custom pH Meter will be a welcome tool.
- The project in **Chapter 8: Two Ballistic Chronographs** is designed to measure the muzzle velocity of various guns from Airsoft pistols and rifles to BB guns and pellet guns. While not intended for conventional firearms, it boasts capability of measuring velocities over 2,500 feet per second. It also introduces some new technology to the stage, including some stand-alone logic, a counter, and a DAC. Two versions of the chronograph are described; the smaller one, Chronograph Lite, measures projectiles with velocities up to about 700 feet per second.
- In **Chapter 9: The Square-Wave Generator**, you build a low cost instrument for generating electronic waveforms. The genesis of the project was to provide a simulator for the Ballistic Chronograph in Chapter 8, but it worked so well that I made a separate project of it. While it falls short of the resolution and flexibility of regular laboratory instruments, you'll find it useful in designing and testing various products—and at a fraction of the price.

- In **Chapter 10: The Chromatic Thermometer**, you create a handy gadget that tells the temperature using a sequence of colored LEDs. While the initial design is simple, the project led to variations that are more complex. You can add a digital readout, a high-accuracy sensor, and a variety of mechanical variants for everything from monitoring a fish tank to a wall decoration.

This book does not cover the basic engineering or programming concepts behind every project in depth, as it assumes you have enough background knowledge to understand those concepts based on a brief explanation. But for the curious reader, the text does provide references where extra information can be found about the design and technology. It also gives background into the history of the project: why I built the project (and why you might want to build it). In all cases, there has been an effort to provide a learning experience at a level the user can appreciate and understand.

Where possible, the projects also suggest alternative approaches that advanced readers can try. To demonstrate why I selected a particular approach, I illustrate how some alternative ways of doing things solve certain problems and cause others. There's a lot of room to personalize, and perhaps even improve on, each project, whether it be packaging, construction technique, or the sketch itself. For example, the Watch Winder can be a utilitarian device or an upscale kinetic sculpture.

WHAT IS MECHATRONICS?

During the course of writing this book, I ran across the term *mechatronics* several times. Being kind of a traditionalist (or just an old fuddy-duddy), I ignored the first several references. However, I eventually took the time to look up the term, and it sure sounds a lot like what we're doing in this book.

Put simply, mechatronics is the process of designing with electronics and mechanical engineering. Tetsuro Mori, the senior engineer of the Japanese company Yaskawa in 1969, coined the term to describe the process of building industrial robots, which requires electrical, mechanical, and computer engineering. A mechatronics engineer unites the principles of mechanics, electronics, and computing to generate a simpler, more economical, and more reliable system.

About the Parts Lists

In the design phase of this book, the selection of parts for projects was often determined by what I might have had lying around the shop. For example, the bearings in the Watch Winder from Chapter 4 were originally bearings I had in my junk bin, but I eventually replaced them with the ones mentioned in the parts list. I have made every effort to make the projects with the tools and materials as described, but I encourage you to use what you have handy.

NOTE *Almost everyone I know who has programmed an Arduino started with the simple "blink" sketch. As a result, many see Arduino as inextricably entwined with LEDs turning on and off. Throughout the book, I attempt to reinforce this association by including LEDs in as many projects as possible. While blinking an LED doesn't even scratch the surface of the Arduino's capabilities, LEDs make the projects more fun and visually interactive.*

Tools and Supplies

Before you begin working through this book, review the following lists of tools and supplies, and note any items you don't have. Not all of these are required for every project, so when you want to build a project, read the required tools and supplies lists for that particular chapter to see if you are missing anything essential. You can purchase most of these items at your local hardware store, but I will indicate where buying online might be a better option.

Drilling, Cutting, and Assembling

Screwdrivers You'll want both Philips and slotted, in multiple sizes.

Dremel tool (or equivalent) A small drill or rotary tool can be very useful for a variety of tasks, from drilling and cutting to etching and polishing. An inexpensive bench attachment turns the Dremel tool into a small drill press, which is really handy, especially for drilling PCBs.

Electric drill Battery operated is preferable. If possible, I suggest a chuck with 3/8-inch or 1/2-inch capacity—the bigger the better.

Drill bit set I recommend purchasing a numbered drill set (that is, with bits labeled #1 through #60) in addition to a fractional drill set.

Pliers I find that a pair of vise-grip pliers, about 6 to 8 inches long, fills many needs for clamping, holding things in place, and tightening. I also recommend getting a good pair of needle-nose pliers.

Saws A simple hacksaw is handy for a variety of tasks. For cutting plastic, there are many options: a keyhole saw works well, or if you don't mind spending some more money, a small variable-speed, hand-held jigsaw (or saber saw) is handy for making a variety of cuts. I use my jigsaw almost exclusively with hacksaw blades. (Practice sawing on some scrap wood if you've never used a jigsaw before. Once you understand how to use it, a jigsaw can be one of the handiest tools in the shop.)

Sharp knife and scissors

Screws and nuts I suggest trying to get a small assortment of both English and metric screws and nuts. There are many such assortments on eBay, if your local hardware store doesn't have a good selection.

Tap and die set A set isn't required for most projects, as individual taps and dies are available, but a set is cheap and handy to have around.

Tapered reamer A tapered reamer is useful in many of the book's projects, and it's a tool I heartily recommend having. I use a set of two inexpensive reamers that I purchased on Amazon, and they work very well on plastic, aluminum, and mild steel. I suggest getting reamers that can create holes up to 7/8 inches in diameter.

Tape I suggest keeping masking tape, double-sided foam tape, and rugged outdoor double-sided tape (3M brand works well) on hand.

Prototyping, Soldering, and Testing

Alligator clips or clip-lead set There are many alligator clips available, and they are very handy when putting together breadboards. Sets are available from multiple sources, including RadioShack and Amazon.

Breadboard and jumper wires These are available from multiple sources, including Pololu and Amazon.

Digital multimeter A broad spectrum of multimeter units is available. You can pay anywhere from under $5 to hundreds of dollars, but low-priced portable units work fine. You will find a multimeter a welcome addition to your household tool collection.

Resistor assortment I suggest checking eBay or Amazon, where you can buy resistors in bulk easily. Some assortments might include 10 units each of 20 or 30 values, while others contain 100 or more resistors per value. These are very economically priced.

Soldering iron and solder Soldering irons are readily available at hardware stores, often for less than $10. Jameco even has an online soldering tutorial (*http://www.jameco.com/Jameco/workshop/learning-center/soldering-basics.html*) that might be worth reading if you're a beginner.

Solder paste This is needed only if you have trouble soldering surface-mount components. While these projects use only a small handful of surface-mount components, you may use more in the future as manufacturers make fewer through-hole versions of newer integrated circuits. I use a lead-free solder paste called Chip Quik. Don't despair, though: you can solder surface-mount components with regular rosin-core solder and a soldering iron as described in "Using SOICs" on page 20.

Solder wick While none of us would ever be careless enough to make a solder bridge across connections, sometimes a gremlin sneaks in and does it anyway. For those occasions, solder wick (a copper braid with a little rosin on it to soaks up solder) lets you remove solder cleanly.

If you like building complete Arduino projects, consider filling out your tool collection with any missing items. Everything described here will surely be useful at some point.

Online Retailers

If you can't find an electronic component or tool at your local hardware store, check one of these online retailers:

- Adafruit (*https://www.adafruit.com/*)
- Amazon (*http://www.amazon.com/*)
- Bitsbox (good in the UK; *http://bitsbox.co.uk/*)
- Digi-Key (*http://www.digikey.com/*)
- eBay (has almost anything you need for this book at low costs; *http://www.ebay.com/*)
- Electronic Goldmine (*http://www.goldmine-elec-products.com/*)
- Farnell (ships globally; *http://www.farnell.com/*)
- Harbor Freight (*http://www.harborfreight.com/*)
- Jameco (*http://www.jameco.com/*)
- MCM Electronics (*http://www.mcmelectronics.com/*)
- Mouser (*http://www.mouser.com/*)
- Newark Electronics (*http://www.newark.com/*)
- Newegg (*https://www.newegg.com/*)
- Pololu Robotics and Electronics (*https://www.pololu.com/*)
- SparkFun (*https://www.sparkfun.com/*)

About the Online Resources

Each project has a Downloads section that lists any sketch, PCB, or template files provided online. Using those files is optional—you can copy the sketch from the book by hand, design your own PCB, and decide where to make holes for components yourself if you prefer. But if you want a place to start, download the resource files from *https://www.nostarch.com/arduinoplayground/*.

0

SETTING UP AND USEFUL SKILLS

This book assumes you have some previous hardware experience, so the projects in it won't hold your hand. That said, if you need a refresher on some basic skills, such as wiring and programming Arduino boards, keep on reading.

This chapter also covers some skills that you will find helpful but that you don't necessarily need to build the projects. For example, in most projects, I provide PCB files that you can use to manufacture a shield PCB, but if you want to make a PCB rather than solder the circuits to prototyping board, read the "Making Your Own PCBs" on page 13. And if you've never assembled a connector yourself or need guidance on working with small-outline integrated circuits, you will find information on that in "Using SOICs" on page 20.

Preparing the Arduino Board

Whether you use an Arduino Nano, a Pro Mini, or one of their clones, there is a good chance your board will arrive with the header pins separate and unsoldered. All of the boards I've purchased came that way (see Figure 0-1).

Figure 0-1: An Arduino Nano clone board with headers and a breadboard, which I use as an aid to soldering.

Before you can use an Arduino or clone, you need to solder the header pins. The strips of headers that come with a processor board usually have more pins than required, and the first step is to trim them to the number you need. The black plastic retainers are grooved to make cutting easy. I use a simple set of diagonal cutters to cut the plastic (see Figure 0-2).

Figure 0-2: The Arduino Pro Mini clone, with the header pins trimmed to length. The 5-pin strip fits on the end of the board.

The next step is to insert the header pins into a breadboard, spaced so the holes in the processor board will fit over the pins. Insert the long end of the header pins into the breadboard, as shown in Figure 0-3. There are four rows of holes left empty between the two rows of header pins—that is, three rows plus the space in the center divide—so the processor board will fit.

The final step is to place the processor board over the short end of the header pins, as shown in Figure 0-4, and solder.

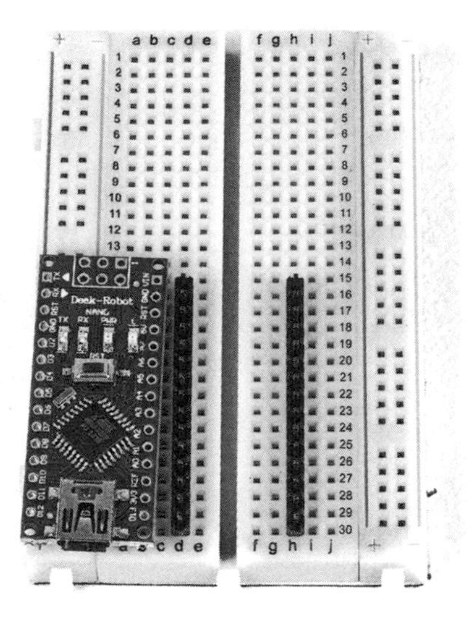

Figure 0-3: The header pins have been inserted into the breadboard in preparation for soldering the Nano clone.

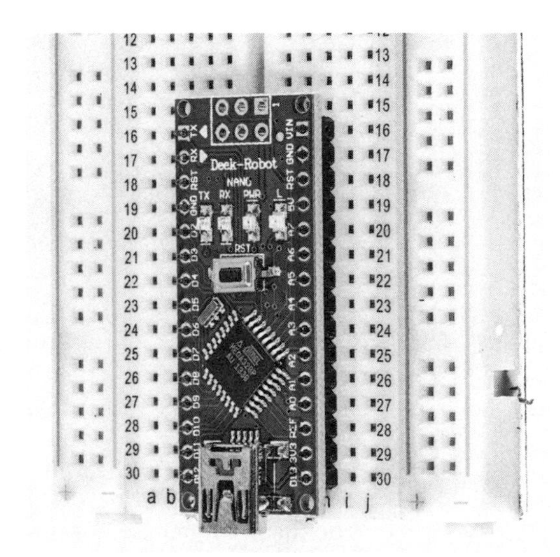

Figure 0-4: The Nano clone in place on the breadboard and ready for soldering

Now your board has all its pins and is ready to be wired up.

Affixing the I^{2}C Board to the LCD

Many projects in this book also use a liquid crystal display (LCD) with an inter-integrated circuit (I^{2}C) interface (see Figure 0-5). The LCDs used in this book can be purchased with or without the I^{2}C adapter board, though I have often had to buy the LCD and the adapter board separately.

If the adapter board isn't already attached to the LCD, connecting the two is about the same as preparing the Arduino board. The adapter board usually comes with header pins installed, so all you have to do is insert them into the display and solder them.

Connecting the display and the adapter usually works without any problems, but in some cases the adapters may have circuitry that almost touches the display board. To avoid connections shorting out, I suggest putting electrical tape on the back of the LCD to insulate it from the connections on the I^{2}C adapter board.

Figure 0-5: A 16×2 LCD and an FTDI module

You may also find that the pins of the header on the I^2C board protrude through the LCD board enough that it causes a problem when mounting the display in a case. Try to solder the I^2C board as far from the LCD board as possible to minimize the amount that the pins protrude through the board. Figure 0-6 shows an adapter board ready to have the pins inserted into the LCD base board.

Figure 0-6: The I^2C board in place, ready for soldering

If soldering the adapter board in such a position proves too awkward, you can insert pins to their limit, solder them, and then trim them with a wire cutter to make them as flush as possible with the LCD board.

Depending on your LCD and adapter board, the I^2C address you need to enter into the sketch may be different. There is a very simple scanner available at *http://playground.arduino.cc/Main/I2cScanner/*. Just follow the instructions to figure out your LCD's I^2C address. 0x27 and 0x30 are common addresses.

Uploading Sketches to Your Arduino

After you've assembled a project's circuit on a breadboard, it's time to load your sketch onto the microcontroller and give it a whirl. I suggest an Arduino Nano, Pro Mini, or clones of those for most projects in this book.

Installing the Arduino IDE

You may already have the free Arduino integrated development environment (IDE) installed on your computer; if not, download the program and install it now. Just visit *https://www.arduino.cc/*, click **Download**, and download the appropriate version of the Arduino IDE for your operating system. The latest version is 1.6.*x*. Then, go to the Getting Started with Arduino page at *https://www.arduino.cc/en/Guide/HomePage/*, and follow the corresponding official installation instructions.

NOTE *If you're not familiar with the IDE, there are a number of tutorials and sample code files on the Arduino website. I strongly recommend that you read them to familiarize yourself with the software.*

Using the Arduino IDE

After installation, open the Arduino IDE. A blank sketch will appear with a name in this format: *sketch_<date>*. To save your sketch, select **File ▸ Save As**. In the dialog that opens, choose where you want to save your sketch and what you want to name it.

You have a choice when creating a new sketch for a project in this book: you can type the sketch into the sketch window, or you can download the sketch file from the resource files at *https://www.nostarch.com/arduinoplayground/* and then copy and paste the code into the sketch window.

I usually like to *verify* the sketch—that is, compile it—before attempting to upload it to the board to make sure no errors crept in as the sketch was typed into the IDE. Verification is easily accomplished by clicking the checkmark in the upper-left corner (see Figure 0-7). The word *Verify* will appear to the right of the five icons on that line when you hover the mouse over the checkmark button.

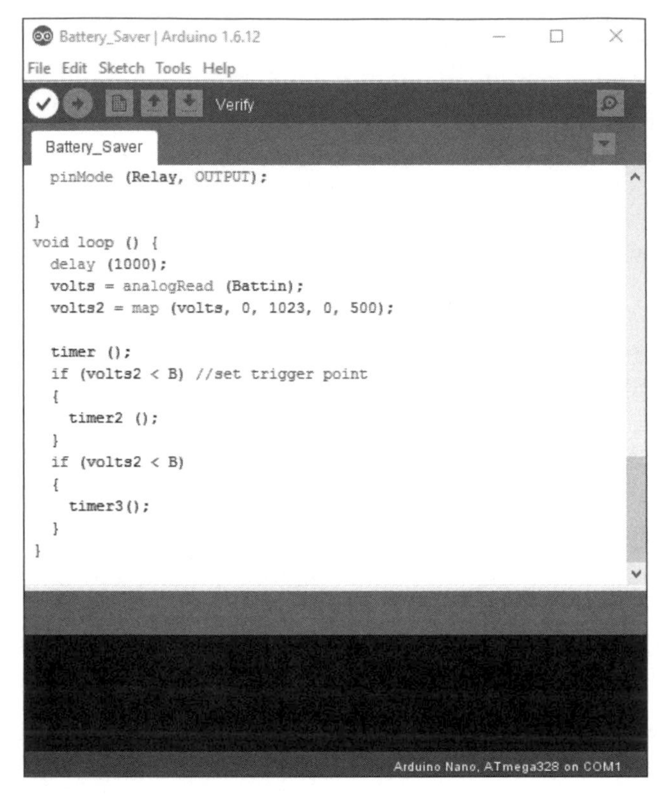

Figure 0-7: The sketch window with the Verify icon clicked at the beginning of the list of icons

If your code compiles correctly, it's ready to upload to your board.

Connecting and Programming an Arduino Nano

After verifying your sketch, you have to connect the Arduino board to your computer. Of the Arduino boards used in this book, the Nano is the easiest to hook up and program, as it includes a built-in USB interface.

For a Nano, find a cable with a USB plug (type A) on one end and a mini-B USB plug on the other; your board probably came with one. Connect the USB end to the computer, and connect the mini-B USB end to the Nano. Select **Tools ▸ Board**, and then select the correct board and microcontroller (see Figure 0-8).

You may also need to select the correct serial COM port for your Arduino, though some versions of the IDE will automatically find a free port and connect to it. Go to **Tools ▸ Port** and select a serial port from the menu that appears. If you have any issues, consult the individual operating system guides at *https://www.arduino.cc/en/Guide/HomePage/*.

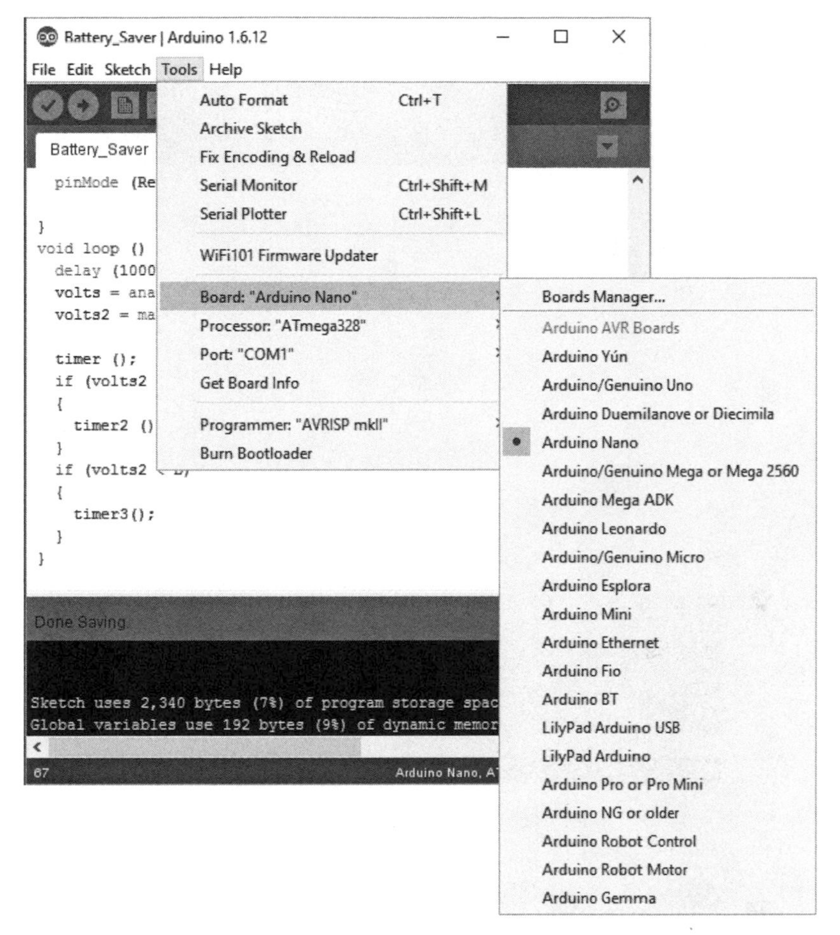

Figure 0-8: The sketch window, with the Tools menu open. I selected Arduino Nano with the ATmega328.

The last step in programming the Nano is to upload the code. First, make sure the board is still plugged into the computer via the USB cable. Then, click the **Upload** button, which looks like an arrow pointing to the right (see Figure 0-9). When you hover the mouse over the Upload button, the word *Upload* should appear to the right of the five main icons.

Uploading code to the Arduino shouldn't take too long, but it depends on the length of the sketch. Afterward, you should be set to power and test your circuit. (Don't forget to unplug the USB before powering it with an external power source.)

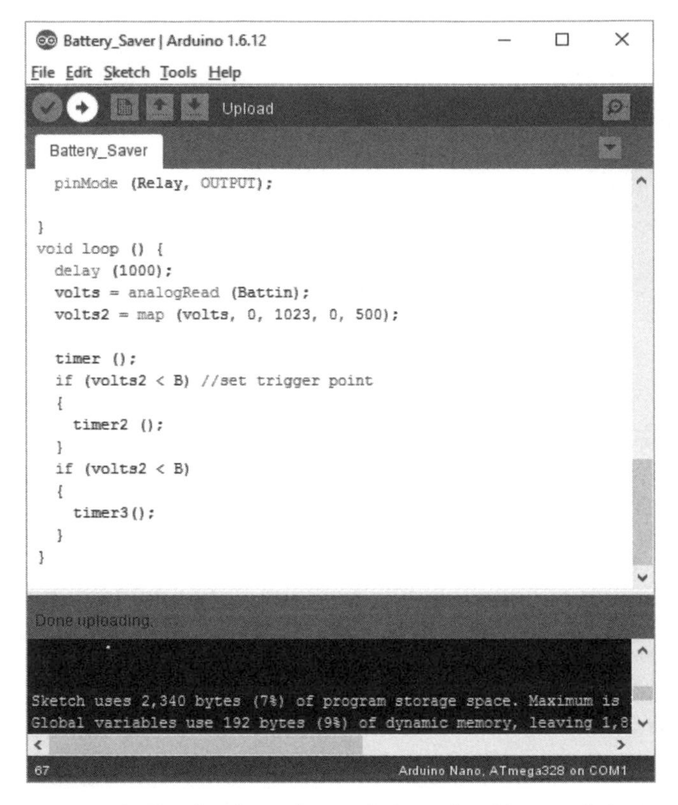

Figure 0-9: The sketch window with the Upload button clicked

Connecting and Programming an Arduino Pro Mini

The Arduino Pro Mini (or clone) works much the same as the Arduino Nano, but it doesn't have a built-in USB interface, opting instead for a transistor-transistor logic (TTL) connection. The easiest way I found to upload code to the Pro Mini was to remove the processor chip from an Arduino Uno, as shown in Figure 0-10, and use the Uno board as a programmer.

The processor-free Uno can be connected to the computer directly via USB, so it can provide power as well as programming signals to a Pro Mini board connected to it. The USB cable for an Arduino Uno is a standard USB cable with a regular (type A) USB connector on one end and a square (type B) USB on the other (see Figure 0-10). More information about USB cables can be found at *https://www.sparkfun.com/pages/USB_Guide/*.

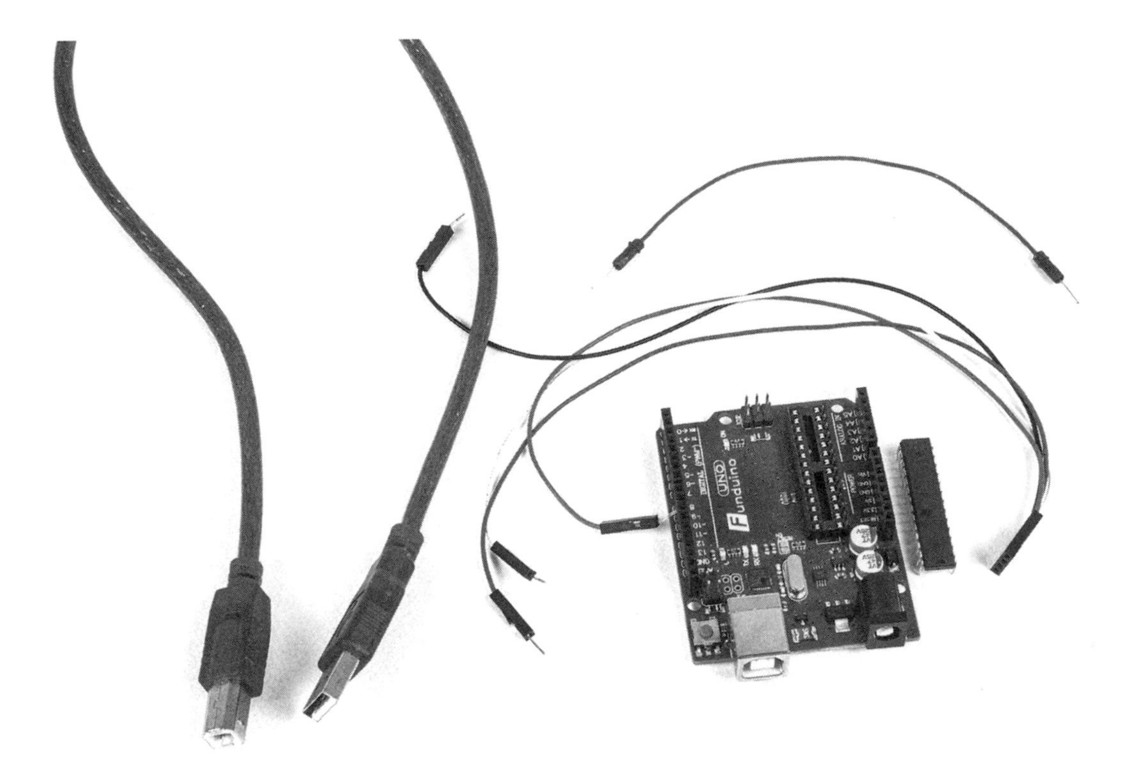

Figure 0-10: An Arduino Uno clone with the processor removed (to the right of the board), a USB cable, the programming cable assembly, and a loose reset wire

Connect the Uno to the Pro Mini as follows:

- Rx on the Pro Mini to Rx on the Uno
- Tx on the Pro Mini to Tx on the Uno
- VCC on the Pro Mini to 5.0V on the Uno
- GND on the Pro Mini to GND on the Uno
- RST on the Pro Mini to RST on the Uno

I made a simple cable to connect the positive and negative voltage supplies as well as the receive (Rx) and transmit (Tx) signals (see Figure 0-11). The individual wires on one end plug directly into headers on the Uno, and a 4-pin plug attaches to the edge headers on the Pro Mini. You could also use separate jumper wires, like those you would use for a breadboard. I have found it easiest to plug the Pro Mini into a breadboard so I can connect the RST signal with a jumper wire, as shown in Figure 0-11.

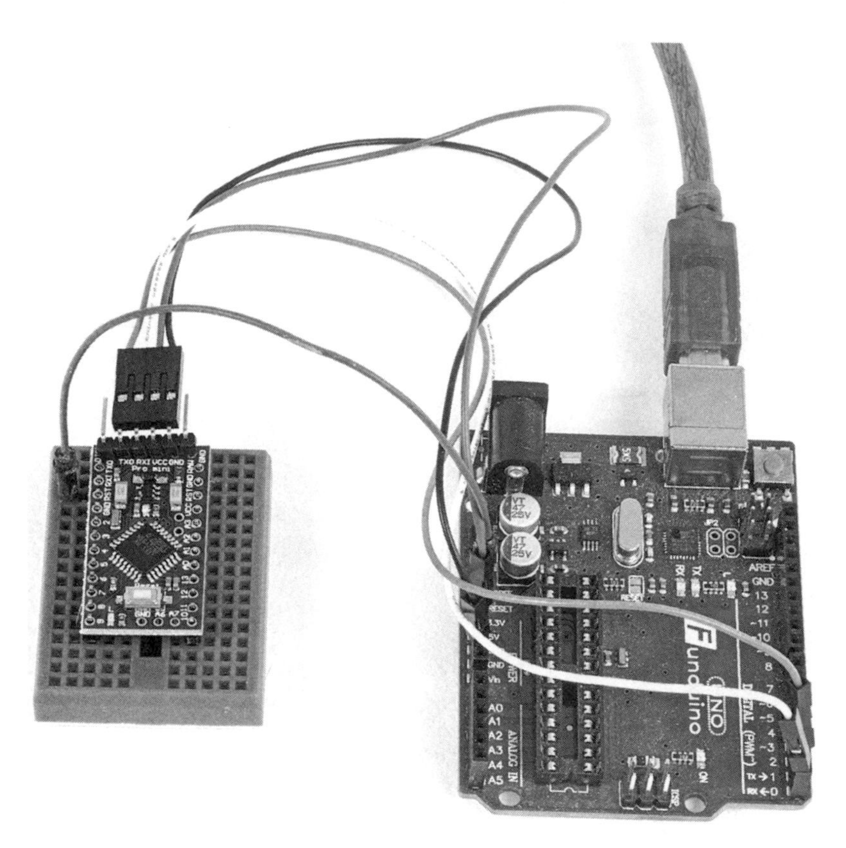

Figure 0-11: The Arduino Pro Mini clone ready for programming with the Arduino Uno clone serving as programmer. The Arduino Uno's USB connection supplies the power.

Make sure that *all* connections line up correctly before plugging the UNO's USB cable into your computer. When you program the Pro Mini, select the proper board from the Board section of the Tools menu; even though you are plugging an Arduino Uno into the computer, you are still programming a Pro Mini. Once this setup is done, you can upload sketches to the Pro Mini just as you would the Nano.

While using an Arduino Uno intermediary is the easiest way to program the Pro Mini, you can also purchase USB-to-TTL devices like the one in Figure 0-12. I purchased several on eBay in the $5 to $12 range, and with a little tinkering (the terminals are sometimes marked differently), they all worked well.

Programming the Arduino is only part of the battle, though. To build a truly permanent project, you need to solder your working Arduino circuit to a board. A custom *printed circuit board (PCB)*, also sometimes called a *printed wiring board*, is the best way to keep your project clean and neat—if you are willing to put in the extra work to make one.

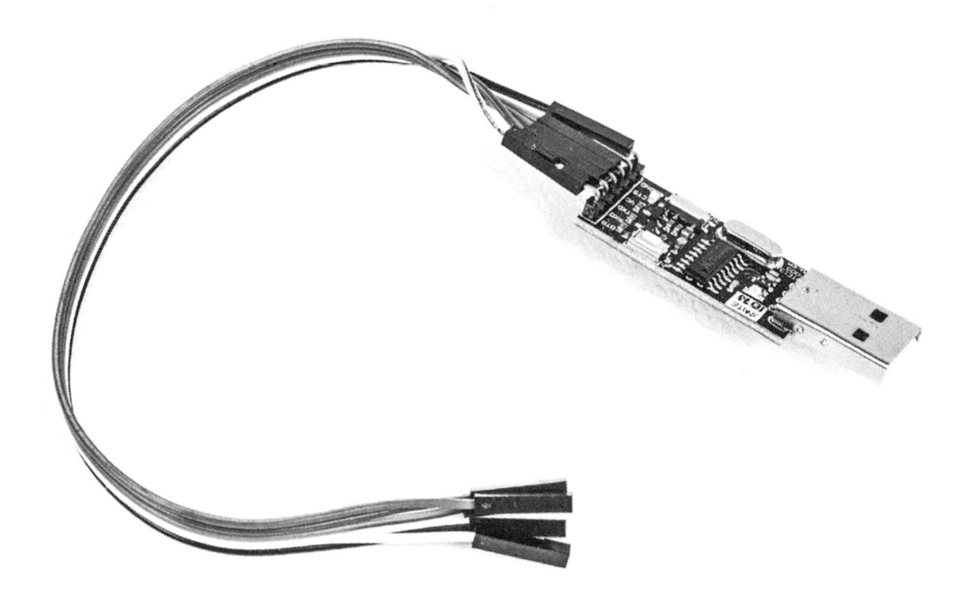

Figure 0-12: A USB adapter for programming the Pro Mini and other controller boards without their own USB interface. This adapter uses a male USB (type A) plug and has a Data Terminal Ready (DTR) pin instead of a reset pin (RST). Most USB-to-TTL devices can be powered with 3.3V as well as 5V, but check before you power them, as some devices operate at 3.3V only.

Using PCB Software

There are many PCB design programs out there, and they vary in complexity and cost. Many are free in order to attract customers to use the company's facilities to make boards. Therefore, there are some hang-ups when trying to use those free tools for DIY boards—for example, the software might have some features locked in the free version. I use ExpressPCB (*https://www.expresspcb.com/*) for both single- and double-sided boards.

NOTE *For double-sided DIY boards, I've had to reverse the image manually. The trick to making double-sided boards is properly aligning the two sides. To greatly simplify the alignment process for many projects, you can make alignment marks and drill alignment holes on the blank copper board before transferring the image. I have also, from time to time, used a drawing program called TurboCAD (similar to AutoCAD) to produce double-sided boards.*

ExpressPCB offers the least expensive solution for making boards that I've been able to find. The company has a MiniBoard service that offers a standard-size board with no frills for a relatively low cost. Further, as the industry creates newer packages, using a software package that includes the newer IC footprints is essential. I have used ExpressPCB to make adapter boards—from SOIC to DIP—and to integrate SOICs into a finished board, as the software works well with the smaller geometries. Even if I want to make a "purpose-built" microcontroller board, which will likely require multiple layers, a ground, and VCC plane, ExpressPCB will probably fill the requirements.

To use the program, simply go to the ExpressPCB website, download the free software, and install it. The ExpressPCB website has several tutorials on using the software, which I recommend you take advantage of. There is a companion free software program, ExpressSCH, which is a schematic capture program for writing your own schematic diagrams. While the features are not as well integrated as they could be, using the programs together has helped with circuit design.

NOTE *All the PCB designs in this book have been prepared using ExpressPCB design software, and they are all available at* https://www.nostarch.com/arduinoplayground/. *To view or change the PCB drawings, you will have to download the software.*

Another advantage of using ExpressPCB is that you can take the same file you develop for making the circuit board yourself and send it out to the company's factory for finishing. I did that for a few of the projects in this book—in most cases, after making my own and wanting to clean up the board. I found the results more than satisfactory. The factory-prepared boards offer plated-through holes—if you build your own double-sided boards, soldering on both sides of the board is necessary. They also include a solder-plate finish and can be made with a solder-resistant coating and silkscreen image printed on the board. Figure 0-13 shows a board I made using ExpressPCB.

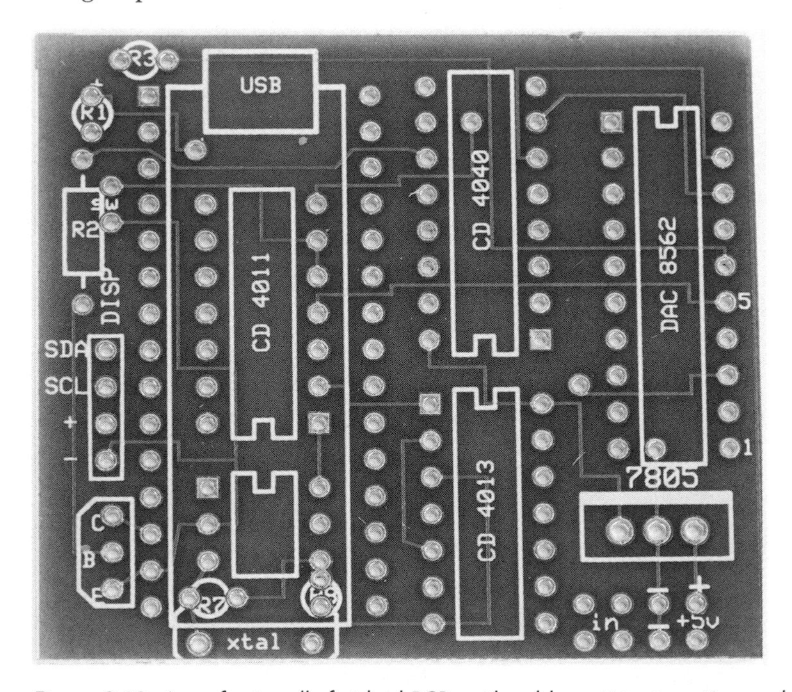

Figure 0-13: A professionally finished PCB, with solder-resistant coating and silkscreening. I used this board to make the Ballistic Chronograph in Chapter 8.

A TIP FOR MANUFACTURING MULTIPLE DIFFERENT BOARDS

You can use ExpressPCB's MiniBoard service to make more than one board for very little cost. The mandatory size for a board to qualify as a MiniBoard—and thus, to get the discount—is 3.8×2.5 inches, and when you bring up the program, a yellow guide box automatically displays an area of that size. In preparing PCBs such as the one in Figure 0-14, I combined several smaller boards into one large "board" by copying and pasting the small boards into the maximum size for the MiniBoard price.

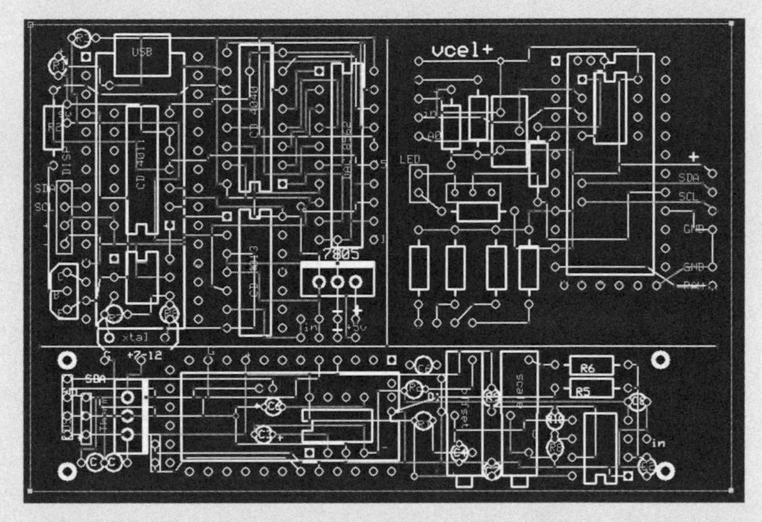

Figure 0-14: Three different boards for one MiniBoard order. The board for the Ballistic Chronograph is on the top left, and the pH Meter's is on the bottom. The top right is for an optical tachometer, which didn't make it into the book. For one price, you get three copies of each board. All you have to do is cut them apart.

Making Your Own PCBs

There are a number of techniques for making PCBs after you design one. As I'll discuss in Chapter 2, the most common method is a subtractive approach, in which copper is selectively removed from a foil-clad phenolic or epoxy/glass board to leave a pattern on the board. The copper can be mechanically milled off, but if you want to make a PCB at home, the most common—and least expensive—approach is to chemically etch the pattern.

When chemically etching a PCB, a circuit pattern is printed on the blank board with a *resist*, a chemical that prevents the copper from being removed by the *etchant* in treated areas. The etchant is an acid that attacks the untreated copper on the clad board. Figure 0-15 shows a copper board's transition into a PCB.

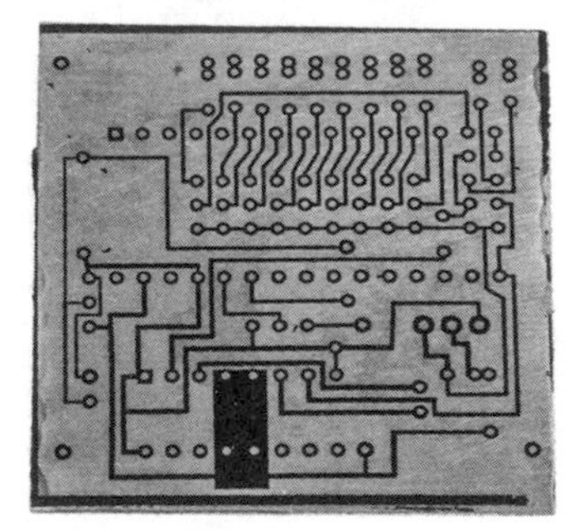

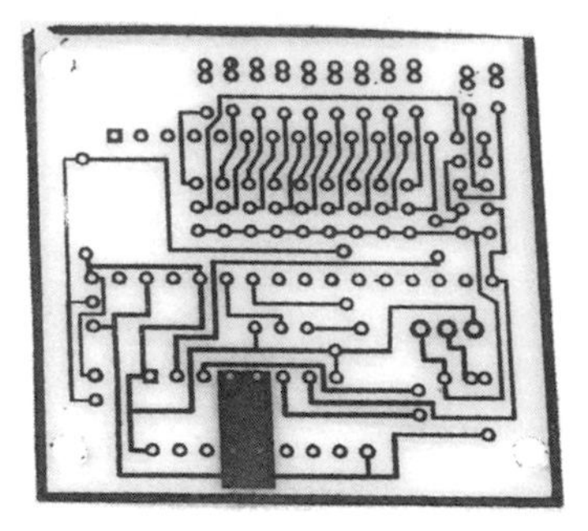

Figure 0-15: From left to right: an untreated, scrubbed copper-clad board; the board with resist printed on it; and the etched board without its holes drilled

In the old days, making a PCB was a tedious and messy job—particularly for a hobbyist. First, you would have to lay out the PCB pattern, which until not long ago, was done with tape on an acetate sheet with a light table. Then, you'd clean the copper-clad board and whirl on the photo resist. This needed to be exposed to UV light and developed with carbon tetrachloride (CCl_4), which is not so good for you, or trichloroethylene (C_2HCl_3), which is not much better. After that, you'd begin the messy etching process with ferric chloride ($FeCl_3$) or ammonium persulfate [$(NH_4)_2S_2O_8$]. With all those steps, you could usually count on spending a good part of a day producing one board.

Today, that's all changed. With today's contemporary PCB software, you can frequently lay out a pattern for a relatively simple single-sided or even double-sided board in less than an hour, depending on its complexity. From there, the process gets even easier.

Applying the Pattern

If you want to learn to etch your own PCBs, go to the PulsarProFX website (*http://www.pcbfx.com/*), which has the tools you need to put an image on a copper-clad board easily. Pulsar's PCB Fab-in-a-Box product is a complete kit that contains all you need to get going and make several boards. One key ingredient is a special paper that you print on with a laser printer and that uses heat to transfer the image to the copper-clad board. The whole fabrication process—before drilling the holes—almost never takes more than an hour, unless you're running low on etchant, which slows the etching time.

Apart from the items in the Pulsar kit, the only tools needed are:

- A laser printer
- A plastic laminator (Pulsar suggests using a GBC laminator, but I've used an Office Depot brand unit for years, and it works fine.)
- A water bath

The procedure for applying a PCB pattern to a copper-clad board is relatively simple:

1. Design the pattern on a computer using a PCB layout program, such as ExpressPCB.
2. Print the image on the special paper provided by Pulsar with a laser printer, not an inkjet. The laser ink is a polymer compound that melts when heated and partially bonds to the paper, leaving your image on the paper. The paper, when reheated on the blank PC board, allows for easy transfer of the image.
3. Transfer the image from the paper directly to a clean copper-clad board, using an inexpensive office laminating machine.

The thermal ink on the paper becomes the resist on the copper-clad board. Pulsar provides an additional, thin film layer thermally bonded to the laser-jet ink, but the ink alone will resist the etchant.

Etching the Board

While copper is not a highly active metal, there are several replacement reactions that etch it effectively. However, many of the resulting byproducts are somewhat toxic, and almost all of them result in materials that have to be discarded or recycled in a special way because they are extremely harmful to the environment. Most copper salts are strong poisons for a variety of plants and animals, including humans.

For a better etching method, I recommend an Instructables page called "Stop Using Ferric Chloride Etchant (A Better Solution)," which you can find at *https://www.instructables.com/id/Stop-using-Ferric-Chloride-etchant!--A-better-etc/*. Read the environmental and personal safety warnings in this tutorial carefully before mixing your etchant.

The system described in this tutorial uses standard household chemicals: hydrogen peroxide (H_2O_2) and muriatic acid (essentially hydrochloric acid, HCl). The process is far more environmentally friendly than the old ferric chloride or ammonium persulfate techniques. You can also regenerate the described solution without having to discard the old solution because it actually uses copper—that is, copper chloride in aqueous hydrochloric acid solution—to dissolve the copper.

In addition to the etchant, you will need a vessel or container to etch the board in. For very small- and medium-sized boards, it's possible to use a cylindrical container, such as the beaker in Figure 0-16.

With your etchant in a safe container, all that's left is to put the PCB in the etchant and take the PCB out when the unwanted copper is gone. In the scenario in Figure 0-16, the circuit board is dipped in and out of the etching solution. Note that the beaker is sitting on a hot plate. Heating the solution accelerates the etching process, but be sure to keep the temperature between 100 and 120°F.

For larger boards, some kind of pan can be used. If you tip the pan, the etchant flows over the board, as illustrated in Figure 0-17.

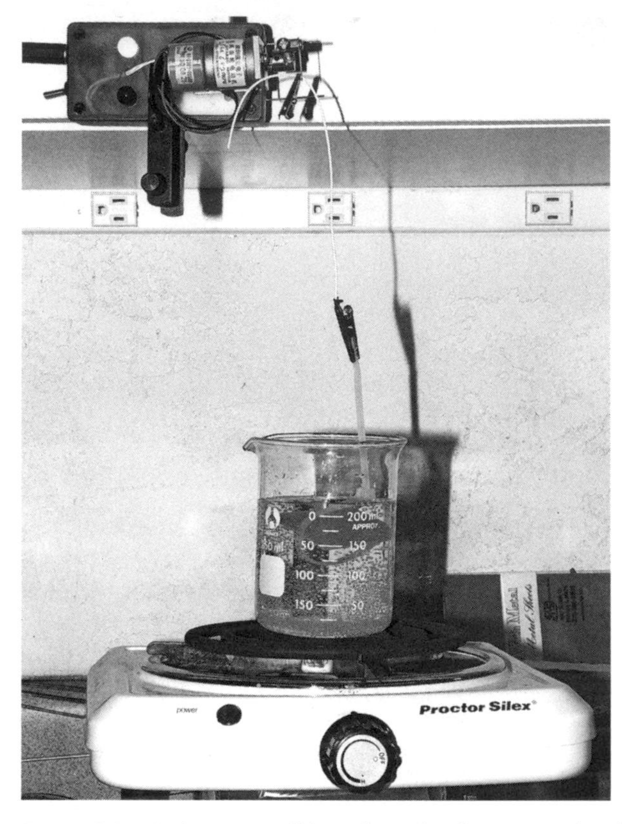

Figure 0-16: Etching a small board in a beaker agitated with the Automated Agitator for PCB Etching from Chapter 2. The board is held using a plastic wire tie.

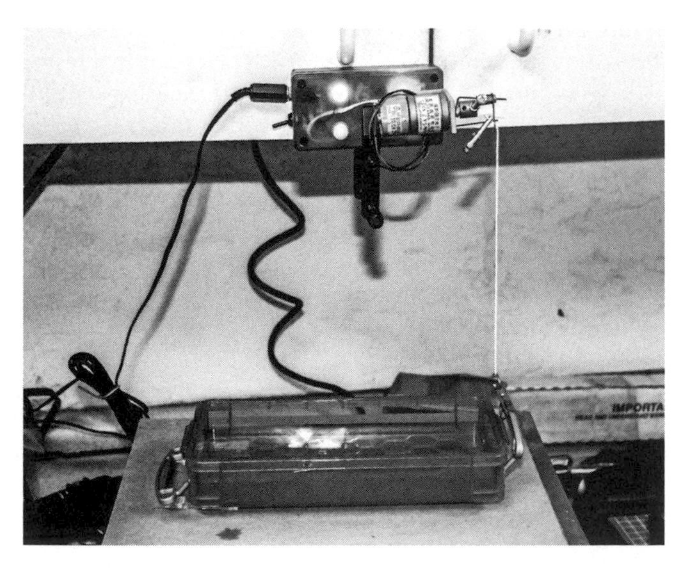

Figure 0-17: Etching larger boards in a container that is tipped by the Automated Agitator for PCB Etching

The pan can be glass or plastic. In the past, I've use a glass baking pan, but the container in Figure 0-17 is plastic. Just be careful with the heat if you use a plastic container. Temperatures in the recommended area should be safe.

In both the etching situations shown, an Arduino-based agitator (see Chapter 2 to build the project) makes etching go even faster. The agitator provides a simple, mechanical way to agitate the etchant and speed up the process.

Drilling the Board

Etching only removes copper, so unless you are making a single-sided, all surface-mount board, you will have to manually drill holes for your components. Drilling the PCB can be tedious depending on your equipment and the number of holes.

In my early days making boards, I used a Dremel tool freehand with a #66 drill. For small projects, hand-drilling is fine, but it's easier to use a drill press for larger ones.

If you already own a Dremel tool, then you're in luck. For under $40, depending on where you buy it, a Dremel drill press accessory works well for circuit boards, as well as hundreds of other tasks. If you don't already have a Dremel tool, you can get one of those for around $30 or less if you shop around. There are a variety of other high-speed drill and drill press combinations available at relatively low prices, too. Check with Harbor Freight and other suppliers of imported products on the web.

If you plan to produce a number of PCBs, I suggest a dedicated drill press rather than a drill press attachment. There are inexpensive units that run on low-voltage supplies as well as high-priced units, such as the Electro-Mechano pictured in Figure 0-18, which is designed exclusively for drilling small holes in jewelry and PCBs. Just pick one that suits your needs.

Figure 0-18: The small Electro-Mechano drill press I use for drilling PCBs

You'll need small drill bits to drill the PCB. I recommend an assortment of 10 tungsten-carbide drills with 1/8-inch shanks, which are available from Electronic Goldmine (part #G15421). A similar assortment is available from Amazon, and I have purchased several sets of these at very modest prices.

Connectors Used in This Book

Throughout this book, I've tried to simplify the use of connectors and minimize the number of different connectors used. But whether you make your own PCBs or not, you will always need some way to interconnect modules like LCDs, I^2C adaptors, sensors, and so on; and sometimes you will have to assemble your own connectors.

The connectors I use quite frequently are a family of connectors on 0.100-inch centers, a standard that works both for male and female headers on PCBs and for stand-alone connectors for some cable assemblies. While the units I use in this book were purchased from Pololu Robotics and Electronics (*https://www.pololu.com/*), the same or similar units are available from many other suppliers, including Jameco, Newark, Mouser, Digi-Key, and so on.

Figure 0-19 shows a few basic connector configurations I've used.

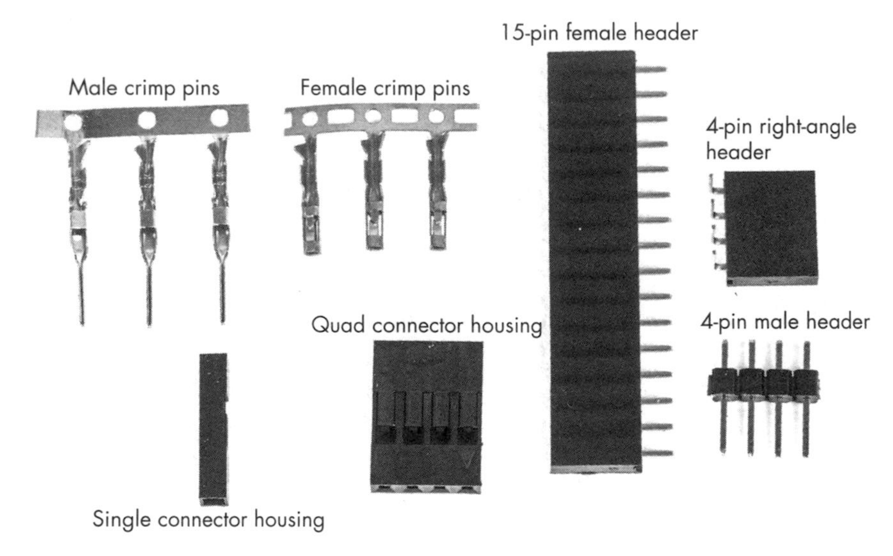

Figure 0-19: A few basic connectors

The male and female crimp connectors are the workhorses in most cables I make. However, these connectors must be crimped onto the wire they connect. To crimp a pin, you can use a professional crimping tool (see Figure 0-20), which results in a nicely finished crimp (see Figure 0-21).

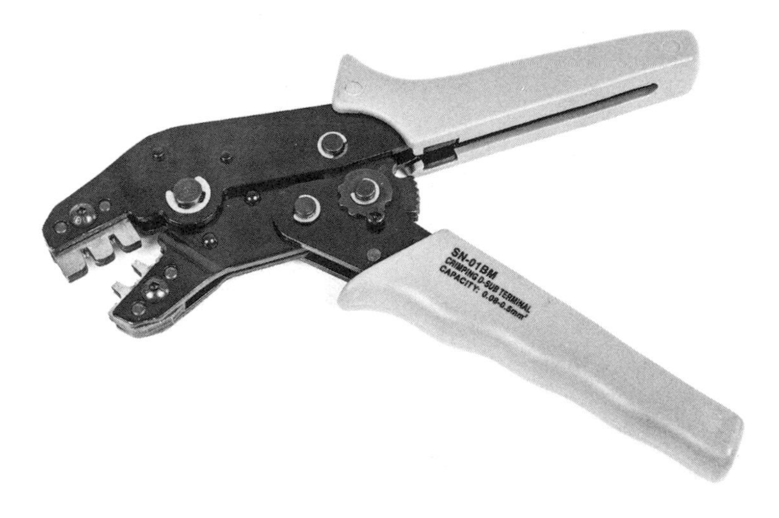

Figure 0-20: Crimping tool used to crimp 0.100 crimp connectors to 26-, 28-, and 30-gauge wire

Figure 0-21: A male crimp connector properly crimped using the crimping tool

The crimper that Pololu sells for its crimp connectors is relatively easy to use and makes a nice solid crimp, but it is a bit pricey at around $30. If you don't want to buy a crimping tool, you can crimp the connectors with a small pair of pliers. The resulting connection may not be as pretty, but it should work just as well. Figure 0-22 shows a cable I crimped using pliers.

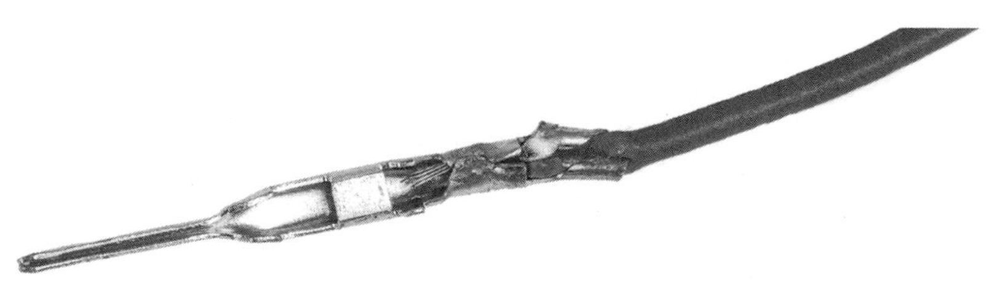

Figure 0-22: A male connector identical to the one in Figure 0-21 that was crimped by hand with a pair of pliers. Both fit snugly in the connector housing and work well.

You can create cables that plug into male headers with female crimp pins. These are useful for connecting parts of a PCB with a cable and for connecting an Arduino board to a shield.

Headers and housings are available in sizes from a single-pin wide up to 10 pins, 15 pins, and beyond. Most projects in this book that involve handmade connectors use 2- and 4-pin connectors.

Using SOICs

Making connectors for through-hole headers is fine, but through-hole integrated circuits with pins on 0.100-inch centers are becoming harder to get. While manufacturers continue to make many ICs in the older format, new designs are often available only as surface-mount components. These new packages are known as *small-outline integrated circuits (SOICs)*. Figure 0-23 shows two SOIC components next to an 8-pin DIP IC, for a size comparison.

Figure 0-23: A standard DIP package (top) compared to two tiny SMD ICs, a 5-pin Linear Technology LTC1799 in a TSOT-23 package (middle) and a 3-pin Maxim MAX7375AUR in a SOT-23 package (bottom), next to a dime for scale

What Are SMT Devices?

The two non-DIP ICs shown in Figure 0-23 are *surface-mount technology (SMT)* devices. SMT devices are soldered directly to the surface of a PCB instead of with their pins protruding through holes in the bottom. The advantage of SMT is that the devices can be made a lot smaller and placed in close proximity to each other, resulting in a more compact device. Many SMT components have 0.95 mm (0.0374-inch) centers or smaller, which doesn't match up to the 0.100-inch centered parts discussed so far.

Using SMT components also reduces wiring lengths, which can be critical at high frequencies. Many circuit boards have multiple layers (projects in this book have a maximum of two layers), and connections between layers were formerly made with holes for the pins on the ICs. These connections are now more commonly made with *vias*, which are small, plated-through holes in the board. Automated pick-and-place equipment is now concentrating on SMT devices, too. One day, resistors, capacitors, inductors, LEDs, fuses, and so on will likely be available only in surface-mount configurations, but it will probably take a while.

The SMT packages that I talk about in this book are *leaded*—that is, the package itself has leads protruding from it, even though the leads are not designed to go through holes in the PCB. Leaded ICs come in a variety of configurations with pins at different spacing, from relatively sparse leads—like the two SMT components in Figure 0-23—to ICs with hundreds of leads.

NOTE *This book avoids certain SMT packages, such as the ultra-small packages with direct-connect patches, where the chip connects directly to the PCB (called chip-on-board), and ball-grid arrays, where the connection is a controlled-collapse solder bump on the bottom of the package.*

The Solder Paste Method

Using leaded SMT chips leaves you a couple of options. One is to design a PCB with the correct pads for the SOIC footprint and solder the IC directly to the PCB. Soldering an SOIC component involves applying solder paste and heating the board itself. While this is a viable approach (and a fair amount of tutorials on the web cover it), populating the board can be difficult—particularly if the board has both through-hole and SMT devices. Unless you have a stencil for depositing the solder paste, the paste has to be applied manually, usually with a syringe and sometimes with a sharp toothpick or dental pick. Figure 0-24 shows a set of tools I have used for this process.

Several online suppliers offer solder paste in syringes at reasonable prices. Most of the solder paste compounds have a melting point between 300 and 470°F (some less), so boards can be soldered in a toaster oven or in a container on a hot plate.

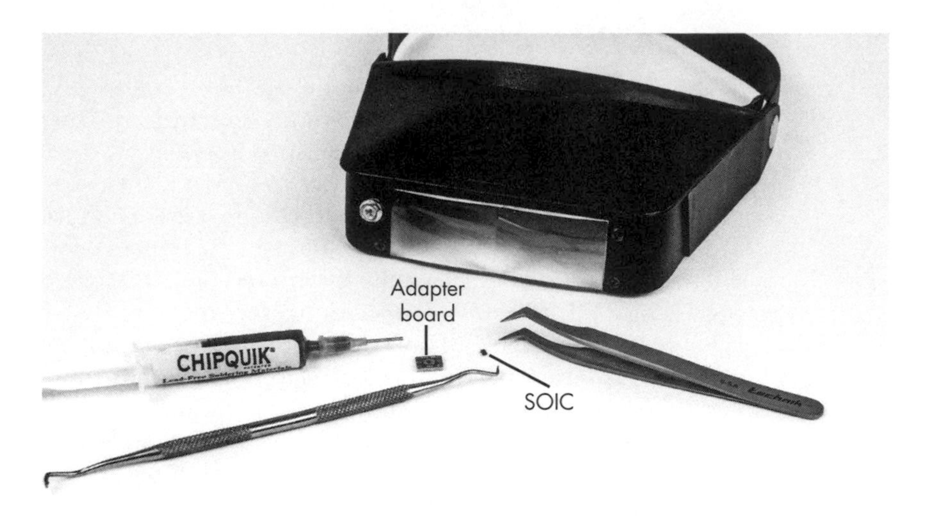

Figure 0-24: Chip Quik solder, a head-mounted magnifier, a dental pick, tweezers, an adapter board, and an SOIC (the speck next to the tweezers), ready for mounting

The solder paste that I use is Chip Quik. It's relatively inexpensive, comes in its own syringe with a tip, and has a melting point of only 138°C, or 281°F. While the tip could be a little smaller, it has worked for most applications. Chip Quik's low melting point makes soldering the board in a toaster oven or on a hot plate easy but could conceivably be a problem otherwise: in a high-current application, a solder joint could heat up enough to melt the solder. But with the voltage, current, and signal levels in this book, I don't expect this to be a problem.

After the solder paste is applied, the components can be carefully placed on the paste with a pair of tweezers and a steady hand. I also use a head-mounted magnifier so I can see the connections. When the chip is set in place, all that remains is heating the assembly to the melting point of the solder, and voilà—the chore is done.

If you opt to heat the board in a standard toaster oven rather than buying a specialized SMT oven, just don't use the same toaster oven you use to cook food. Many solders still contain lead, which has been deemed not-so-good for you. Flux materials (present in the solder paste to make the solder flow more easily) and binders also contain certain volatile compounds that may be unhealthy if ingested.

You can also put smaller boards in a small, clean metal can. Then, place the can on a hot plate, and set a small scrap piece of steel (aluminum will also work) on top of the can to hold the heat in. When the solder paste has melted, remove the heat. The process usually takes only a few minutes.

For someone like me with fat fingers, 0.95 mm is pretty small, whether applying solder paste, placing a component, or soldering a leaded SMT component directly. The process has, on occasion, taken me several tries. If you're not quite ready to try the solder paste solution directly on your main PCB, consider buying an adapter board to convert SMTs to conventional 0.100-inch center through-hole mounting.

Several vendors offer small adapter boards that convert from an SOIC package to DIPs with 0.100-inch centers. The adapter board in Figure 0-25 is from Futurlec (*http://www.futurlec.com/*). Futurlec adapter boards go for all of $0.28 each, so I ordered a variety, including those for 8-, 14-, 16-, and 18-pin SOICs.

Figure 0-25: Futurlec 6PINSO23 adapter board

The best soldering solution, even with the adapter, is to use solder paste and an oven (or a can, as I just described). But if you don't have access to the materials for that technique, you can always solder the SOIC component directly.

Soldering Directly

Soldering an SOIC component directly is a little tricky. Doing so requires a soldering iron with a fine tip, though I used the tip I use for everything else. (I believe mine is 0.7 mm.) Here's how this approach works:

1. First, place male headers in a breadboard with the adapter on top, and solder them to make a stable platform (see Figure 0-26).

Figure 0-26: Adapter board with stakes installed and plugged into a breadboard for soldering the IC. The particles shown are residue from the solder and flux, which I later removed using alcohol and a Q-tip swab.

2. Even though the adapter has solder plate on the copper, it's not thick enough to secure the leads of the IC. More solder is needed, so carefully melt a thin layer of solder on only one pad. (Often, I place too big a blob of solder and have to remove it with solder wick, but that too works out fine, as it still leaves a thin coating of solder on the pad itself.)
3. Place the component on the adapter board, hold it down securely (I apply pressure with a dental pick), and put the hot iron on the lead that has the solder under it.
4. Once the first leg is secure, it holds the device in place, and you can carefully solder the other terminals with the iron.

Figure 0-27 shows a board I soldered this way. It may not look real pretty, but it works.

Figure 0-27: A completed adapter board with the male headers installed, the chip soldered, and a decoupling capacitor soldered across two of the pins. This is the cleaned-up version of Figure 0-26.

A completed adapter board like the one in Figure 0-27 can then be mounted on a conventional through-hole board with holes on 0.100-inch centers. I used this technique on the Ballistic Chronograph (Chapter 8) and Square-Wave Generator (Chapter 9) projects in this book.

Closing Thoughts

With the knowledge in this chapter and some previous electronics experience, you are ready to tackle any project in this book. I will cover other important techniques and information on an as-needed basis throughout.

1

THE REACTION-TIME MACHINE

In this chapter, I will show you how to build a time machine—that is, a Reaction-Time Machine. I'd love to say that this project will bring you "back to the future," but alas, it won't. The "time" it's looking at is the time it takes you to react to a stimulus, which makes for a fun game. This project is designed to accurately measure an individual's reaction time and provide an area for comments on the level of the individual's performance (see Figure 1-1). There is also plenty of room to personalize the game to make it even more fun for you, your friends, and your family.

Figure 1-1: Completed Reaction-Time Machine

Required Tools

Soldering iron and solder

Drill and drill bits

Mounting tape

Wire cutters

Parts List

This project has one of the smallest parts counts of all the projects in this book, but don't let that attenuate its value for you. My family and friends have enjoyed playing the game repeatedly, and it's portable, so you can take it with you to get-togethers and other events.

Here's what you'll need:

One Arduino Nano or clone

Two SPST momentary switches (preferably one with a red button and one with a button of a different color)

One SPST toggle switch

One red LED

Two 10-kilohm resistors

One 470-ohm resistor

(Optional) One audible annunciator, Mallory Sonalert or similar

One 4×20 LCD

One I^2C adapter, if not included with the LCD (see "Affixing the I^2C Board to the LCD" on page 3)

NOTE *I purchased a 16×2 LCD and its external I^2C board separately and soldered the two together. However, many online vendors offer the same display and I^2C adapter already soldered for about the same price or less than the two boards separately. Check eBay in particular.*

One 9V battery

One 9V battery clip

One 3.5 mm jack (if remote switch is used)

One Hammond 1591 BTCL enclosure

28-or 30-gauge hookup wire

22-gauge solid conductor wire

Downloads

Before you start this project, check the following resource files for this book at *https://www.nostarch.com/arduinoplayground/*:

Sketch file *Reaction.ino*

Drilling template for case *ReactionEnclosure.pdf*

Reaction vs. Reflex

People often confuse reactions and reflexes, so I will start by defining both. *Reflexes* are involuntary, automatic responses to a stimulus. In a reflex action, the stimulus bypasses the brain and travels from the source of the stimulus to the spinal cord and back to the receptor that controls the response, without any cognitive acknowledgment. (Though I know many people for whom almost all stimuli—and information—seem to bypass the brain, often just getting lost instead.) Think of the doctor hitting your knee with a patellar hammer to trigger your knee-jerk reflex.

Reactions, on the other hand, take the stimulus to the brain to be processed, and then a return reaction travels to a receptor to result in some motor action. This process takes somewhat longer than a typical reflex, though some athletes are said to have reaction times so fast that it's possible their response is more similar to a reflex than a reaction.

NOTE Sports Illustrated *has done interesting work in this area, with eye-opening articles on baseball players and other athletes who have what appear to be exceptional reaction times.*

How Does the Game Work?

The Reaction-Time Machine game measures how long it takes an individual to press a button in response to a visual stimulus—in this case an LED. With a minor modification, you can add an auditory stimulus to the game: simply replace the LED with an audible annunciator, such as a Mallory Sonalert. Reaction time is measured in milliseconds or seconds (your choice), and it is the time between the moment the stimulus is activated and the moment the participant presses the button.

HISTORY OF REACTION-TIME DEVICES

Over the years, there have been many devices to measure reaction time. One of the simplest I remember from years ago required you to keep your fingers on either side of a ruler held by another person in mid-air. When the ruler was dropped, you would see how far it traveled before you could grasp it. The distance was translated to time using the algebraic equation

$$S = \frac{1}{2}AT^2,$$

where S is the distance traveled, A is the acceleration due to gravity, and T is the reaction time. After you build this project, try both the ruler test and the Reaction-Time Machine to see how close your times are between devices.

Measuring Time with the Arduino Nano

While there are many ways to measure elapsed time, this project takes advantage of the Arduino Nano's ability to keep accurate time. Microcontrollers keep time exceptionally well, and they measure the time that elapses between one input and another with a minimum latency. In addition to timing your reactions, the Nano shows the result on an LCD.

The Nano does almost all of the work in this project; the other components are basically passive. After testing some early builds, I added features to the sketch to make the game more interesting and accurate. For example, I initially used a simple pushbutton to reset the Nano and start a counter. The participant would press the red stop button as soon as the LCD indicated so, and the Nano measured the time between pressing the reset and stop buttons. I found, however, that the player could anticipate the reset button being pushed and come up with some amazing reaction times.

To prevent the player from anticipating when the stimulus is about to occur, I had the Nano start the timer on a delay instead. The version in this book generates a random delay from when the reset button is depressed, activates the stimulus after the random delay, and counts the time from the stimulus to the moment the participant responds by depressing the stop button. That solved one problem.

Then, one of the participants tried to jump the gun and get an early start by holding down the stop button. I solved this problem by setting a minimum reaction time in the sketch. Any time under that minimum throws an error, and the LCD displays "Jumped the Gun" to indicate that the player pressed the button too soon.

I used a relatively large display—4 lines with 20 characters each—so there would be enough room to display the reflex time and some commentary on the relative prowess of the player. You can make your commentary as funny or serious as you want, but it must not exceed 60 characters in length—that is, three lines of 20 characters each. While I leave the commentary up to you, the sketch for this project includes some ideas that I used when putting it together. You can always edit the commentary and reload the sketch to show comments specific to a set of users, like friends or relatives.

Expected Speed Ranges

Most individuals' reaction times seem to vary greatly, based on the small sample I tested. Interestingly, age doesn't seem to be a factor. The average reaction time was around 200 milliseconds, and that is the average reaction time identified by many researchers.

The fastest response of anyone I sampled was 105 milliseconds; however, the individual was not able to repeat that performance. Several individuals scored between 105 and 125 milliseconds, but not consistently. Significantly lower reaction times may well be anomalous or the result of an individual actually anticipating the stimulus. My players' failure to repeat extremely fast reaction times would tend to bolster that idea. (I wouldn't want to accuse anyone of successfully pre-guessing the release moment.)

The Schematic

While the display could have been wired directly, using the I^2C interconnect made it a lot simpler and reduced the interface to only four wires: positive, ground, data, and clock (see Figure 1-2).

The only components needed are the Nano, three switches (one toggle switch for power and two momentary pushbutton switches for activate and reset), an LED, the display, and three resistors. Despite the relatively sparse parts count, the project performs elegantly.

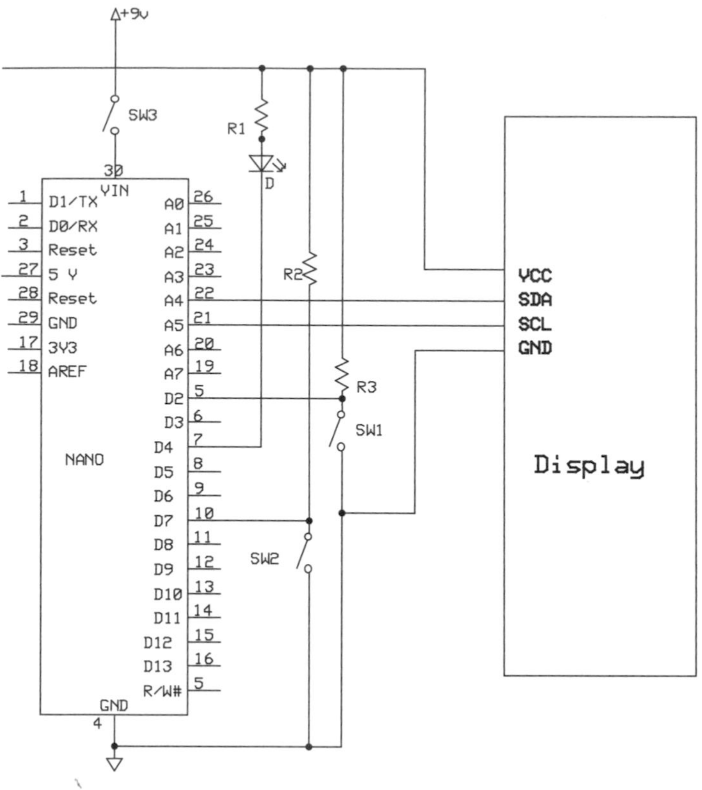

Figure 1-2: Schematic diagram of the Reaction-Time Machine

The Breadboard

As is the case for most of my Arduino projects, the first step is to prepare a breadboard to prove the concept and test the sketch. Here's how to wire up the breadboard:

1. Connect the red positive rails on the breadboard together.
2. Connect the blue negative rails on the breadboard together.
3. Insert the Arduino Nano (or clone) in the breadboard, leaving two rows on one side and three on the other. (If the Nano does not come with stakes soldered in, see "Preparing the Arduino Board" on page 2.)
4. Connect the 5V terminal on the Nano to the red positive rail on the breadboard.

5. Connect the GND terminal on the Nano to the blue negative rail on the breadboard.
6. Connect the negative wire from the battery connector to the blue negative rail. Remember that the breadboard has no switch, so you must disconnect the battery to turn it off.
7. Connect the positive lead from the battery connector to VIN on the Nano. (Do not connect the positive terminal of the battery to the red positive rail—it could permanently damage the Nano.)
8. Attach 5-inch wires to two normally open momentary pushbutton switches. (I use 22-gauge solid conductor wire so it can plug in to the breadboard directly.)
9. Prepare a wire harness for the LCD (see "Affixing the I^2C Board to the LCD" on page 3).
10. Connect the red wire from the LCD to the red positive rail on the breadboard (5V) and the black wire from the LCD to the blue negative rail.
11. Insert the yellow wire from the display (SDA) to pin A4 on the Nano.
12. Insert the green wire from the display (SCL) to pin A5 on the Nano.
13. Connect one side of each pushbutton switch to the blue negative rail.
14. Connect the other side of the red reaction switch (SW2) to pin D7 on the Nano.
15. Connect the other side of the yellow reset switch (SW1) to pin D2 on the Nano.
16. Connect a 10-kilohm resistor from pin D7 on the Nano to the red positive rail.
17. Connect a 10-kilohm resistor from pin D2 on the Nano to the red positive rail.
18. Connect the anode side of the LED (the longer leg) to the red positive rail and the cathode side to an empty row on the breadboard.
19. Connect a 470-ohm resistor from the cathode side of the LED to pin D4 on the Nano.

Upload the *Reaction.ino* sketch to the Arduino Nano (see "Uploading Sketches to Your Arduino" on page 5), and you should now be ready to go. Figure 1-3 shows the breadboard laid out with the switches dangling from their wires.

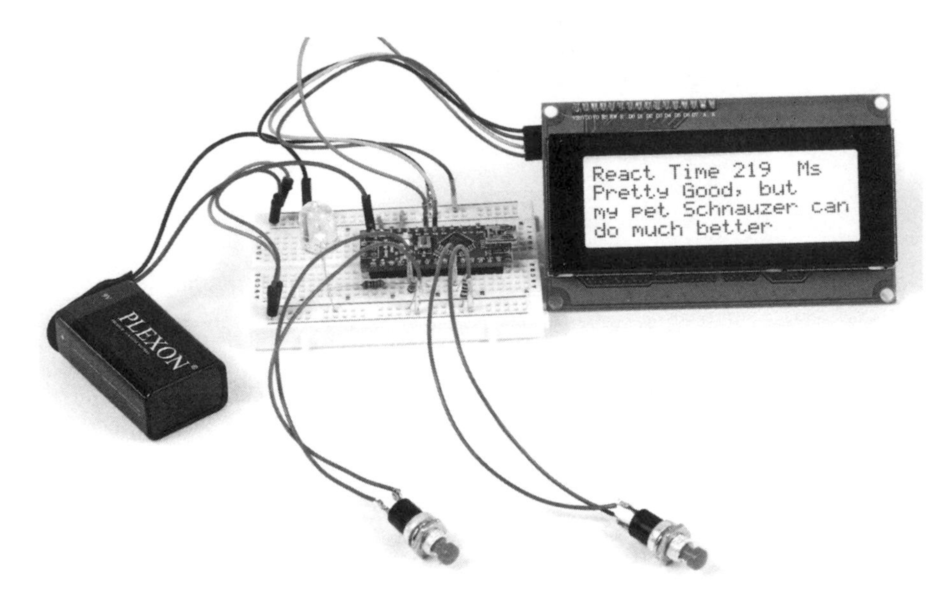

Figure 1-3: The breadboard setup for the Reaction-Time Machine. Because there is no on/off switch, you have to disconnect the battery to shut it off.

The Sketch

The sketch is the actual computer program that tells the Arduino what to do and when to do it. It is written in a language of its own that comprises structures, variables, arrays, functions, and so on, which represent a recipe for the microcontroller to follow. This language is converted into a sequence of zeros and ones that are routed to various parts of the controller and can perform storage, timing, comparison, arithmetic functions, and more.

The process of converting a computer language to a sequence of zeros and ones is called *compiling*. The compiling routine in the Arduino Integrated Development Environment (IDE) is activated when you click the Verify and Compile buttons in the upper-left side of the Sketch window.

The sketch gets pretty long because of all the messages that can be inserted when it checks the score; however, the basic operation uses only a handful of code lines. You can use the scoring function as is, modify it, or copy and paste it to make a new scoring function. As you'll see in my messages options, I've had fun with it.

The following code has been truncated to minimize the number of lines. However, you can simply go to *https://www.nostarch.com/arduinoplayground/* to download the entire sketch, which includes a number of messages.

```
/*
Includes score function, random number generation, false start
"jump the gun" indicator, and multiple messages spaced 10 ms apart

Mod for "jump the gun" gives response if time <70 ms
*/
```

```
#include <Wire.h> //Libraries included
#include <LiquidCrystal_I2C.h>

int start_time = 0;
int stop_time = 0;
int reacttime = 0;
int x;
int R;
int randnumber1;
int z;

LiquidCrystal_I2C lcd (0x3F, 20, 4); //Initiate LCD

void setup() {
  Serial.begin (9600);
  pinMode(2, INPUT);
  pinMode(4, OUTPUT);
  pinMode(7, INPUT);
  lcd.init();
  lcd.backlight();
}
//Begin function "score"
void score() {
  lcd.clear();
  lcd.print("Reaction Time ");
  lcd.print(reacttime);
  lcd.print(" ms");
  lcd.setCursor(0, 1);

  if((reacttime >= 105) && (reacttime < 135)) {
    lcd.print("Approaching Superman");
    lcd.setCursor(0, 2);
    lcd.print("but you can still do");
    lcd.setCursor(0, 3);
    lcd.print("a lot better");
  }

  if((reacttime >= 135) && (reacttime < 180)) {
    lcd.print("Superhero Status");
    lcd.setCursor(0, 2);
    lcd.print("but not yet");
    lcd.setCursor(0, 3);
    lcd.print("Superman");
  }

  if((reacttime >= 180) && (reacttime < 225)) {
    lcd.print("You are trying ??");
    lcd.setCursor(0, 2);
    lcd.print("but not hard enough");
    lcd.setCursor(0, 3);
    lcd.print("still a loser");
  }
```

```
  if(reacttime > 225) {
    lcd.print("Lost your touch");
    lcd.setCursor(0, 2);
    lcd.print("If you ever had it");
    lcd.setCursor(0, 3);
    lcd.print("on the border of wimpy");
  }
}

//Begin main program
void loop() {
  digitalWrite(4, HIGH);
  lcd.clear();
  lcd.print("System is Armed");
  delay(1000);
  lcd.setCursor(0, 1);
  lcd.print("      READY   ");
  lcd.setCursor(0, 2);
  lcd.print(" Push Red Button");
  lcd.setCursor(0, 3);
  lcd.print("When Red lamp lights");

  randnumber1 = random(5, 25); //Generate random number between 5 and 25
  R = randnumber1;
  for(x = 0; x < R; x++);
  delay(5000);
  if(x == R) {
    digitalWrite(4, LOW); //Turn on start lamp
    start_time = millis(); //Initiate timer
    lcd.clear();
    lcd.print("Mash React Button");
    lcd.setCursor(0, 1);
    lcd.print("          ");
    lcd.setCursor(0, 2);
    lcd.print("          ");
    lcd.setCursor(0, 3);
    lcd.print("          ");

    while(digitalRead(7) == 1); //Wait for response

    stop_time = millis(); //Complete timing cycle
  }

  reacttime = stop_time - start_time;

  if(reacttime < 70) { //Jump the gun indicator
    lcd.clear();
    lcd.setCursor(0, 0);
    lcd.print("Too anxious. You");
    lcd.setCursor(0, 1);
    lcd.print("(Jumped the Gun)");
    lcd.setCursor(0, 3);
    lcd.print("Could be Fatal!");
  }
  score();
```

```
  Halt:
  while(digitalRead(2) == 1);
}
```

The `#include` lines initiate the libraries: the I^2C library, *Wire.h*, establishes the rules for I^2C communications, and the LiquidCrystal library allows the Arduino to control LCDs. Then, we define the seven variables used to calculate reaction time. Next, `setup()` sets up the serial communication—in case you want to adjust the code and view it on the serial monitor—and defines various pins as inputs and outputs. Inputs are required for the reset and stop buttons, and an output pin is defined for the LED that tells the player when to press the stop button.

Customized Reaction Commentary

One of the most entertaining aspects of this project is the chance to get creative when displaying the player's reaction time. After `setup()`, the sketch shows a function called `score()`, which lists different comments that could be displayed on the LCD based on the participant's response speed. A function may not necessarily be the most efficient approach (a look-up table or other approach could also have been used), but it works well enough. I used only a single scoring function in this iteration; however, you could easily define as many as you like and change your sketch to select one. For example, you might write a second function called `score1()` that could include a different set of comments and timing. Then, to switch from one function to the other, you'd have to change only the line that calls `score()` to call `score1()` instead.

To customize the sketch to include comments that could refer to your own friends or family members, you can simply enter your comments in place of the ones that are in my sketch. Don't forget to keep the text you want to print to the LCD in quotes so the Arduino recognizes the printable characters.

A word on the reaction time itself: each comment is for a range of reaction times of either 5 or 10 milliseconds. I selected these ranges arbitrarily. After you play with the Reaction-Time Machine for a bit, you may wish to change these ranges based on the fact that users' responses may cluster around a particular area, such as from 195 to 225 milliseconds. I found that many reaction times were in the 190 to 250 milliseconds range, but your friends and family may be different. In that case, you can separate the comments by as little as 1 or 2 milliseconds so players don't keep getting the same comment.

You can add as many comments as you wish, up to one comment per millisecond. If you accidentally overlap the times, the sketch may not compile.

NOTE *You can find reaction-time measurement tools on the web if you want to see how your game's measurements compare. However, their accuracy is suspect because of the latency in the PC itself.*

ON WRITING CODE TO SET UP LCDS

There are a few points to note about the setup of the LCD. The sketch uses a LiquidCrystal library, *LiquidCrystal_I2C.h*. If this library is not included in your Arduino IDE, you can easily download it using the instructions provided in the reference section on the Arduino website (*http://www.arduino.cc/reference/*).

In addition, each I^2C device comes with its own I^2C address. This allows several I^2C devices to be used on a single serial line. Usually the device documentation provides the address—in the case of the I^2C LCD I used, the address was 0x3F. Thus, when the sketch initiates the LCD, the code looks like this:

```
LiquidCrystal_I2C lcd (0x3F, 20, 4);
```

However, different displays come with different addresses. If you have an I^2C device that you do not have an address for, you can easily find the address by hooking up the device to an Arduino, downloading a scanner sketch from *http://playground.arduino.cc/Main/i2cScanner/*, and running the sketch. The scanner sketch should display the I^2C address on the serial monitor.

Many projects in this book use similar code to work with an LCD, so refer to this box any time you need a refresher on how that code works.

What Happens in the Loop

Now let's look at the sketch's loop. After `void loop()` initiates the start of the program, the program calls `digitalWrite(4, HIGH)` to turn off the active light. Then, the LCD screen is cleared, and text is written to the LCD to indicate that the system is armed and ready for a player to push the reaction button as soon as the red LED illuminates.

Next, a random number between 5 and 25 is generated, and the program calls `delay(5000)` to count every five seconds from zero to the random number. As soon as the random number is reached, three things happen: first, the annunciator lamp illuminates; second, an internal timer is started in the Nano; and third, the display then changes to read "Mash the React Button."

NOTE *A wider range of random numbers might make this game even more interesting for players. You can easily experiment by changing the random number count, the delay, or both.*

The Nano is then instructed by `while(digitalRead(7) == 1);` to wait until the reaction button is depressed. After the button is depressed, the Nano calculates the reaction time with `reacttime = stop_time - start_time`. This time will be displayed on the LCD and used to select the appropriate comment in the `score()` function. Also, if the player's reaction time is less than

70 milliseconds at this point, then the conditional statement looking for a participant to be "jumping the gun" displays appropriate wording for the LCD. The system is then halted and ready to be reset.

Otherwise, the serial print block is included in case you want to adjust the code and view it on a serial monitor. It also helps for debugging purposes.

Finally, the `score()` function is invoked, followed by the `Halt` command, and the system is ready to have the reset button depressed.

Construction

Building the Reaction-Time Machine can be as simple or as complex as you want. Initially, I placed all the components in the vinyl package that a flexible wrist brace came in. I cut a hole for the display connectors with an X-ACTO knife and punched the holes for the switches and LED with a paper punch, followed by a tapered reamer. The result was somewhat crude, as shown in Figure 1-4.

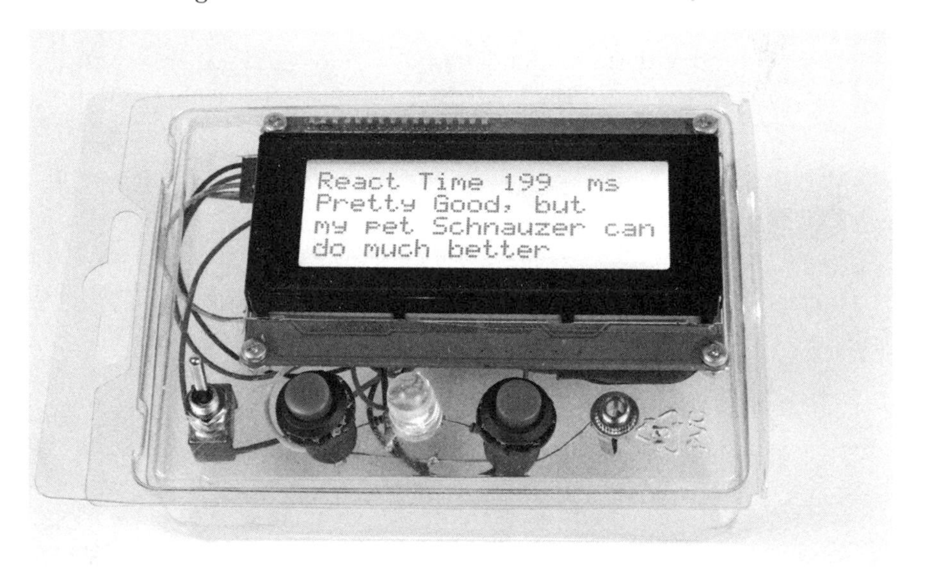

Figure 1-4: This was the Reaction-Time Machine's original, primitive package, which worked but turned out to be too flimsy. The vinyl was only 0.018 inches thick.

Preparing a Sturdy Case

Of course, a real case makes the game much sturdier, which is important when you have competitive players mashing those buttons. To keep things as simple as possible, I employed one of the clear ABS plastic cases from Hammond (1591 BTCL). The clear top of the case allowed me to place the display behind the cover rather than machining out a hole for the display to protrude through. To mount the components, I simply drilled holes in the cover according to the drawing in Figure 1-5.

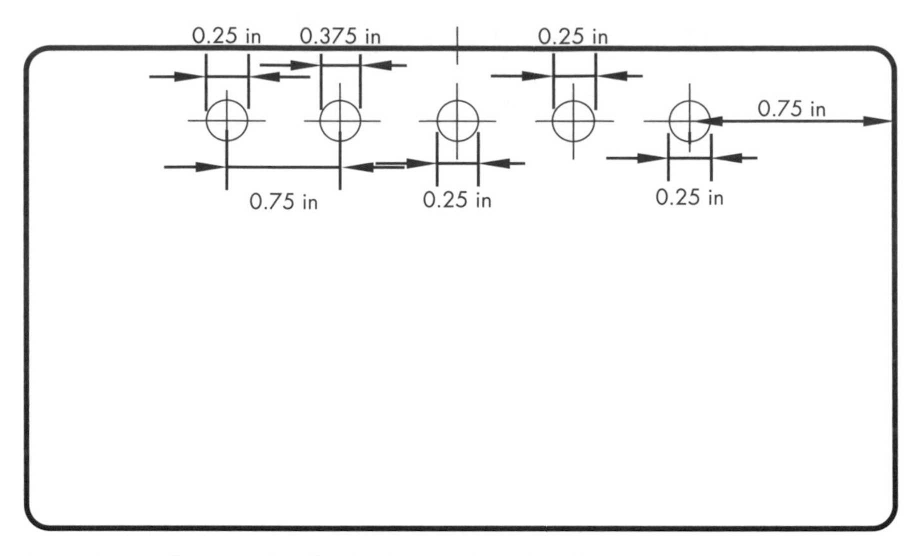

Figure 1-5: Drilling template for the Reaction-Time Machine

Quarter-inch holes work well for the momentary pushbutton switches, as well as for the toggle switch and 3.5 mm jack. For the 10 mm LED, I used a 3/8-inch drill and then reamed the hole out to make a tight fit. No other mounting hardware for the LED was necessary.

NOTE *The 3.5 mm jack is wired in parallel to the execute switch. If you want to use an external stand-alone switch, it can simply plug in to the jack. I abandoned the effort, however, as most participants preferred to hold the box in their hands.*

Mounting the Hardware

To mount the display to the case, I used two-sided 3M Indoor/Outdoor Super Heavy Duty mounting tape. I cut two sections the size of the LCD display's end bezel sections and bonded the display directly to the cover. The tape is difficult to remove, so make sure to place it right the first time. I used the same tape to mount the Nano and the battery holder to the back of the display. When mounting the display, I also used wire cutters to carefully cut off the corners of the display circuit board so it would fit far enough into the case without hitting the cover mounting pylons. See Figure 1-6 for the finished product, viewed from the underside.

Once all the components are in place, all that remains is to solder the components together, inserting the resistors where required. Take particular note of the I^2C adapter, which is the black paddleboard just below the switches and LED. While I could have bent the connectors and used a header to wire that up, the case may not have closed, depending on how carefully I crimped the connectors. Instead, I elected to solder the wires directly. It was only four wires, and it worked without much trouble. Finally, I printed out and attached labels from a Brother label maker. Figure 1-7 shows the completed unit.

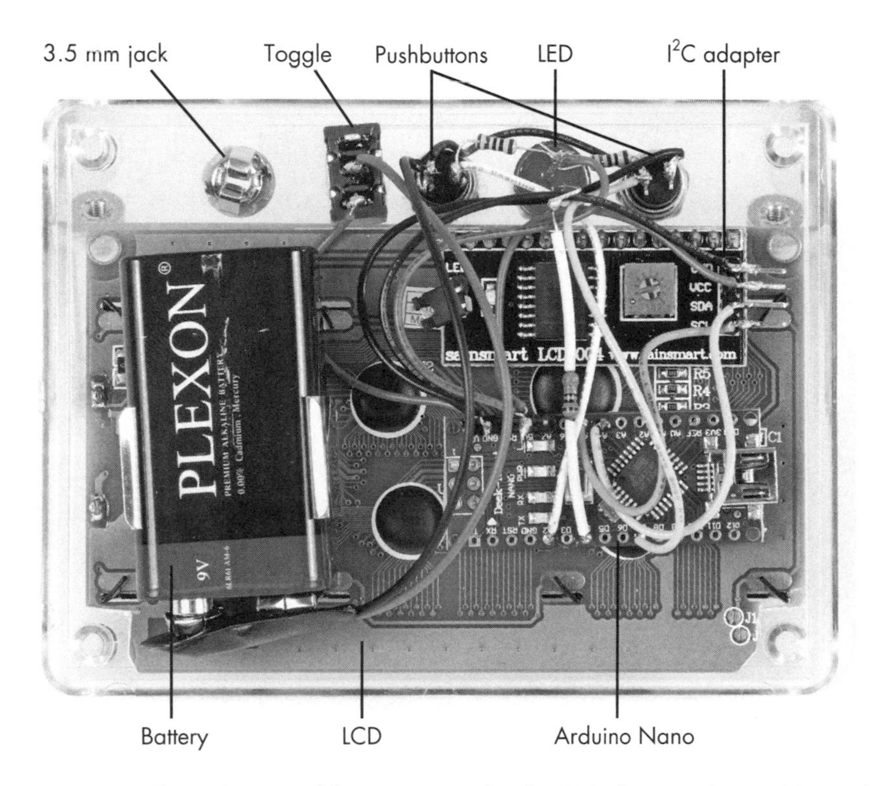

Figure 1-6: This is the rear of the unit mounted in the ABS plastic enclosure. Notice that the corners of the display (lower left and right) have been clipped off to fit around the top mounting pylons. The 3.5 mm jack is not wired, as I decided not to use it in this implementation.

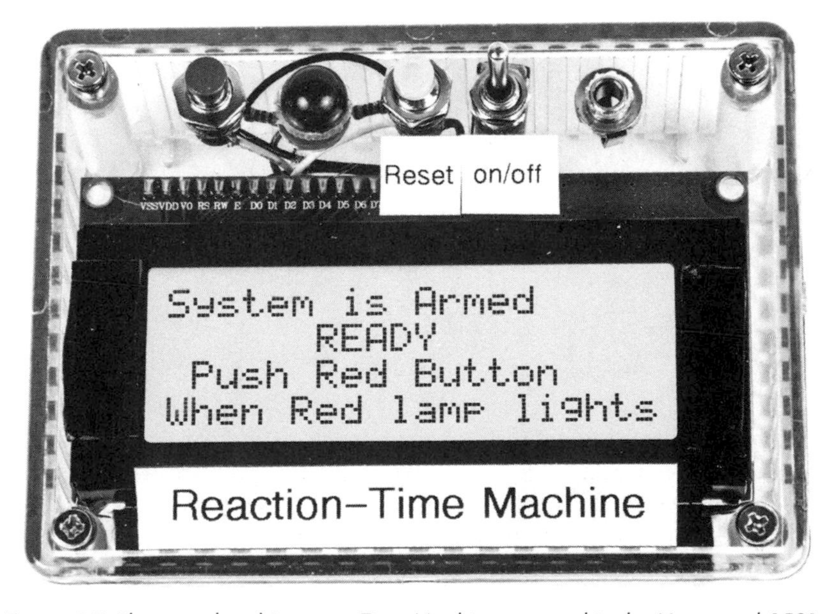

Figure 1-7: The completed Reaction-Time Machine mounted in the Hammond 1591 BTCL clear plastic enclosure

Ideas for Customization

There are many variations you could implement to increase the versatility and enjoyment of the Reaction-Time Machine. For example, as I developed it, I connected a Hall effect switch to one of the analog inputs and modified the sketch to automatically decrease the reaction time by a percentage when the Hall effect switch is activated. Then, I taped a small magnet to my finger that sat opposite the Hall effect switch so as I grabbed the box, it activated the switch. When I played, my reaction time was reduced by around 20 percent, while others had an actual reading. Far be it from me to suggest that readers try to hoodwink their adversaries, of course!

There are other modifications that can be made, such as incorporating a tone sound, or beep, as the sketch counts up to the random number. This can easily be accomplished with the addition of an annunciator and a few lines of code. If you're ingenious, there are other sound effects you could add, such as a vulgar sound that plays when poor scores are achieved.

You can also exercise your brain and add code to the sketch that will average scores after, for example, three tries before you reset it. I experimented with many variations as I played with the device, but I would caution that you can spend a great deal of time for minimal advantage. Put the game together and enjoy.

2

AN AUTOMATED AGITATOR FOR PCB ETCHING

This project uses the Arduino microcontroller to sense change in a motor's current drain (the rate at which the motor uses electricity) and then reverse the direction of the motor. There are numerous applications for the measurement and use of current drain, and this project provides an example method that can prove useful in the development of future electronics projects.

"Making Your Own PCBs" on page 13 illustrates different ways to design and make circuit boards at home for a very modest cost using readily available and environmentally safe household products. Part of this process includes etching the copper off a clad board. The process is more efficient when the board is agitated in the etching solution, resulting in a laminar flow of liquid across the surface of the board in both directions. Depending on the chemical activity of the etchant and thickness of copper to be etched, this process can take anywhere from 10 or 15 minutes to well over half an hour! Standing there stirring the pot is pretty boring, but you can create a device that dunks the board in and out of the solution for you (see Figure 2-1).

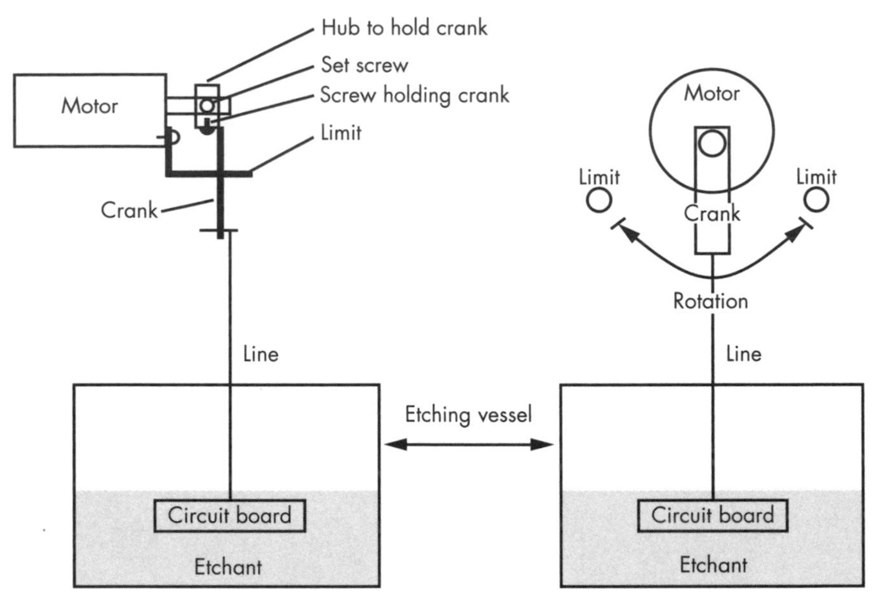

Figure 2-1: Illustration of the motor, crank, and etching vessels set up to dip a circuit board in and out of the etchant. While there are many ways to agitate a circuit board, dipping it into and out of the etching solution works well, especially for small boards.

In this project, the Arduino is measuring the current from the motor. When the motor's rotation reaches the limit pin, it begins to stall, increasing the current drain. The Arduino reacts to the increase in current by reversing the motor.

INSPIRATION BEHIND THE AUTOMATIC MOTOR REVERSAL PROJECT

This project has its roots in a problem my friend had with a model train set accessory. The accessory included a tramway to take make-believe skiers up and down a miniature mountain. The original mechanism failed, so I created a little circuit to drive a DC motor that moved the skiers up and down. My idea was that when the tramcar reached either the top or bottom of its run, the motor would slow down or stall, resulting in an increase in current drain. That excessive current drain would reverse the motor by changing the polarity and thereby send the car back the other way. To date, the skiers are still at the bottom of the mountain because my friend and I never installed the board, but the core circuit works well and promises other interesting applications.

The ability to receive an input, process the information, and produce an output is *the* fundamental function of any microcontroller. In this case, the Arduino starts the motor turning, waits until it detects the motor drawing more current than usual, and then reverses the motor's rotational direction. This simple function has a number of different applications: you could use the voltage drop to provide a safety turn-off for an overloaded motor, create a system to limit motion, and more.

Required Tools

One 6-32 tap

Drill and drill bits

Needle-nose pliers

Parts List

One Arduino Nano or clone

One SN754410 quad H-bridge IC, with socket if desired (Note that if you use the socket, you lose whatever value the PCB offers as a heat sink.)

One printed circuit board (PCB) or perf board

One current-limiting resistor (You should have a selection available for experimentation, from 1 ohm to 10 ohm. A 1/8 W resistor will work for smaller motors, but get a 1/4 or 1/2 W resistor for larger loads.)

Two 330-ohm, 1/8 W resistors

Two LEDs, one red, one green

One LM7805 voltage regulator

One plastic enclosure (I recommend the Hammond 1591 XXATBU.)

Two 2-pin female headers to connect the motor to the shield

Four 4-pin female headers to plug the Nano into

One small solder lug

One 3.5 mm, 2-conductor jack and plug

One SPST toggle switch

One plug-in wall adapter with an output of 5 to 12V at 200 mA or better

One gear head motor (I used a 6V motor, the Amico 20 RPM 6VDC.)

Two M3×0.5 mm screws with threaded spacers

Limit wires, preferably 0.039 piano wire or spring wire

Scrap brass or aluminum

One 4-40 or 6-32 screw

Downloads

Before you start this project, check the following resource files for this book at *https://www.nostarch.com/arduinoplayground/*:

Sketch *Reverse.ino*

Shield *Reverse.pcb*

Template *MotorMount.pdf*

How Automatic Motor Reversal Works

The Arduino is perfect for this project because it can control the whole system, and it simplifies the problem of accommodating different motors with different current requirements. Implementing the project in discrete components would require several more components than the equivalent Arduino circuit. Further, changing values for different motors or different reversal thresholds would mean changing a lot of hardware, but with Arduino, you just have to make a simple program change. The Arduino also provides the flexibility to add delays at each end of the run if desired.

The motor circuit you'll connect to the Arduino uses a resistor between the power supply and the motor (see Figure 2-2). When the motor slows or stalls, the current increases, creating a voltage drop across the resistor.

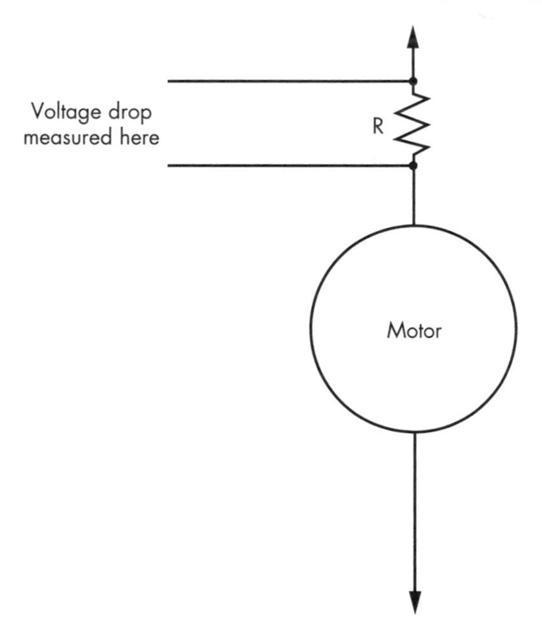

Figure 2-2: A voltage is created across the resistor between the positive supply and the input to the motor. It is this voltage that triggers the operation of the circuit.

The voltage drop across resistor R is the real-world input to the microcontroller. In this project, that voltage drop is fed to the Arduino Nano's two analog input pins that straddle the dropping resistor. The microcontroller digests this input and creates an output designated by your program.

NOTE *You could implement the circuit with only a single analog input, but that would curtail some of the flexibility of the circuit—particularly if you use motors that run at different voltages.*

The Schematic

The agitator circuit feeds the voltage that appears across resistor R1 into two of the Arduino's analog input pins, A0 and A1, setting up the real-world input (see Figure 2-3).

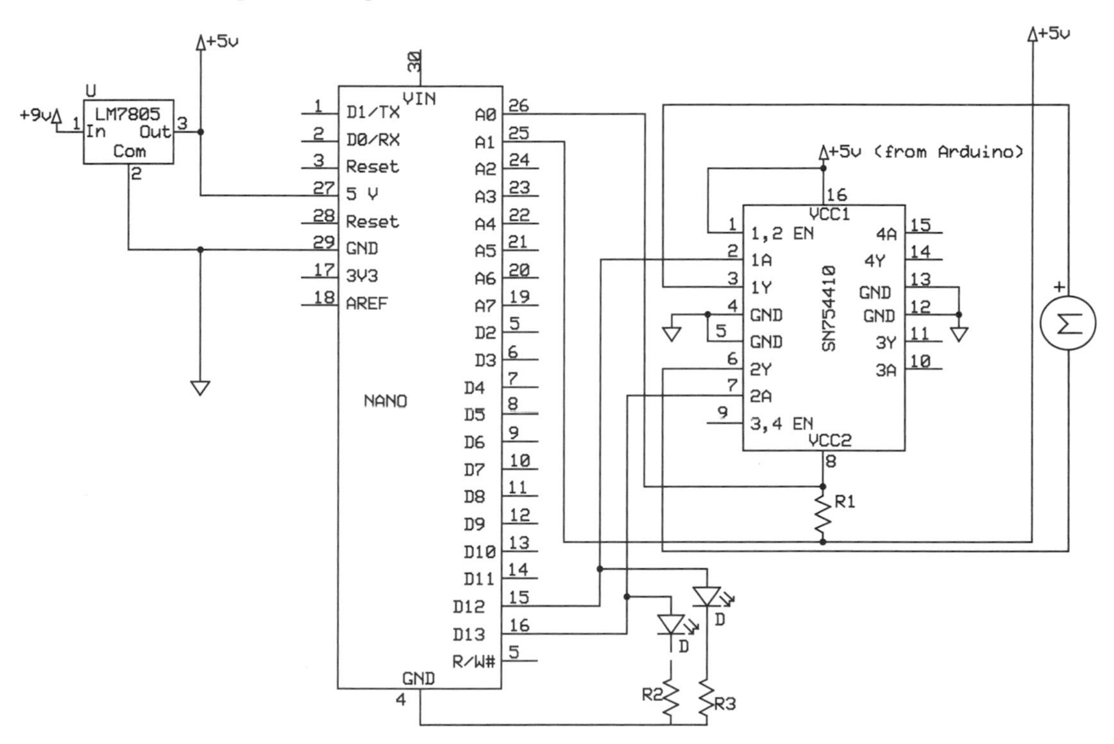

Figure 2-3: The completed schematic for this project shows the 5.6-ohm voltage-drop resistor (R1), the two LEDs (D), the 330-ohm current-limiting resistors (R2 and R3), and the quad H-bridge (SN754410), of which half is used.

All grounds in this circuit are connected together, and the voltage across pins A0 and A1 is the voltage your program will use to decide when to reverse the motor's direction. Note that this voltage is not referenced to either the positive or negative rail, but it must be between 0 and 5V to prevent damage to the microcontroller. If you get stuck on wiring the H-Bridge, see "Using an H-Bridge" on page 48.

The analog-to-digital converter (ADC) behind each analog pin provides 10 bits of resolution, which means the converter can deliver up to 1,024—that is, 210—different values, from 0 to 1,023, depending on the input.

Thus, if the power supply is 5V, each increment is roughly

$$5\text{V} \div 1023 \approx 0.0048\text{V} \,.$$

Determining the Reversal Threshold

In order to write a program that tells the Arduino when to reverse your motor, you have to determine that point yourself, with some math and a little bit of faith.

First, determine the current drain of the motor you're using. It's usually printed on the motor's label. The motor I used has a current drain of about 40 milliamps (mA), or 40 thousandths of an ampere (see Figure 2-4). Now we get into the heavy math. You're going to have to use a formula known as *Ohm's law* to determine the voltage threshold to set in the sketch.

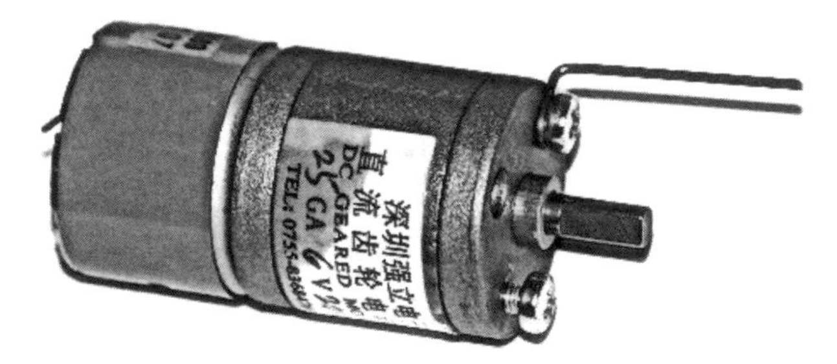

Figure 2-4: I used an Asian import motor, shown here with one limit pin installed, that has demonstrated reliability and performance. The screws are M3×0.05.

I used a 5.6-ohm resistor in series with my motor circuit. Using Ohm's law, which states that voltage equals current times resistance ($V = IR$, with voltage in volts, current in amperes, and resistance in ohms), we're able to calculate that 40 mA times the resistance of 5.6 ohm is about 0.224V:

$$\frac{40\ \text{A}}{1000} \times 5.6\ \Omega = 0.224\text{V}$$

Now, go back to the ADC. It has 1,024 units to represent 5V, so each unit represents 0.0049V. A little arithmetic reveals that the 0.224V dropped represents about 46 units out of the 1,024:

$$\frac{0.224\text{V}}{0.0049\text{V per unit}} = 45.71\ \text{units}$$

There are some estimates you have to take on faith—at least until you confirm with a test. This is one. As a motor is slowed or stalled, the current drain increases. Depending on the motor, the increase in current is typically somewhere between two and four times the normal current drain, but possibly more.

NOTE *With no load (or minimal load), current drain on the motor is minimal. With a usual running load, current can be four to five times the no-load current. With a heavy load, current can be as much as 10 times that, depending on the motor design.*

So according to our good-faith model, a good place to start setting the threshold for reversing the motor would be in the area of 90 to 100 units of the ADC's 1,024 units.

Alternatively, you could use a digital multimeter to measure the exact current drain first (see Figure 2-5). To use a multimeter to measure current drain, set its indicator to 200 mA to start; you may need to set it as high as 10 A if the motor doesn't move when you build the circuit described here.

Figure 2-5: Multimeters are handy for many projects and useful to have around the house. They're available from a variety of sources at a range of prices. I use this cheap one from Electronic Goldmine, but if you plan to do high-voltage experiments, invest in a really good multimeter.

Build the circuit as shown in Figure 2-6, and then connect the red lead of the multimeter to the power supply. Connect the black lead of the multimeter to the motor lead to complete the circuit. If the reading is negative, reverse the red and black leads of the multimeter. Depending on your power supply voltage and the motor's voltage requirement, you may also need to connect the motor to power through a voltage regulator circuit, as described in "The Voltage Regulator" on page 58.

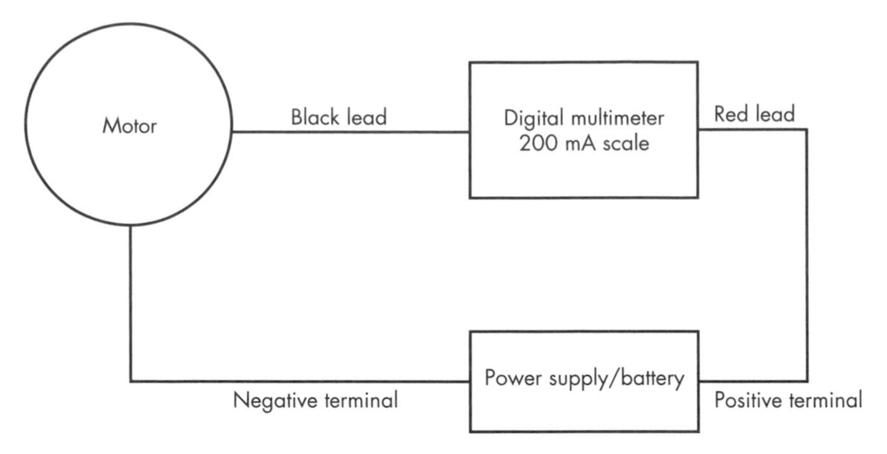

Figure 2-6: Connection diagram for measuring the current drain of the motor

To check the current drain, hold the shaft of the motor to slow it, and watch the readout on the multimeter. You can get an accurate indication of the number of ADC steps by plugging your readout in to Ohm's law, calculating the voltage, and converting into steps, as I did.

NOTE *In the sketch, I use a value of 100 as the threshold for reversing. You could also calculate the absolute value of the voltage drop by multiplying 100 by 0.0049V:*

$$100 \text{ steps} \times 0.0049\text{V per step} = 0.49\text{V}$$

Remember, the exact threshold depends on the type of motor you use. Different motors will have different current capabilities and may even require a different value resistor. Also, note that the value of current drain is not precise. The nature of permanent magnet motors is such that the current drain under load will be a range, not an exact number.

As the current increases, the voltage drop increases until it reaches the point where the microcontroller is instructed to do something. At that point, the difference in analog voltage that appears between A0 and A1 is above the preset threshold, which will set the Arduino into action. Once the threshold is reached, the Arduino tells the H-bridge to reverse the current to the motor.

Using an H-Bridge

You'll likely encounter an H-bridge driver in future projects because it's a very versatile part and can serve numerous functions. There is quite a selection of H-bridge chips available, but I've been using the Texas Instruments SN754410 quad H-bridge. It's popular because it operates over a wide voltage range and is extremely flexible—and inexpensive. The logic operates at a 5V level, while the drive can be as much as 36V with a continuous output of 1 A (and a peak output of 2 A), making it capable of driving a wide variety of hobby motors, solenoids, and even relays. It comes in a standard 16-pin dual inline package (DIP). The DIP package was a longtime standard but is slowly being replaced by newer types (see "Using SOICs" on page 20). It's the conventional centipede-looking circuit.

Figure 2-7 shows the pinout for the SN754410 H-bridge, and Table 2-1 shows its function table. You'll find more information in Texas Instruments' data sheet at *http://www.ti.com/lit/ds/slrs007b/slrs007b.pdf.*

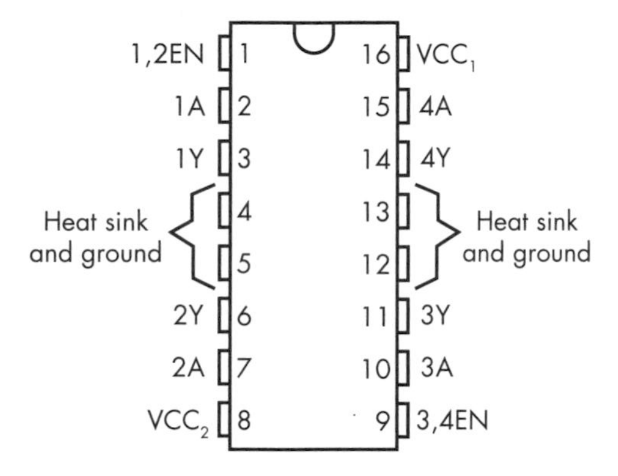

Figure 2-7: The pinout for the SN754410 quad H-bridge chip used in this project. Note that pin 1 is in the top-left corner of the chip when viewed from the top with the notch pointing up.

Table 2-1: Function Table for the SN754410

Inputs		**Output (Y)**
A	**EN**	
H	H	H
L	H	L
X	L	Z

According to the data sheet, in this function table, *H* stands for *high level*, *L* stands for *low level*, *X* means the level is irrelevant to the circuit behavior, and *Z* indicates high impedance, which turns the motor off.

The H-bridge is an elegant motor-control solution for several reasons. It allows you to reverse the polarity from a single supply, and it provides for different logic and control voltages. In addition, if both inputs of the dual H-bridge are either high or low, there will be no output. The sketch takes advantage of that in a function written to stop the motor. Other projects in this volume also use this capability.

The Breadboard

For most Arduino projects, I suggest building the circuit on a breadboard first to make sure you're going in the right direction and to prove your initial hypothesis. Use a standard breadboard and the plug-in wires that are sold as accessories for the breadboard (see Figure 2-8).

Figure 2-8: Typical small breadboard and plug-in wires

Before you begin building the circuit on the breadboard, look over your Arduino. Many Arduino boards come complete with the male headers already soldered in place. However, that's not always the case; some Asian suppliers include the headers loose with the processor board. If your board lacks headers, see "Preparing the Arduino Board" on page 2 for complete instructions on attaching them.

Most breadboards include a red and blue stripe on the entire length of each side of the board; the holes next to these stripes are used for power (+) and ground (–), respectively. Before you hook up the circuit, use a wire to connect the red column on the right to the red column on the left. Connect the blue columns to each other, too.

WARNING *Do* not *connect the red column to the blue column! This will cause a short circuit and will burn out the electronics.*

Figure 2-9 shows my breadboard for this project, and the schematic from Figure 2-3 lays out the connections.

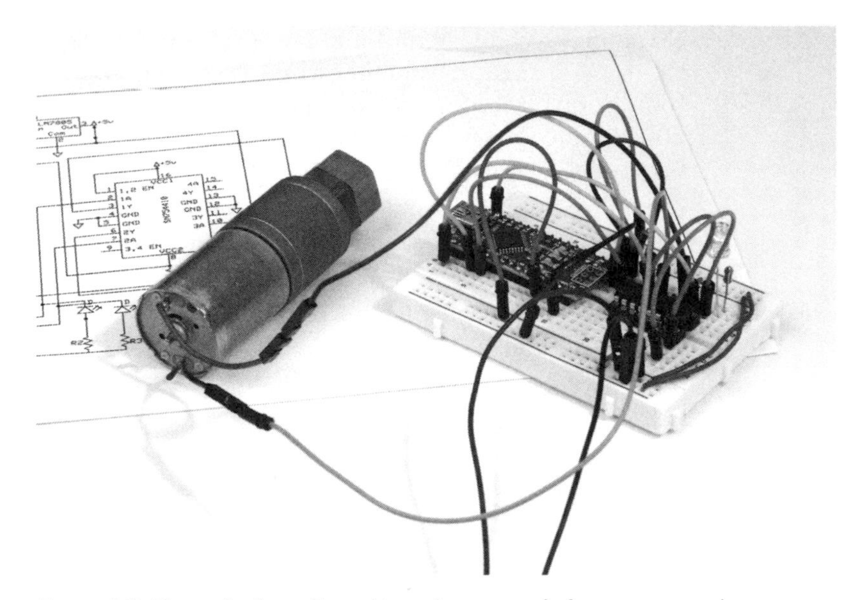

Figure 2-9: This is the breadboard I used as a proof-of-concept to make sure everything worked as anticipated.

WARNING *Don't plug the Arduino in to the computer while it is actually receiving power from the voltage regulator. This could burn out the Arduino.*

I suggest prototyping your circuit as follows:

1. Insert the Nano board into the breadboard, leaving a couple of rows of holes at one end.
2. Place a wire from the pin labeled *5V* on the Nano (pin 27) to the red positive rail on the breadboard.
3. Place a wire from GND on the Nano (pin 29) to the blue negative rail on the breadboard.
4. Find three consecutive holes on the board where they will not connect to anything and insert the three leads of the LM7805 into them.
5. The input lead of the LM7805 will go to the 9V power supply, the ground of the LM7805 will go to the blue negative rail, and the output of the chip will go to the red positive rail. (See Figure 2-10 for the LM7805 pinout.)

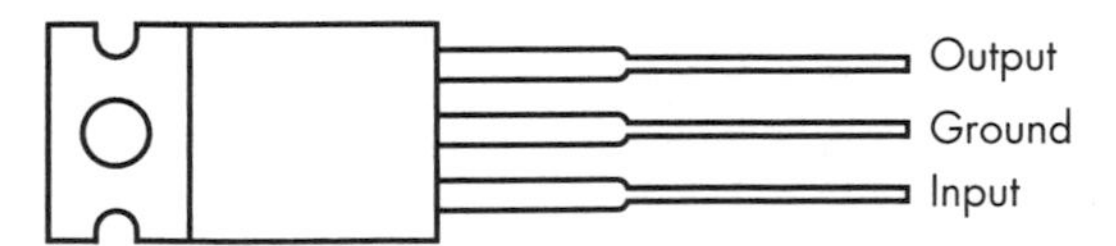

Figure 2-10: Pinout of the LM7805 regulator

6. Insert the H-bridge into the breadboard with the notch facing the Nano, and leave a couple of rows between the H-bridge and the Nano.

7. Use a wire to connect pin 1 and pin 16 of the H-bridge together (see Figure 2-7). Then, use another wire to connect pin 1 to the positive connection on the breadboard. This connection from pins 1 and 16 provides the voltage to run the logic on the H-bridge and also to enable the section of the H-bridge used.
8. Use a wire to connect pins 4 and 5 of the H-bridge, and then connect them to the negative terminal on the breadboard. Running a wire from either pin 4 or pin 5 to ground will do the trick.
9. Similarly, connect pins 12 and 13 of the H-bridge together, and connect them to ground.
10. Use a wire to connect one side of the motor (it doesn't matter which) to pin 3 of the H-bridge, and connect pin 6 of the H-bridge to the other side of the motor.
11. Connect digital pin D12 of the Nano to pin 2 of the H-bridge.
12. Connect digital pin D13 of the Nano to pin 7 of the H-bridge.
13. Connect one side of the 5.6-ohm resistor (R1) to pin 8 of the H-bridge.
14. Connect the other side of resistor R1 to the red positive rail on the breadboard.
15. Insert a wire from pin 8 of the H-bridge to analog pin A0 of the Nano.
16. Insert a wire from the positive (red) connector to analog pin A1 of the Nano.
17. Insert the positive side (long lead) of one LED to D12 of the Nano.
18. Insert the negative side of the LED into an empty row on the breadboard.
19. From that row with the negative side of the LED, connect a 300-ohm resistor (R2) to the blue negative rail.
20. Insert the positive side (long lead) of the second LED to D13 of the Nano.
21. Insert the negative side of the second LED into an empty row on the breadboard.
22. From that row with the negative side of the second LED, connect a 330-ohm resistor (R3) to the blue negative rail.

The VCC2 supply drives the output to the motor. It goes from the positive side of the supply—the output pin of the regulator in the schematic—through resistor R1 to pin 8 of the H-bridge. VCC2 becomes the low-voltage side of resistor R1; it will have a lower voltage as the load on the motor increases because the other end of the resistor is attached to the positive of the power supply. The VCC2 supply voltage can be anywhere from the 5V that the logic uses to the 36V limit of the H-bridge. For this project, I simply tied the voltage-drop resistor directly to the 5V supply, which worked well with a 6V motor.

The Nano's D12 and D13 output pins drive the A inputs of the H-bridge, while A0 and A1 inputs straddle the voltage-drop resistor, R1.

It's this voltage-drop value that tells the Arduino to change the outputs to instruct the H-bridge to reverse the motor. When output D13 is high and D12 is low, output pin 2Y on the H-bridge becomes positive while 1Y remains negative. When D12 is high and D13 is low, the reverse happens, and 1Y becomes positive while 2Y stays negative. When both pins have high or low output, they are at the same potential (or voltage), and the motor is not driven. (Refer to the function table in the H-bridge chip's data sheet, or see Table 2-1.)

The Sketch

The following sketch is written so that when the motor reaches its limits in one direction, both outputs go low, and when it reaches its limits in the other direction, both outputs go high. When both outputs are either high or low, there is no potential across the motor and it is stopped for a specified delay time. After the delay is satisfied, the motor starts in the other direction. Because LEDs are wired to pins D12 and D13, you'll also get a visual indication. Both LEDs are illuminated when the motor pauses in one direction, and both LEDs are off whe the motor pauses in the other direction.

```
/* Sketch for the Automatic Motor Reversal Project
*/

//Identify pins that will not change
const int ledPin1 = 12; //LED1 in schematic
const int ledPin2 = 13; //LED2 in schematic
const int analog0 = A0;
const int analog1 = A1;
int analogValue0 = 0; //Identify variables for analog inputs
int analogValue1 = 0;
int analogdifference = 0;
int threshold = 100; //The threshold value calculated to stop the motor

int reading;
int state;
int previous = LOW;
int count = 0;
int numberstops = 250;
int time = 0; //The last time the motor reversed

//Amount of time to wait to get rid of the jitters when the motor reverses
int debounce = 400;

❶ void setup() { //This is the setup routine
//Initializes pins as input or output
  pinMode(analog0, INPUT);
  pinMode(analog1, INPUT);
  pinMode(ledPin1, OUTPUT);
  pinMode(ledPin2, OUTPUT);
```

```
  Serial.begin(9600); //Was used in setting up the parameters
}

❷ void loop() { //This begins the processing section
  //Enter an endless do-nothing loop after the counter reaches the limit
  while(count > numberstops) {
    digitalWrite(ledPin1, LOW);
    digitalWrite(ledPin2, LOW);
  }

  analogValue0 = (analogRead(analog0)); //Read the analog values
  analogValue1 = (analogRead(analog1));

❸ //Setting up the analog difference
  analogdifference = analogValue1 - analogValue0; //This is the voltage drop
  //analogValue1 will be greater than analogValue0

  //These were added to view what was happening on the serial monitor
  Serial.print("count =    ");
  Serial.println(count);
  Serial.print("analogdifference =      ");
  Serial.println(analogdifference);
  Serial.println();
  Serial.print("numberstops =     ");
  Serial.println(numberstops);

  //This comparator looks at the difference or drop across the resistor
❹ if(analogdifference > threshold) {
    reading = HIGH;
  }
  else {
    reading = LOW;
  }

  //Toggles the output and includes the debounce
❺ if(reading == HIGH && previous == LOW && millis() - time > debounce) {
    if(state == HIGH) {
      state = LOW;
    }
    else {
      state = HIGH;
    }
    //Increments the counter each time the motor reverses
❻   count++;
    time = millis();
  }

  //Writes the state to the output pins that drive the H-Bridge
  digitalWrite(ledPin1, state);
  digitalWrite(ledPin2, !state);
```

```
  previous = reading;
}
```

This sketch sets up human-understandable aliases for the pins the project uses and adds convenient constants and variables for referencing analog inputs and other key values. After the sketch defines and initializes the input and output pins at ❶, it starts the main loop at ❷.

Inside the main loop, the sketch finds the voltage drop across the resistor in terms of analog steps ❸. At ❹, the sketch determines whether the reading was high or low. Threshold values from 100 to 120 work reliably for the 6V, 20 RPM motor I used, but you may need to experiment to find the right value for your motor. See "Determining the Reversal Threshold" on page 46 for more on how to estimate the threshold value. The reading at ❺ dictates whether to reverse the motor.

THE DROPPING RESISTOR IS KEY TO SENSING CURRENT

I've tried this reversing circuit with several similar motors, and I've only ever needed to make a slight adjustment to the threshold value in the sketch. But for a motor with extremely high or low current drain, you may need to anticipate a much different value for `analogdifference` and/or use a different dropping resistor, which was R1 in the schematic. You might need to reduce the value of the dropping resistor to something like 2.2 ohms, which then requires a reduction in the value you compare `analogdifference` to.

For most small motors, the lower the value of the dropping resistor—which is usually between 1 and 10 ohms—the better, as the analog difference tends to be more stable. For other motors, experiment to find the resistor value that works best.

When the sketch checks `reading` to see whether the motor needs reversing, it also uses the `debounce` value to assure that a high reading wasn't caused by electrical noise created by the motor's commutator or brushes during a legitimate reversal. I set `debounce` to 400, but you may have to adjust that for different motors. For larger motors specifically, this may need to be set a little higher.

This sketch also includes a few functions that aren't strictly necessary to reversing the motor but are helpful when using the motor as a PCB agitator. These aspects of the project may appeal to you in other applications, too, so let's look at them in more detail.

One of the things that I added was a counter to track the number of times that the motor reversed. In the sketch, the count increment appears at ❻ as `count++`. In the project, when a certain value of `count` is reached, the motor stops (if `count = numberstops`). If you wanted to set off an alarm, such as an audible noisemaker, to tell you it's finished, that can easily be accomplished by adding a line to write to one of the digital outputs. I set a maximum `count` value in the sketch, using `numberstops = 250`, so the motor will reverse 250 times and then stop. That provides a little more than 15 minutes of etching time with the motor I've selected running at 5V, which should be enough to etch most circuit boards.

When the maximum count is reached, the sketch enters the `while` loop at the beginning, stopping the agitation. This basically stalls the processor, and you have to hit the power switch to restart, or reset, the agitator. The placement of this loop near the beginning of the software is just a reminder that it's there.

The thinking behind the count, optional alarm, and stop capabilities is that a reminder to check on your board is helpful. If the board has completed etching, continued agitation would speed undercutting of the traces, which is not a good thing because it weakens (and can break!) small copper traces. On the other hand, if it fails to etch in a reasonable time, you might need to refresh the etchant.

MOD: ADJUSTABLE STOP AMOUNT

If setting a fixed stop maximum in a sketch doesn't leave you satisfied, try connecting a potentiometer between power and ground with the adjust pin, which is usually the center pin on the potentiometer, to the A2 input pin of the Arduino. Then, set `numberstops` equal to the value of A2, which should range from 0 to 1,023, depending on the position of the potentiometer wiper.

Here's how the sketch would differ. First, change

```
int numberstops = 250;
```

to

```
int numberstops = setNumber;
```

Then, add the following:

```
int setNumber;
int analogPin2 = A2;
int analogValue2;
setNumber = analogRead(analogPin2);
```

Because the timing is relative, you could use a 270-degree rotation linear potentiometer and make some rough markings on the enclosure to indicate the number of counts.

The Shield

For this project, I recommend making a small PCB *shield*, which is basically a host board designed to plug into the Arduino Nano. With a shield, your motor reversal project can remain compact, and you can design and build it with a minimum of effort.

PCB Layout

You could just solder the parts for your project directly to a piece of perforated project board, but I believe creating and populating the shield takes less time than putting the parts on a perforated board and wiring them by hand. You'll also gain invaluable experience by preparing, etching, drilling, and assembling your own PCB. And in the end, some projects are complex enough that wiring by hand just won't be an attractive option. (See Figure 5-13 on page 148 for an example.)

To make my printed circuit layouts, I use a free software program called ExpressPCB. Figure 2-11 shows my layout of the PCB.

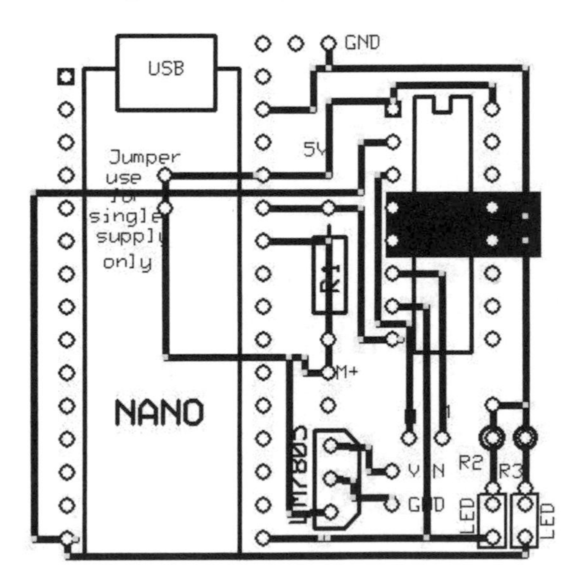

Figure 2-11: This is the actual PCB pattern I used in the project. The Arduino Nano can be soldered directly to the board or can plug in if you use header connectors.

If you don't want to lay out your own PCB but still want to make the board, download the *Reverse.pcb* file from *https://www.nostarch.com/arduinoplayground/* and follow the directions in "Making Your Own PCBs" on page 13. When you've made your PCB, just solder all the components to it in the right places, and you'll be done with the shield.

Shield Design Notes

If you lay out your own shield, there are a few design factors you should definitely keep in mind.

Analog Inputs

Be certain to connect the A1 and A0 inputs to the correct sides of resistor R1, according to the schematic in Figure 2-3. A1 should attach to the power supply side and A0 to the H-bridge side. In the sketch, to compare the analog values, we take the difference as `analogdifference = analogValue1 - analogValue0`, with `analogValue1` as the input at the high end of the resistor. In this case, `analogValue0` is A0, and `analogValue1` is A1.

Grounding and Heat Sink

Pins 4, 5, 12, and 13 are ground on the H-bridge, and they are also a heat sink to keep the chip from overheating. A small area on the proposed shield is included to increase the heat sink area. If you're using a relatively small motor—such as the 6V, 20 mA unit—no more heat sinking is required. If you're using a much larger motor or driving a heavy load, consider using the second side of the PCB as a heat sink.

The Voltage Regulator

This project uses its own 5V regulator to supply power to the Nano. A 9V, 200 mA plug-in wall adapter is connected to the voltage regulator LM7805 on the shield, which reduces the voltage from about 9V to 5V. An external regulator is included so a more powerful regulator than the one built into the Nano can be used. Make sure to connect the pins of the regulator correctly (see Figure 2-10).

You could feed a 7.5V DC or 9V DC wall supply directly to the VIN pin of the Nano and use the onboard regulator, which worked with my motor. But if you use a larger motor—or higher-current LEDs—it might tax the onboard regulator and could conceivably burn it out.

The higher the voltage of the power supply, the more work the regulator has to do to bring it down to 5V. Overtaxing the regulator could cause it to heat up and fail. For example, feeding the regulator 12V is probably at the high end for 5V regulation. A 9V input is better, and a 7.5V input is better yet. If the regulator chip gets warm, add a heat sink to the tab. A small piece of aluminum is often sufficient, but a regular heat sink can be used. And while it's good to have the supply voltage as close to the output voltage as possible, remember that the regulator needs at least 1V above the regulated output to work, so it must be fed with at least 6V, which is a 5V-regulated output plus 1V. Input voltages above 12V are feasible, too, but just be sure not to exceed the limits of the device.

MOD: USING A HIGHER VOLTAGE

If you use a higher-voltage motor for this project, it will turn faster, have more torque, and so on. But you *can't* simply connect the higher voltage to the high end of the dropping resistor connected to pin 8 of the H-bridge. That would cause the voltage between both A0 and A1 and ground to exceed 5V, which is hazardous to the health of the ATmega328 microcontroller on the Arduino. (This is the only time that the voltage referenced to ground is important.) Thus, a modification is required. Look at R1 in the schematic in Figure 2-12. The supply first goes to resistor R2; R2 joins with resistor R3, which goes to ground.

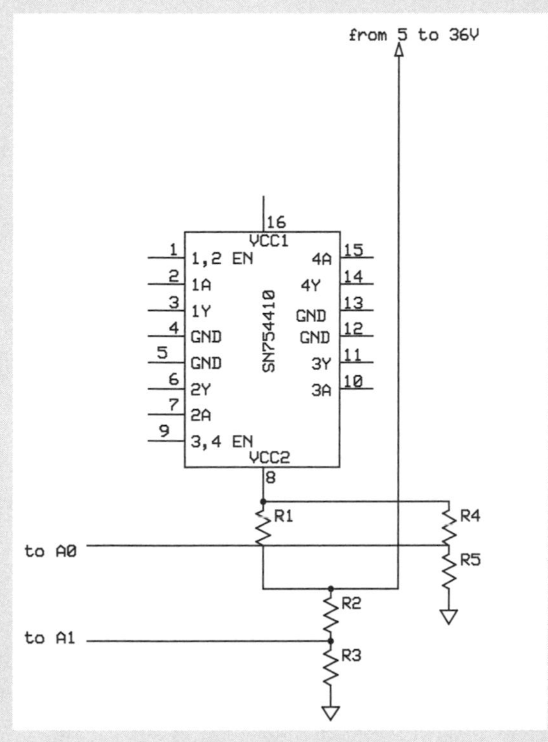

Figure 2-12: If you elect to use a higher voltage and drive a faster motor, you will have to modify the circuit by adding voltage dividers in front of both the A0 and A1 inputs.

To avoid damage to the Nano processor, you will want to keep the voltage that appears at that joining point under 5V, referenced to ground. The easiest way to do this is to use a voltage divider. Two resistor pairs divide the higher voltage: the first pair is R2 and R3; the second is R4 and R5. The value of these resistors should be such that the output at the joining of each pair—R1 and R2, and R4 and R5—is somewhat less than 5V for whatever value of input voltage you use.

(continued)

Use this formula:

$$V_{out} = V_{in} \times \frac{R2}{R1 + R2}$$

and the schematic in Figure 2-13 to determine the values of the resistors to use in a voltage-divider circuit.

For example, if you start with 9V and arbitrarily select a 10-kilohm resistor in series, you would have to shunt it with a 12.5-kilohm resistor to ground, according to the calculator. The closest resistor I had was 12 kilohm, and it worked fine. If you can't find a standard resistor to fit your needs, you can also combine two standard values in parallel to achieve the value you want with this formula:

$$R\ total = \frac{R1 \times R2}{R1 + R2}$$

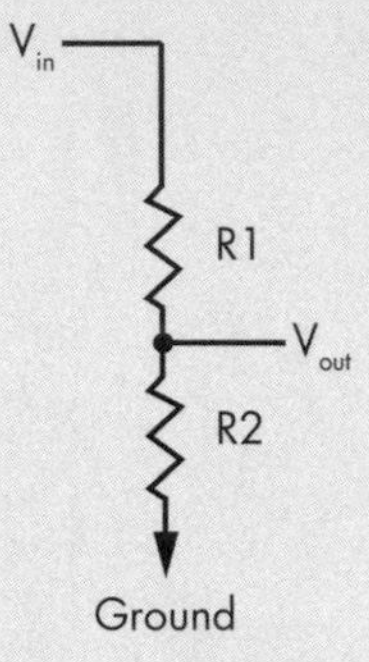

Figure 2-13: A basic voltage divider. To find the resistors you should use, plug the values from your own divider into the formula as if your divider were this circuit.

If you don't want to do the algebra yourself, you could use one of the convenient online voltage-divider calculators such as *http://www.sengpielaudio.com/calculator-paralresist.htm* or *http://www.raltron.com/cust/tools/voltage_divider.asp*. SparkFun also has an excellent tutorial on voltage dividing, with a calculator of its own: *http://learn.sparkfun.com/tutorials/voltage-dividers/*.

Directional LEDs

Of course, what Arduino project would be complete without blinking LEDs? As you'll see in the schematic and on the shield PCB, I included two LEDs: a red one for clockwise rotation and a green one for counterclockwise rotation. But which direction belongs to which LED is your choice: simply reverse the motor leads to change the LED status.

Construction

For this project, you'll use the motor-reverse technique to create an agitator that accelerates the etching of PCBs. To do this, you'll suspend a PCB from an Arduino-driven motor over etching solution, as shown in Figure 2-1. A small enclosure will contain the Arduino Nano, the shield, the motor with limit wires, direction LEDs, a power switch, and the power jack.

After assembling the box, you just have to mount it somewhere above your etching setup and attach the reverser, either directly to the PCB or to a tray. I clamped my box to a cabinet door above my workspace, with a place for the etching vessel below (see Figure 2-14). The entire system can be assembled and disassembled quickly.

Figure 2-14: For larger PCBs, try etching in a tray for a more conventional approach. Just attach the motor reverser to your tray to agitate the board rather than using the reverser to dip the board in and out of the solution.

Construction of the rest of this project takes a little bit of patience and perhaps some ingenuity in scavenging some of the parts required. You will need a couple of M3 screws to mount the motor to the motor plate—in this case, a small aluminum L bracket—and some limit wires, preferably made of 0.039 piano or spring wire. You'll also need a small block of scrap brass or aluminum—round or rectangular, doesn't matter—to attach to the motor shaft and crank, a long 4-40 or 6-32 screw to act as the crank, and an M3 spacer and solder lug to attach the agitator line to the crank. Figure 2-15 shows the nearly-finished, unmounted product.

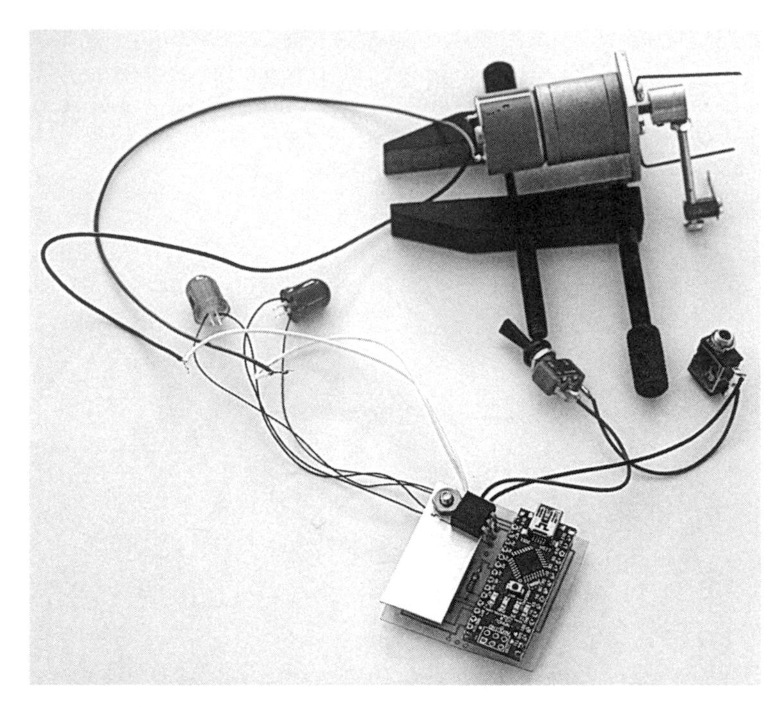

Figure 2-15: Wire up your components and lay them out for a final test before you put them in an enclosure. For the test, I held the motor in a clamp so the crank was free to move. The regulator heat sink obscures much of the shield.

The Limit Wires

The limit wires will create resistance to the motor's rotation by essentially bumping into the motor crank. The point in the rotation where they strike the crank is the limit of rotation. When the crank runs up against the limit wire, the wires prevent the motor from turning and initiate the reversal.

I recommend piano or spring wire to provide a little spring as the crank hits it at the extent of rotation. Use a pair of needle-nose pliers to bend two pieces of the limit wire into shape (see Figure 2-16). These wires will fit on the motor mount screws outside of the motor mounting bracket. You can change the limit of rotation by loosening the screw and rotating the wire.

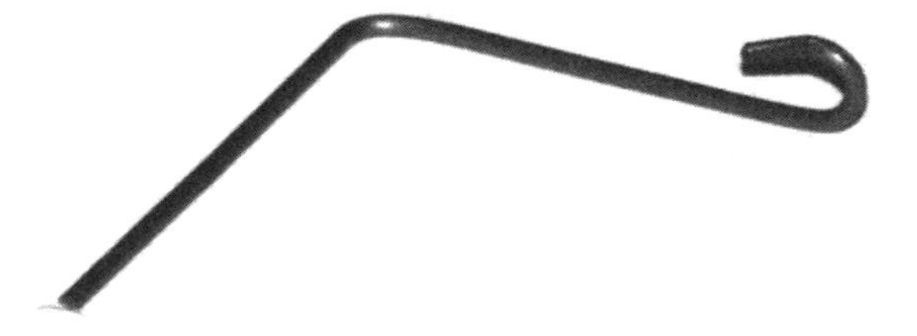

Figure 2-16: This is how the limit pins are formed. A good pair of needle-nose pliers does the trick.

The Crank Bushing

The crank bushing is simply what transfers the rotation of the motor to the crank. Figure 2-17 details the construction of the bushing, the spacer, and the solder lug.

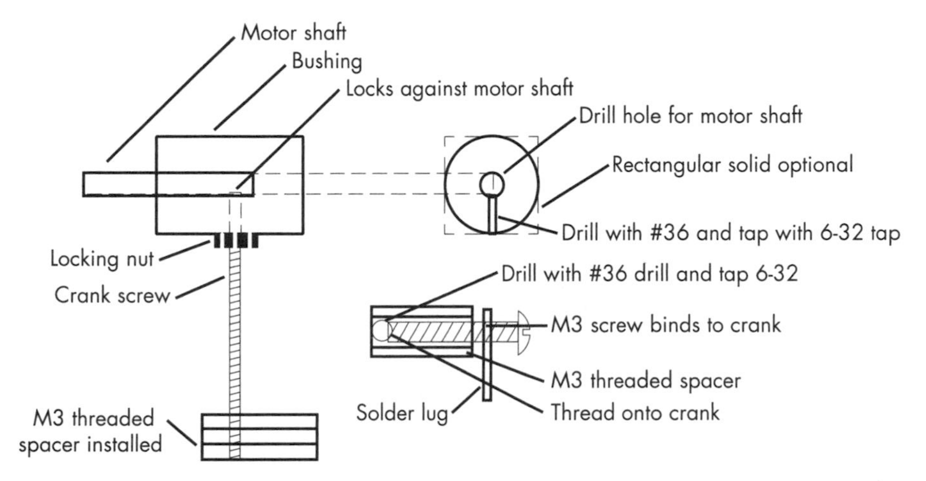

Figure 2-17: The detail of the drive mechanism that transfers the rotation of the motor to the lifting motion of the agitator

While there can be a number of different variations in your approach to assembling this part of the project, here's the sequence I used to put it together:

1. Drill a hole for the motor shaft through the center of the bushing, which can be a small piece of brass or aluminum round stock about 0.5 inches in diameter and 0.75 inches long. A rectangular piece will work just as well. Use a drill that is as close to the size of the motor shaft as possible. For example, if your motor shaft is 0.157 inches in diameter like the one I used, then a 11/64-inch drill bit is close enough. It isn't important to get the hole exactly on center—just close.
2. In the bushing, perpendicular to the motor shaft hole, use a #36 drill to drill a hole. Then, tap the hole you drilled so a long 6-32 screw can serve double duty as a setscrew and crank. You can also use a separate setscrew to move the crank farther from the motor, as I did in Figure 2-18.
3. Thread the crank screw into the bushing so it bears tightly against the motor shaft, and use a locking nut to hold the screw in place (see Figure 2-18).

Figure 2-18: A photograph detailing the head of the crank. Note the solder lug used to hold the wire and the alligator stop clip on the left side.

4. At the end of the crank, you are ultimately going to attach the line that will pull the PCB in and out of the etchant. This fitting can be just a nut, or even an alligator clip, attached to the crank. However, in the detail, I used an M3 hex female-female spacer that was 7 mm long. I drilled clean through the spacer to one side, starting on one of the flat surfaces with the same #36 drill. I then tapped the hole with the 6-32 tap and threaded it onto the crank.
5. Take an M3×0.5 mm machine screw and put it through the solder lug (see Figure 2-19 for the lug itself and Figure 2-18 for the lug in place). Screw it into the standoff all the way so it binds on the crank screw.

Figure 2-19: The solder lug used to hold the wire that holds the etching board. If you can't purchase something similar, you can easily make one with a piece of scrap metal or plastic.

My local Ace Hardware store had all of the accessories I needed, with the exception of the M3 spacer, which I got from eBay. You should be able to find the same items at Home Depot or Lowe's.

Packaging

The shield and Nano fit in a standard plastic enclosure (see Figure 2-20). Drill holes in the enclosure for the 3.5 mm power jack, the SPST switch that serves as a power switch and reset, the indicator LEDs, and the motor wires.

Figure 2-20: Completed enclosure with motor, limit wires, direction LEDs, power switch (reset), and power jack. The LEDs light up, with one for each direction. When the motor pauses in one direction, both LEDs turn on; when it pauses in the other direction, both LEDs turn off.

Most 3.5 mm jacks use approximately a 1/4-inch hole, which is the same sized hole as the switch. If you want a tight fit, 15/64 inches is closer. Whether you use a 5 mm or 10 mm LED will dictate the size of the holes required for those. It's been my experience that different brands tend to have slightly different diameters, so you might want to try a smaller drill first and test whether the LED fits. The arbitrary English-sized drill bits for the 5 mm and 10 mm LEDs are 3/8 inches and 3/16 inches, respectively. If you have a set of tapered reamers, you can start with a smaller hole and ream it out to make a tight fit for the LEDs.

Mount the motor on a small piece of aluminum angle, readily available at most hardware stores. I purchased a 1-inch section of 1.5×1.5–inch aluminum angle and cut it down to size with a hacksaw. If you're using the motor I use, you can copy the template in Figure 2-21 or download and print it from *https://www.nostarch.com/arduinoplayground/*, cut it out, tape it to the aluminum angle

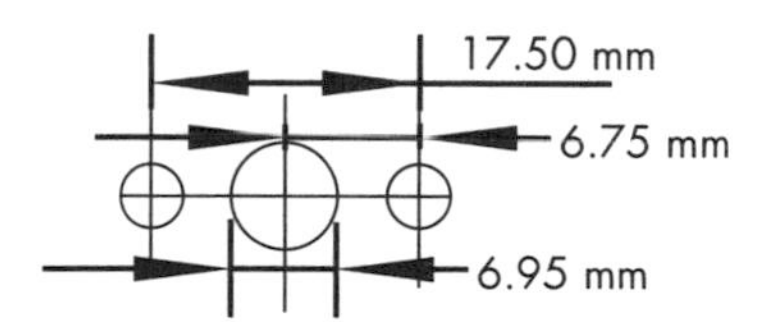

Figure 2-21: Template for the motor mount

bracket, and carefully mark the hole positions on the bracket with a center punch or nail. Now, drill the holes—1/8 inches for the motor mount and 5/16 inches for the center hole. If you use a different motor, you will have to measure and mark out the mounting holes.

Just use some double-sided foam tape to secure the shield to the enclosure if you think you'll want to use it in another project. Otherwise, attach it to the inside with standoffs and screws in any size you like.

The Etching Process

There are a number of techniques for making PCBs. The most common is a subtractive approach, which involves starting with a copper clad board, or a copper foil bonded to an electrically insulating substrate, from which the copper is selectively removed to leave a pattern on the board. While the copper can be mechanically milled off, the most common approach is to selectively etch the pattern on the board chemically.

In the chemical etching process, the circuit pattern is printed on the blank board with a chemical resist so that the copper is removed by the etchant in the areas not treated with the resist. The etchant is a chemically active material that attacks the untreated copper on the clad board, leaving you with only the copper you need for your circuit. I describe how to etch circuits step-by-step in "Making Your Own PCBs" on page 13, and this project makes that process easier.

Our goal is to suspend an unetched circuit board over the etchant in the vessel and keep it in the etchant for the maximum time as the agitator goes up and down, resulting in a laminar flow of etchant across the surface of the circuit board. I suggest using a nylon cable tie to hold the circuit board during the etching process, as nylon is relatively impervious to the etchant. You could attach the tie, in turn, to the motor shaft with an alligator clip so the board is easy to remove (see Figure 2-22).

I used a 250 mL beaker as an etching vessel. For very small boards, this works extremely well. For larger boards, I recommend a large measuring cup, such as a 2 qt Pyrex cup. A 600 mL beaker works for intermediate-sized boards. For even larger boards, you can use a tray, as illustrated in Figure 2-14.

The switch and power input are located on the left-hand side of the enclosure. To hold the board being etched, I suspended a wire through the solder lug and attached that wire to the board with a small alligator clip. On the back of the lug, you can either tie a small knot in the wire or attach a clip of some sort to make sure the wire doesn't fall through the lug and into the acid. In my setup, a clamp (behind the motor in the photo) holds the enclosure to an overhanging door.

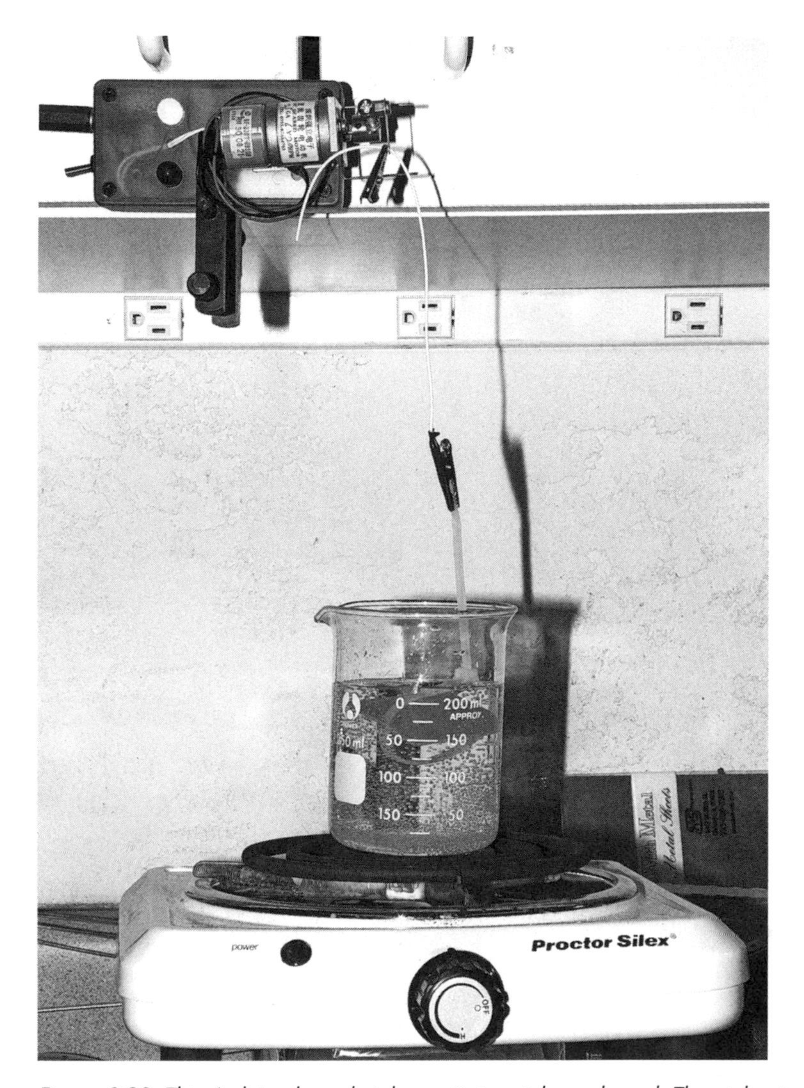

Figure 2-22: This Arduino-based etcher-agitator etches a board. The etchant should turn emerald as the copper is etched. The board is held by a wire tie that is attached to a wire by an alligator clip. The wire goes through a hole on the crank and is held in place with another alligator clip. One of the LEDs is lit.

Note that the etching vessel is sitting on a hot plate. Though etching will occur at room temperature, it's accelerated somewhat by heating. Be careful not to get the etchant too hot: if you set the hot plate on low to keep the liquid at about 100 to 120°F, it will speed etching without softening the resist.

3

THE REGULATED POWER SUPPLY

Whether you use a standard bench power supply or run your Arduino off the USB port of your computer, sooner or later you're going to need a stand-alone, regulated power supply capable of providing a variable voltage. This project shows you how to make exactly that, using only a handful of inexpensive parts. A variable power supply is one of the most frequently used tools in many workshops. This one is easy and fun to build, and you will find that you end up using it over and over again.

When set for 5V or 3.3V, the Regulated Power Supply can power most Arduino projects with ease. You can also use it to power some ancillary piece of equipment, to vary a particular voltage in a system while the main power is fixed, or simply to test a lamp circuit, LED, motor, or other device.

The circuit uses the extremely versatile LM317 regulator chip. If you ever find yourself in need of a precision voltage regulator with some unusual demands, look up the LM317 on the web. The JavaScript Electronic Notebook has a particularly good article titled "LM 317 Voltage Regulator Designer" by Martin E. Meserve, which can be found at *http://www.k7mem.com/Electronic_Notebook/power_supplies/lm317.html.*

Required Tools

Soldering iron and solder

Drill and drill bits (3/8 and 1/4 inches)

Hacksaw or keyhole saw (nibbler or other)

Phillips head screwdriver

(Optional) Tapered reamer set

(Optional) Crimping tool

Parts List

The Regulated Power Supply is capable of providing an adjustable voltage from 1.25V to about 12V at up to 1.5 A, depending on the fundamental power you use. It uses the LM317 single-chip voltage regulator to set the voltage. To build this project, you will need the following parts:

One Arduino Pro Mini or clone

One LM317 voltage regulator

One LM7805 voltage regulator

Two 2.2-ohm, 5 W resistors

Three 10-kilohm, 1/8 W resistors

Three 6.8-kilohm, 1/8 W resistors

One 68 μF tantalum capacitor

Three 0.1 μF ceramic capacitors

One 1 μF tantalum capacitor

One 16×2 LCD

One I^2C adapter, if not included with the LCD

Four 4-40 screws

Eight 4-40 nuts

One 5 mm LED (for power indicator)
One SPST switch
One 470-ohm, 1/8 W resistor
One 10-kilohm, 1/8 W potentiometer
One Hammond panel/case (#1456CE3WHBU)
Two banana plug jacks
One 3.5 mm jack
One 12V 2 A AC adapter
One power adapter jack
One PCB/shield
Six 1×4 headers
Four 1×4 housings
Four 1×2 headers
One heavy-duty TO-220 heat sink
One medium-duty TO-220 heat sink
Four male crimp connectors
Four female crimp connectors
30-gauge hookup wire
One knob to cover the potentiometer shaft
Double-sided foam tape

A note on heat-sink selection: there are a wide variety of heat sinks available. The one pictured in Figure 3-2 is the Futurlec TO220ST, which works okay but runs pretty warm as you approach the 1 A range. A larger one that will still fit in an enclosure may be better. Futurlec TO220SMAL, the heat sink for the LM7805, is sufficient for the job.

Downloads

You will find the following files in this book's online resources to help you complete this project:

Templates *PowerSupplyFront.dxf, PowerSupplyFrontBottom.dxf*
Sketch *PowerSupply.ino*
Shield file *VoltageRegulator.pcb*

WHAT THE REGULATED POWER SUPPLY IS AND ISN'T

The power supply in this project is not intended to replace a regular bench power supply. It does not provide any current limiting and is rated up to 1.5 A, due to the current capacity of the LM317. However, for a wide variety of applications—including all of the projects in this book—it works very well. If you're new to Arduino, it will provide a solid power supply and save you a lot of batteries, if that's how you've been powering your projects. If you already have a full-sized bench supply, this project will be indispensable as a second supply. The Regulated Power Supply can be used for several projects in this book that require a secondary power supply.

I have used the Regulated Power Supply in other applications, and at the end of this project, I'll illustrate a more simplified version that can be used as a remote supply. There are times when you just don't have the proper voltage supply for a project, and this build fits the bill.

A Flexible Voltage Regulator Circuit

The basic LM317 regulator circuit in Figure 3-1 is the heart of this voltage regulator. Though it is relatively rudimentary, the chip's simplicity belies what a powerful and versatile tool it can be.

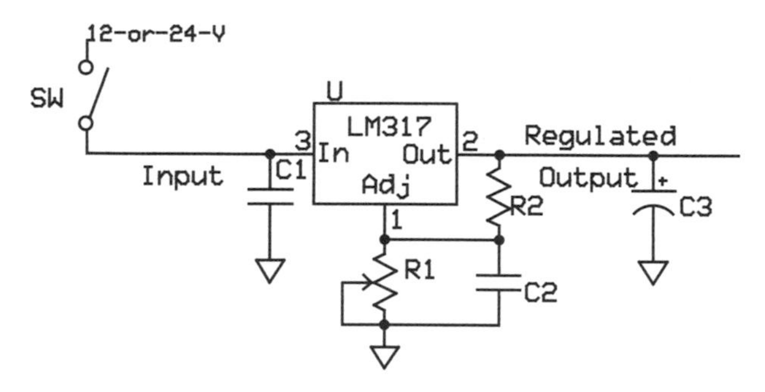

Figure 3-1: This is the schematic diagram for the regulator component of the Regulated Power Supply. The complete schematic with the display is shown in Figure 3-4.

I have used some variation of this circuit for many applications, from a stand-alone variable supply to an integrated part of a larger system, always with good results.

In the shop, I sometimes use kind of a "hair-wired" version, shown in Figure 3-2 (left), for testing LEDs and controlling things like motor speed and lamp intensity. I've also used a breadboard version, shown in Figure 3-2 (right), to implement the regulator circuit. In both cases, I used a trimmer potentiometer, which requires an alignment tool or screwdriver to make adjustments.

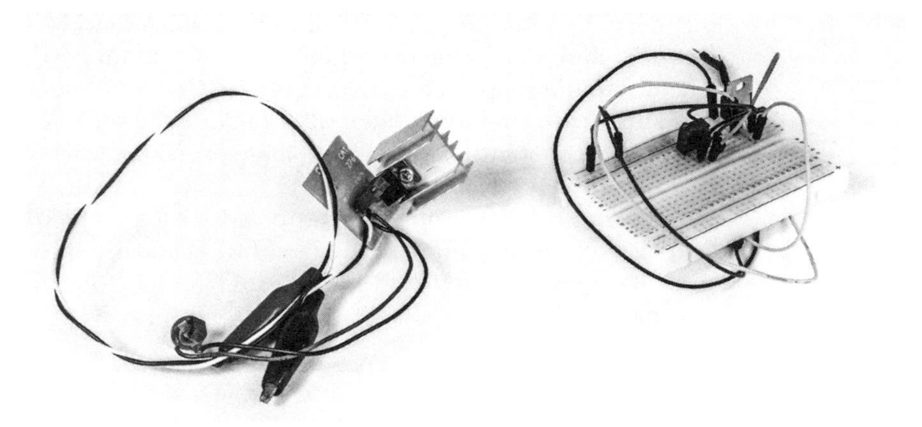

Figure 3-2: The "hair-wired" version of the voltage regulator circuit with a heat sink screwed to the LM317 (left) and a breadboard version of the regulator (right). I used this in an application with minimal power requirements and thus did not include the heat sink.

While those regulator circuits worked, one reason for using the more refined Regulated Power Supply format in Figure 3-1 was to eliminate the awkward trimmer potentiometer and instead use a standard 270-degree potentiometer and knob so that I could make adjustments quickly, easily, and repeatedly. But the main reason I built it was to have a secondary variable-voltage power supply with digital readout readily available on the bench.

How the Circuit Works

The circuit for this project is not overly complex. In essence, it measures the voltage at the output of the LM317 regulator using the onboard analog-to-digital converter (ADC) and compares it to an internal reference voltage. The result is sent to the LCD screen. However, the Arduino Pro Mini 5V version can accept a maximum of only 5V at the analog inputs. We therefore use a *voltage divider* to make sure that the voltage at the analog input pin doesn't exceed 5V. (Make sure voltage supplied to the LM317 is no greater than 12V.)

In this case, a voltage divider comprises two resistors connected in series, straddling the output of the LM317 and the ground rail of the breadboard (see the schematic in Figure 3-4). The voltage coming from the LM317 gets divided across the two resistors, R2 and R3. As you will note in the sketch, converting from the divided voltage back to the original levels for the display is simply a matter of reversing the arithmetic.

As shown in the schematic in Figure 3-4, the output of the LM317 connects to both the R2-R3 voltage divider and to resistor R1. Together, resistor R1 in parallel with R9 and the load, or whatever you want to power with the power supply, can be seen as another voltage divider. The R1-R9 voltage divider has a resistance of only 1.1 ohms, so the voltage drop across it is going to be relatively small. According to Ohm's law ($I = V/R$) the voltage across R1 and R9 is going to be 1.65V for a maximum current drain of 1.5 A (the maximum supported by the regulator IC): 1.5 A = 1.65V/1.1 Ω.

This means that when the LM317 provides about 12V and the load draws 1.5 A, there will be a 1.65V voltage drop across R1 and R9, leaving 10.45V at the power supply's output.

Looking at the schematic in Figure 3-4, we are comparing the voltage at analog inputs A0 and A2 of the Pro Mini. If you use the same values for the voltage divider for A0 and A1, you can eliminate one of the sets of resistors and simply connect A0 to A1.

VOLTAGE DIVIDER RESISTOR VALUES

The values of the resistors needed to achieve a certain voltage are determined using this formula:

$$V_{out} = V_{in} \times \frac{R2}{R1 + R2}$$

The schematic of a typical voltage divider circuit is shown in Figure 3-3.

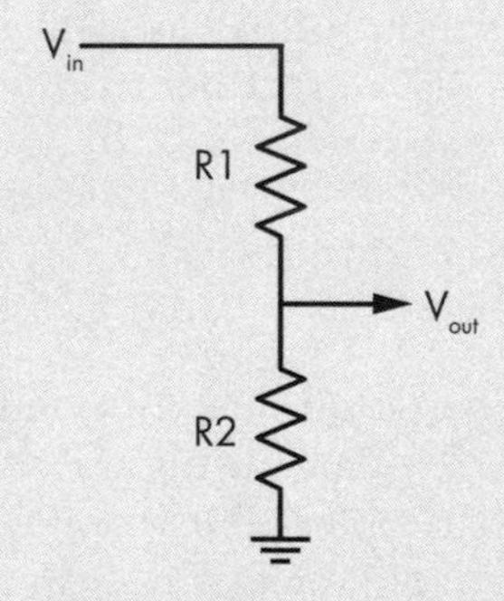

Figure 3-3: A typical voltage divider circuit

You can do the algebra if you'd like, but it's easiest to use an online calculator, such as the one at *http://www.daycounter.com/Calculators/VoltageDivider-Calculator.phtml*. In this project, the objective is to achieve an output of around 5V. We'll start with a 12V input and a 10-kilohm resistor, represented by R1 in the formula and marked R2 in the schematic. Fill in the calculator fields with this information, and the formula will give you a resistor value of 7.1-kilohm for R2 in the schematic and R1 in the formula. The closest standard resistor value is 6.8 kilohm, so the project uses that along with the 10-kilohm resistor in its voltage divider.

But why start with a 10-kilohm resistor? The first reason is to avoid drawing too much current. Even if the entire 12V dropped across the 10-kilohm resistor, it would result only in a nominal drain of 1.2 mA. Second, I have a lot of 10-kilohm resistors in the parts bin, and I am sure you do, too.

I use three sets of voltage dividers in this circuit. The first looks at the voltage at the output of the regulator, which is ultimately displayed on the LCD. The other two divide the voltage in front of and behind the voltage-dropping resistor so that the amperage can be measured according to the formula $I = V/R$, where V is the voltage drop across resistors R1 and R9, and R is the value of those two resistors combined. Could I have eliminated one set of voltage dividers? Yes, by joining A0 and A1 together. I thought, however, that I might want to change those values at some point to increase the accuracy of the ammeter by bringing the value closer to the Arduino reference voltage, so I did not join them in my version of the project.

The Schematic

While I wanted the Regulated Power Supply to be relatively robust, I didn't want it to be overly complex or hard to build. The hair-wired and breadboard versions did well in temporary or emergency applications when used with a digital multimeter (DMM), but I sought to build something more permanent that would have its own voltage and current readout, and that would stay on the workbench or sit on my desk as a regular addition to the tool set. Figure 3-4 shows the full schematic for the Regulated Power Supply.

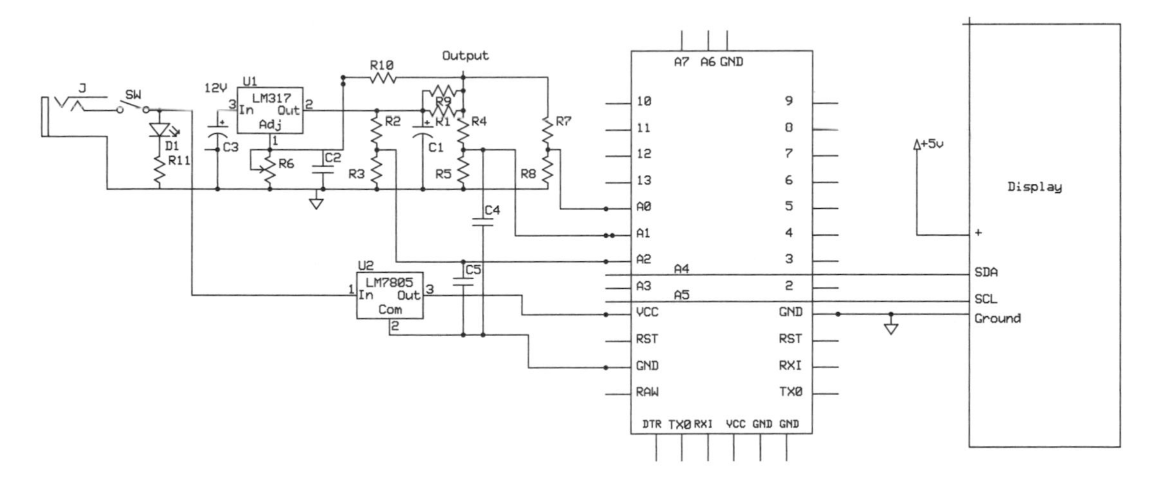

Arduino Pro Mini
16-2 LCD/I2C display
U1: LM317
U2: LM7805
R1: 1.2-ohm, 5 W resistor (two 2.2-ohm resistors in parallel)
R2, R4, R7: 10-kilohm, 1/8 W resistor
R3, R5, R8: 6.8-kilohm, 1/8 W resistor
R6: 470-ohm, 1/8 W resistor
R7: 10-kilohm, 1/8 W potentiometer
C1: 68 µF tantalum capacitor
C2: 1 µF tantalum capacitor
C3, C4, C5: 0.1 µF ceramic capacitor
SW: SPST switch
D1: LED
R11: 470-ohm, 1/8 W resistor

Figure 3-4: Schematic for the Regulated Power Supply

The Breadboard

As in all of my Arduino projects, I began with the standard breadboard. To make life easy, I used a standard potentiometer with pins that would fit into the 0.100-inch-spaced breadboard holes. With a little effort, a standard 16 mm rotary potentiometer (R7 in the schematic) with printed circuit board connectors will just about fit into every other hole in a breadboard. Figure 3-5 shows an overhead view of the finished breadboard before you power it.

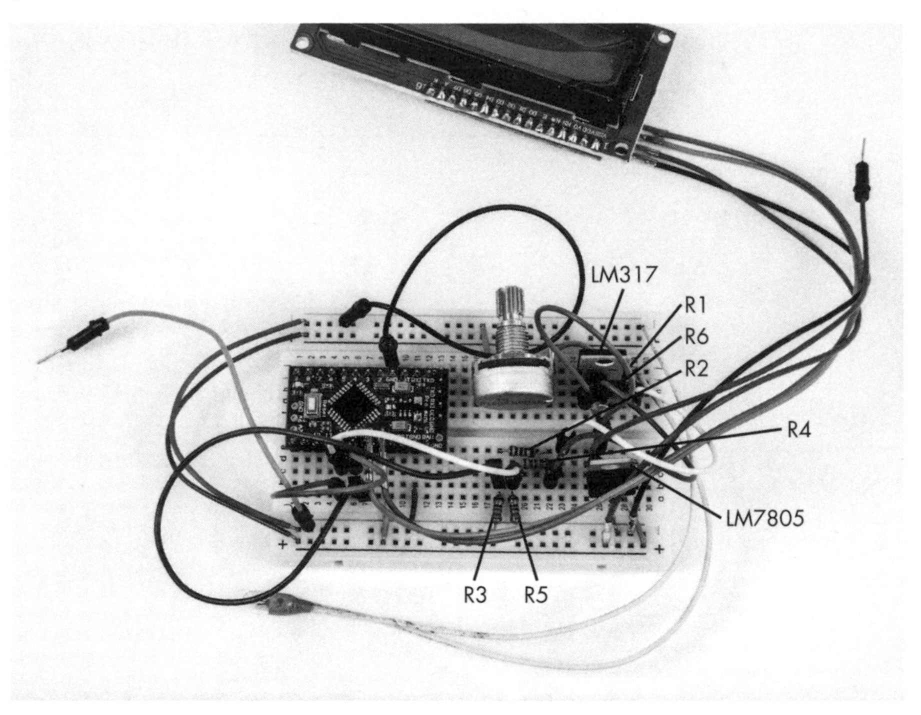

Figure 3-5: The breadboard for the Regulated Power Supply. The capacitors in the schematic—C1, C2, and C3—are not included in the breadboard but should be included in the completed unit.

Preparing the Arduino Pro Mini and LCD

The Arduino Pro Mini may or may not come with the male headers attached. If it doesn't, you'll have to solder them yourself (see "Preparing the Arduino Board" on page 2). Make sure the number of header pins in your strip matches the corresponding holes in the Pro Mini; you may have to cut the strip to the proper number of pins if the included strip is too long. Trim two strips of headers to size and place the long ends of the two header strips into a breadboard, spaced so that the Pro Mini board will fit over them. Put the Pro Mini in place, and solder all the header pins. Then, take two header pins (use the surplus from the longer header or purchase these separately), insert them in the A4 and A5 holes in the Pro Mini, and solder. These are the pins used for the LCD.

Finally, install five header pins on the edge of the board (at the TX0 and RXI end). Some boards come with straight headers, others with the long pins bent at a 90 degree angle. In most of the applications, I have found it easier to work with straight headers. You can use right-angle headers, but it may be more difficult to plug in the connector for programming the board, so I recommend replacing any right-angle pins with straight ones. You also might want to take a 1/2-inch length of 22-gauge wire and solder it to the short end of the RST pin so it sticks up. A female header connector will connect to this during programming.

You will now have to get the LCD/I^2C assembly ready. If you purchased the display and adapter separately, you will have to assemble them. Go to "Affixing the I^2C Board to the LCD" on page 3 for instructions. If you purchased the display with the I^2C adapter, it's ready for assembly.

Building the Breadboard

Here's the step-by-step guide to putting together the breadboard:

1. Insert wires to connect the two positive (red) rails together.
2. Insert wires to connect the two negative (blue) rails together.

WARNING *Be careful not to cross the two and connect the positive rails to negative rails. That could cause a short circuit and damage the hardware.*

3. Insert the 10-kilohm rotary potentiometer into the breadboard.
4. Insert the LM317, with or without heat sink attached, near the potentiometer, as shown in Figure 3-5. (See Figure 3-6 for the pinout of the LM317.)

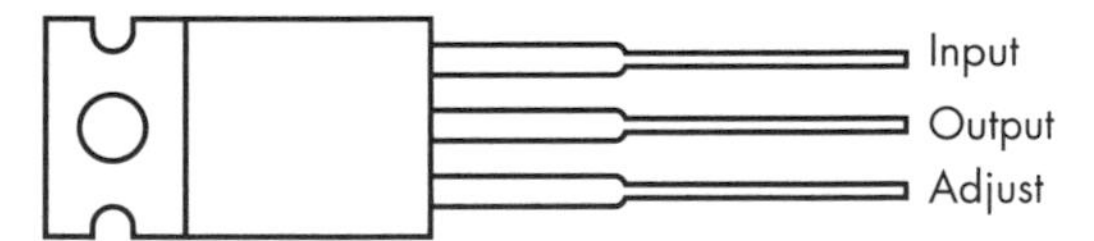

Figure 3-6: Pinout of the LM317 regulator

5. With the potentiometer shaft facing you, connect the leftmost pin and center pin of the potentiometer together, and then connect both to the adjustment (ADJ) pin of the LM317.
6. With the potentiometer in the same orientation, connect the rightmost pin to the blue negative rail (ground).
7. Connect a 470-ohm resistor from the output pin to the ADJ pin on the LM317.

8. Connect a 1.2-ohm resistor (R1 in Figure 3-4—I used two 2.2-ohm resistors in parallel) from the output pin of the LM317 to a load of your choosing. For test purposes, I used a 1/8 W resistor and connected a 5V, 30 mA incandescent indicator lamp for the load. (You may want to use the actual R1 and R9 resistors that you will use in the finished unit, so you can adjust the sketch before completing the unit.)
9. Connect the input pin of the LM317 to the 12V system input voltage. This is the wire going from the LM317 to the upper alligator clip in Figure 3-5.
10. Connect the blue negative rails to the negative side of the input power (probably a wall plug).
11. Insert the LM7805 into the breadboard, as shown in Figure 3-5. (See Figure 3-7 for the pinout of the LM7805.)

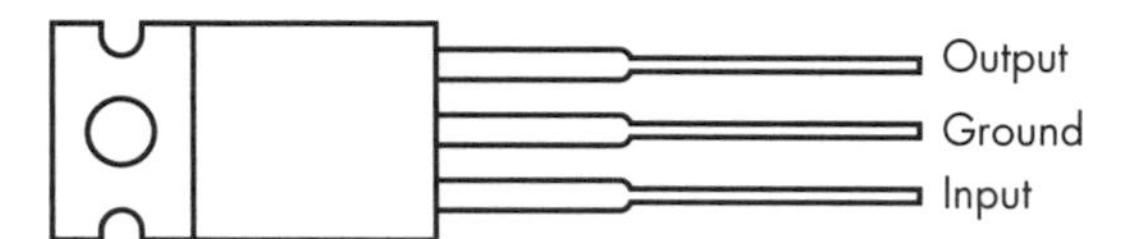

Figure 3-7: Pinout of the LM7805 regulator

12. Connect the output pin of the LM7805 to the red positive rail of the breadboard.
13. Connect the input pin of the LM317 to the input pin of the LM7805. This is the point at which the input voltage from the power source will be connected.
14. Connect the ground pin of the LM7805 to the blue negative rail of the breadboard.
15. Insert a 6.8-kilohm resistor, and connect one side to the blue negative rail. This is resistor R3; the other side will connect to resistor R2. See Figure 3-5 for a top view of the breadboard.
16. Insert a 10-kilohm resistor (R2) into the breadboard with one side connected to the LM317 output pin and the other side connected to resistor R3 from step 15.
17. Connect resistor R1 from step 8 from the output pin of the LM317 to a blank hole in the breadboard. This row will be the output of the regulator.
18. Connect the voltage divider: first, insert a 10-kilohm resistor (R4) into the breadboard. Then, connect one side of resistor R4 to the same row as R1 (you'll have to use a jumper wire) and the other side to an empty row on the breadboard.
19. Insert a 6.8-kilohm resistor (R5) into the board with one side connected to the open side of R4 and the other side of R5 connected to the blue negative rail.
20. Insert the Arduino Pro Mini in the breadboard so that it straddles the center break, as shown in Figures 3-4 and 3-5.

21. Use a jumper to connect the VCC terminal of the Pro Mini to the red positive rail.
22. Use a jumper to connect the GND pins of the Arduino Pro Mini to the blue negative rail. (There are at least two to choose from—one is located between RST and D2, and the other is located between RAW and RST on the other side. Take your pick.)
23. Use a jumper wire to connect the joining point of R4 and R5 to pins A1 and A0 on the Arduino Pro Mini.
24. Find the junction point of R2 and R3, and use a jumper to connect that junction point to the A2 terminal on the Arduino Pro Mini.
25. Load the sketch onto the Arduino Pro Mini. (I often remove the Pro Mini from the circuit completely to program it. It's a little less confusing.)
26. Connect the LCD/I^2C display by connecting VCC and GND to the red positive and blue negative rails on the breadboard, respectively.
27. Connect the SDA to analog pin A4 on the Arduino Pro Mini and SCL to analog pin A5.
28. Connect the input of the LM7805 voltage regulator to some pin where the +12V will be attached.
29. Connect the output of the LM7805 to the red positive rail, and connect the ground to the blue negative rail.

Once all of those connections are in place, you're set to go. Upload the sketch and test the circuit.

The Sketch

The Regulated Power Supply sketch is about as simple as I could make it. The only difficulty is that although I use 1% tolerance resistors throughout, I've found some variation in resistance value. So be aware that you may need to make an adjustment to the sketch to accommodate for this.

Here is the sketch:

```
// Regulated Power Supply with volt and current read

#include <Wire.h>
#include <LiquidCrystal_I2C.h>

LiquidCrystal_I2C lcd(0x27, 16, 2); //Check your library for specific LCD
                                    //code both here and in setup.

float low_side_res = A0;
float volt_two;
float volt_three;
float volt_disp;
float low_side_res_2 = A1;
float hi_side_res = A2;
float volt_drop_1;
```

```
float amp;
float amp_3;
float amp_4;
float amp_disp;

void setup() {
  lcd.init();
  lcd.backlight();
}

void loop() {
  volt_two = analogRead(low_side_res);
  volt_three = (volt_two*5)/1024.0;
  volt_disp = volt_three*(10000+6800)/6800; //Actual voltage reading
  amp_3 = analogRead(low_side_res_2);
  amp_4 = analogRead(hi_side_res);
  amp = amp_3 - amp_4;
  amp_disp = amp *5/1024*(10+6.8)/6.8/1.22*.9; //Calculation of amperage I=V/R
  //*0.9 = adjustment for random error in ref voltage in pro mini

  lcd.setCursor(1,0);
  lcd.print("Volt    ");
  lcd.setCursor(12, 0);
  lcd.print(volt_disp);
  lcd.setCursor(1, 1);
  lcd.print("mA      ");
  lcd.setCursor(11, 1);
  lcd.print(amp_disp*1000,2);
}
```

First, this sketch imports some libraries and sets up the LCD (see "On Writing Code to Set Up LCDs" on page 36). It then defines a series of variables, all floats, to use when setting the voltage, reading from the analog pins, calculating values to display on the LCD, and so on. The `setup()` loop is very short: it has only two lines for initializing the LCD. The main `loop()` reads the battery voltage and current, and performs the necessary calculations to display on the LCD. The `volt_disp` value is the voltage to be displayed on the LCD.

The Shield

While the circuitry is not overly complex, using a shield will simplify many of the connections for driving the LCD and constructing the voltage dividers, and it will make the assembly of the Regulated Power Supply easier than point-to-point wiring. Figure 3-8 shows the shield I designed, though you could also design your own, of course. The PCB file is available at *https://www.nostarch.com/arduinoplayground/*.

While I used two layers to construct the shield, with a little effort and a slightly larger board, the circuitry could be accommodated on a single layer.

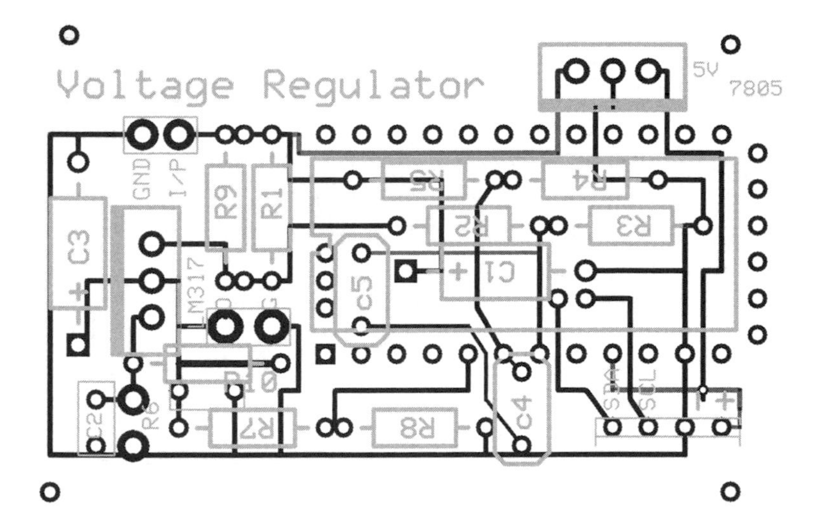

Figure 3-8: The PCB shield used in the Regulated Power Supply. Black is the top layer, dark gray is the bottom layer, and light gray is the silkscreen layer.

The shield doesn't need to be populated in any particular sequence, but some components will be easier to fit before others. I suggest soldering in this order:

1. First, insert 2.2-ohm, 5 W resistors R1 and R9 into the PCB. These are voltage-dropping resistors that create the voltage for the ammeter (mA), which provide a total resistance (R1 in the schematic) of 1.1 ohms at 10 W. The resistors are a little longer than the configuration on the board, so you'll have to bend the leads to make them fit. When the Regulated Power Supply is running close to its maximum rating, expect these resistors—and the LM317 itself—to get a little warm.
2. Capacitor C1, a 68 µF tantalum capacitor, will be a tight fit for the holes. To make sure it doesn't interfere with the LM317, install the capacitor first. Then, install the LM317, making sure to leave room for the heat sink. Remember that the heat sink is likely to get pretty warm.
3. Make sure to install the LM317 with the pins correctly oriented according to the pinout in Figure 3-6 and the schematic in Figure 3-4. If you use the provided shield files, the thick line on the LM317 silkscreen corresponds to the metal tab on the IC. If you insert the part the wrong way, the system won't work, and the part could burn out. It would also be a pain to remove.
4. Install the LM7805 regulator in the upper-right section of the PCB, and make sure it matches the pinout in Figure 3-7 and the schematic in Figure 3-4. You can use the heavy line in the silkscreen image of the PCB as a guide.

5. Try both voltage regulator ICs in the shield with the heat sinks installed (at least temporarily) to make sure they can fit without touching any active components. Remember that the heat sinks are active: the heat sink (tab) of the LM7805 is at ground potential, and the heat sink (tab) of the LM317 is at the output potential.
6. Install resistors R2, R3, R4, and R5. It's easier to place those before installing the female headers for the Arduino Pro Mini.
7. Then, solder in C1, C2, and the wires that will connect the Arduino Pro Mini to potentiometer R6, which you will mount on the chassis. You can leave the wires a little long and trim them when you install the shield in the enclosure. Install capacitors C4 and C5 as indicated in the schematic in Figure 3-4.
8. Next, solder the female headers that comprise the mount for the Pro Mini and the connector for the LCD. The LCD connections are the SDA, SCL, –, and + connections in the bottom right of the PCB. I used male stakes and a female-to-female connector cable to connect from the LCD to the shield. (To learn how to make the custom connector, see "Connectors Used in This Book" on page 18.)

For the LCD and Arduino Pro Mini, I usually insert the headers into the board, solder just one pin, and then, with my finger on the top, heat that one pin and push on the connector to make sure it fits flush against the board. For the pins of the Pro Mini, I use only female headers for those that are active—that is, that have copper traces going to them. I also like to place one pin (I usually use a 1×4 pin header) right at the last pin on the Pro Mini to make alignment easy. This would translate to pins RAW, GND, RST, and VCC. In addition, I like to place at least two headers diagonally for mechanical stability. This would correspond to pins D8 and D9 on the Pro Mini. The male headers on the Pro Mini for A4 and A5 are located just above pins A2, A3, and VCC on the main row of connections.

Construction

When the Regulated Power Supply is all soldered together, you will need to prepare an enclosure and mount the circuit inside. I selected a nice-looking, powder-coated metal enclosure, approximately 2 1/4 × 3 1/4 × 4 3/4 inches. Figure 3-9 shows the completed unit driving a couple of incandescent panel lamps.

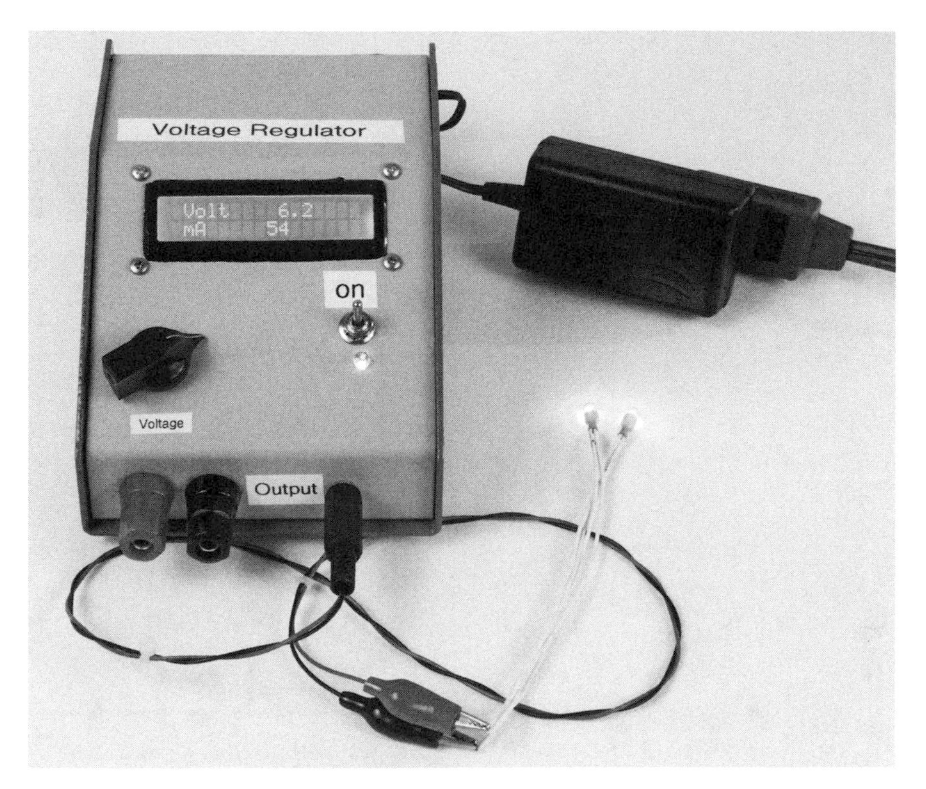

Figure 3-9: The completed Regulated Power Supply

Bear in mind, though, that while the case is not delicate, the paint is easy to scratch, so be careful. It's also a little pricey—coming in around $20—but as I will have it on my workbench all the time, I thought it was worth it. Of course, you could also use a different enclosure of your choosing and modify the templates provided with this book accordingly.

Preparing the Enclosure

If the front panel of the enclosure is sloped, you will need to put a piece of scrap wood behind the areas you need to center punch and drill to help hold it in place. Make sure to measure, center punch, and drill holes carefully.

The templates for this project are shown in Figures 3-10 and 3-11, and they can be downloaded from *https://www.nostarch.com/arduinoplayground/*.

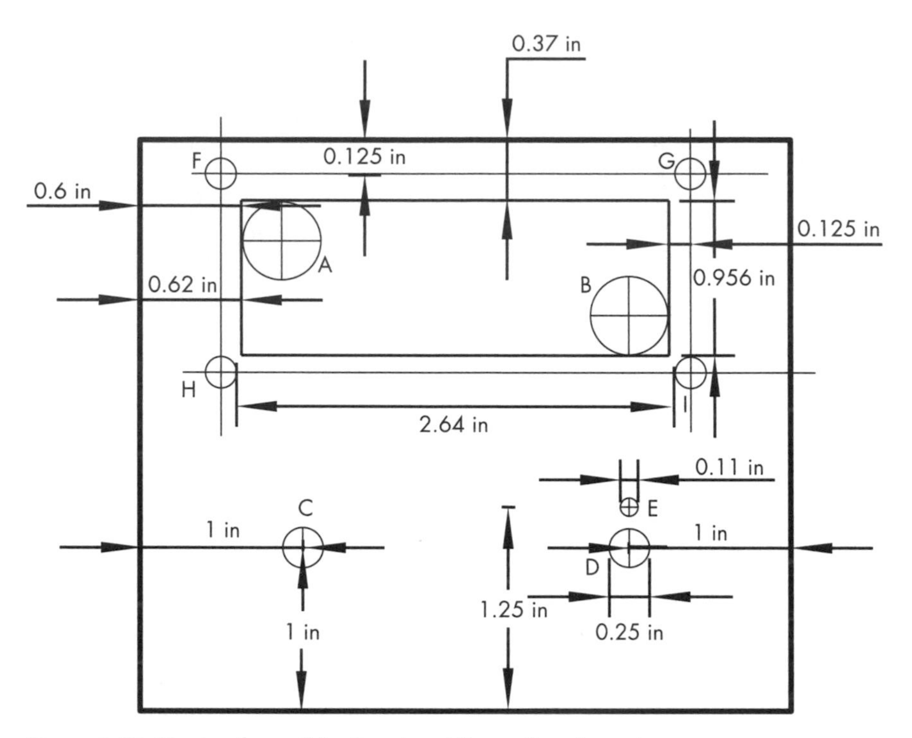

Figure 3-10: Sloping face of the Regulated Power Supply enclosure

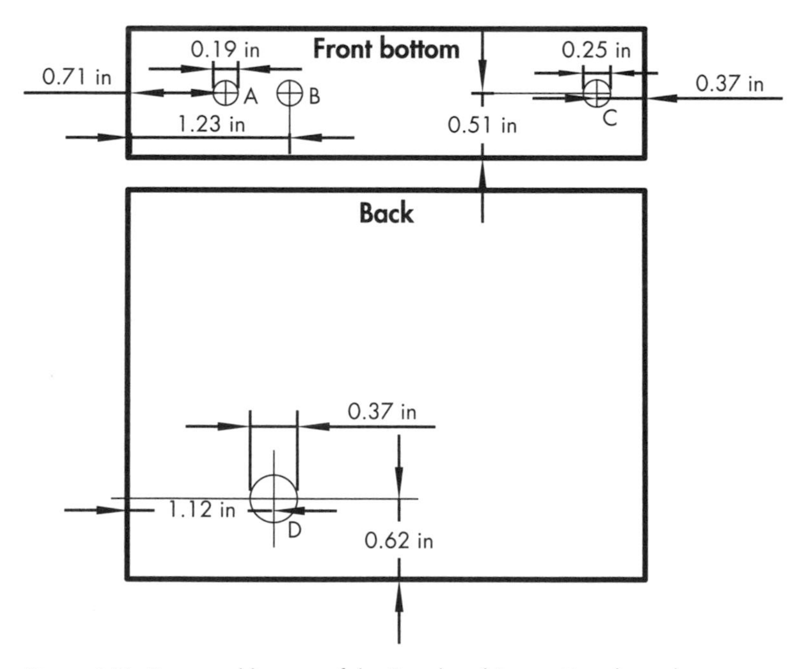

Figure 3-11: Front and bottom of the Regulated Power Supply enclosure

Here is how I suggest you prepare the enclosure:

1. Center punch and drill holes for the potentiometer, on-off switch (1/4 inches), and power indicator LED. See Figure 3-10 for the front panel dimensions.
2. Center punch and drill the hole for the power input jack in the rear of the panel (see Figure 3-10).
3. Drill holes for the output binders and 3.5 mm jack on the front of the case, as shown in Figure 3-11.
4. Carefully measure and mark the cutout for the LCD, as shown in Figure 3-10. Center punch and drill 1/2-inch holes in the corners of the LCD screen area to help initiate saw cutting. You can eyeball this based on the diagram in Figure 3-10 or download a PDF file of the template. Either trace the image onto your enclosure with carbon paper or simply mark the corners with a center punch and connect the punch marks.
5. Carefully cut out a hole for the LCD. There are a variety of tools you can use to do this. I first drilled holes A and B and then used a keyhole saw with a fine hacksaw blade (available at local hardware stores) to cut between the holes. Remember that the cutting occurs on the outward thrust, so you needn't keep pressure on the blade on the return stroke. You can clean up the burrs with a file.
6. Carefully fit the display into the window, and file where necessary to get a secure fit. The backlight protrudes on one side of the display, so in order to avoid crushing the backlight, you can use nuts as spacers to keep it separated from the panel.
7. Drill holes F, G, H, and I, and fasten the display in place. As you do so, check carefully that the spacer nuts are wide enough (4-40 nuts come in different dimensions), and, if necessary, use two nuts or a nut and a washer to space out for the backlight.

Mounting the Circuit Board

Once you have assembled and tested the shield and mounted the LCD, install the potentiometer, on-off switch, LED, and power jack. Then it's time to mount the shield in the enclosure. Originally, I drilled four holes in the board so that I could screw it into the enclosure, but I've found that 3M double-sided mounting tape also does the trick. I mounted the entire board, heat sinks and all, with a 1 1/4-inch length of the 3/4-inch wide double-sided adhesive. (The manufacturer claims it will hold 2 pounds.)

The adhesive is relatively aggressive, so before applying it, make sure you plan carefully where you want to mount the board so it will not be in the way of other components. I mounted the board upside down on the top section of the case so all components and connections were on the same platform (see Figure 3-12).

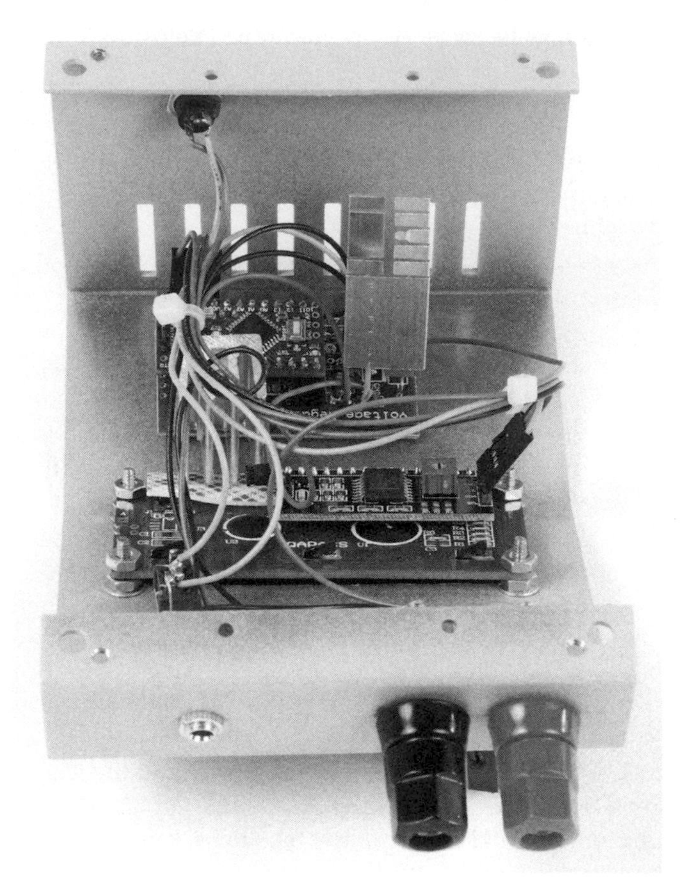

Figure 3-12: The completed (upside-down) assembly with the board, LCD, output connectors, potentiometer, on-off switch, LED, and output power jack all mounted on the inside of the top of the enclosure.

The final step is to connect the on-off switch, LED pilot lamp, LED current-limiting resistor, potentiometer, power input jack, and output connectors to the PCB according to the project schematic. I used 28-gauge hookup wire to tie everything. The two binding posts/banana plug jacks and the 3.5 mm jacks are wired in parallel. Figure 3-12 illustrates how I mounted the shield. Note that I used small wire ties, which are optional, to keep the wires neat.

Before closing up the enclosure, make sure there are no areas that might result in a short circuit. For example, I placed a small piece of insulating tape between the LCD screen and the shield. They might not be touching at the time of assembly but could touch when you close the enclosure. You can apply insulating tape to prevent this kind of short.

After testing for short circuits, all that remains is to put the two halves of the case together. The connections for the output—the binding terminal/banana jack and 3.5 mm jack—are in parallel. The final step is to screw the two pieces together, and you're off and running. I put a pointer knob on the potentiometer even though there are no markings on the case. You can add a dial if you want, but I find the digital readout sufficient.

And, the coup de grâce: remove the protective paper from the adhesive on some rubber feet, and install them on the bottom of the enclosure. Voilà!

The full Regulated Power Supply project will take up a decent amount of space on your workbench, but you will find the digital readout and banana jacks invaluable. If you're feeling adventurous, however, you can build a smaller version of the same circuit, minus the jacks and LCD, shown in Figure 3-13.

Figure 3-13: This is the quick-and-dirty power supply with the top off. I made it before biting the bullet and making the full-fledged Regulated Power Supply.

You can find instructions on how to build the Mini Regulated Power Supply at *https://www.nostarch.com/arduinoplayground/*.

4

A WATCH WINDER

If you're a collector of automatic, or self-winding, watches, you're probably familiar with watch winders and what they do. But why have a watch winder in a book on Arduino microcontroller projects? The answer to that will become increasingly clear as we look at the technology in this project. Further, over the course of this project, we'll take a quick look at some automatic watch lore and how these seemingly anachronistic devices have survived and prospered in the digital age. Even if you do not collect such treasured timepieces, this project may just inspire you to start your own collection.

Why a Watch Winder?

Because, as a collector, you own more than a single automatic watch, you might want to think about keeping the watches that aren't currently on your wrist wound. If you read up on mechanical watches and winders, you will find many pros and cons (probably more pros) of using a watch winder. One big pro is that multifunction watches can take a long time to set if they run down. There are also arguments that if a mechanical watch sits in one position and doesn't run, the lubricant tends to migrate to a low point. Regular motion from a watch winder or from being worn keeps the lubricant distributed and in the bearings where it belongs. While many subscribe to this viewpoint, there is no real evidence either way.

There is yet another compelling argument for a watch winder. As a collector, it's nice to be able to display more of your collection than just one-at-a-time on your wrist. Many of the commercial winders available come inside exotic wood cases to show off the watches. But inexpensive winders tend not to be reliable, and the expensive ones are, well, expensive.

I took a chance and bought one of the more economical models and put two of my mechanical watches—a real and a faux Rolex—in it and figured I was done. But after less than six months, the winder failed. I took it apart, and it appeared to be very poorly designed and made. Even if I replaced the failed motor, the rest of the mechanism would probably not be reliable. While using the winder, the faux Rolex did not wind all the way and did not keep good time.

At that point, the question was whether to dig deep in my pockets for the $400 or $500 winder (there are even models that sell well in excess of $1,500, $2,000, or more) that promises reliability or to try to do better. So the gauntlet was, metaphorically, thrown down. The challenge was to design and build a reliable watch winder that would provide both a showcase for my watches and have the flexibility and control over timing that I wanted in a robust mechanical format. Arduino was the obvious choice for controlling the frequency of watch turns, and the mechanics went around that.

As you build your Watch Winder, you will find a lot of room for personalization both in the mechanical construction and the sketch. While a watch winder is a utilitarian device made to keep your watches wound, this version provides an elegant display platform for your timepieces—and it is itself a work of art, a kinetic sculpture. You can see the final result in Figure 4-1.

Because I selected Arduino as the logical timing element, I had to plan the other electronics and software around that. We will revisit the H-bridge circuit from the PCB Etcher (see "Using an H-Bridge" on page 48) to drive the motor in both directions, and we'll use transistors for increased drive for the high-output LEDs. We'll also use a Hall effect sensor to measure the rotation of the watches.

Figure 4-1: The finished Watch Winder. Unfortunately the black and white image doesn't do it justice: the brightly colored LEDs illuminate the device using the acrylic as a light guide to transport the various colored LEDs.

The sketch developed for this project uses functions and arrays to flash the LEDs in repeating patterns. The sketch also instructs the controller to read the state of the Hall effect sensor, which is either zero or one. Knowing this state allows the controller to decide when to wind the watches and to keep count of the number of turns to ensure that the watches don't get over- or underwound.

THE MYSTIQUE OF THE AUTOMATIC WATCH

The automatic watch was invented in the early 1920s and was commercialized several years later. Over the next several years, many improvements were made until it reached the level of sophistication of today's instruments. Automatic watches operate by using a pendulum attached to a ratchet assembly: the ratchet assembly winds the watch's mainspring as the pendulum swings. A built in slip-clutch mechanism prevents overwinding. See Figure 4-2 for a look inside one of these watches.

Figure 4-2: An automatic watch with the back removed, exposing the pendulum and the fulcrum (the screw in the center), which combine with a ratchet assembly to wind the watch's mainspring

Automatic watches from just about all watch manufacturers enjoyed broad success for several decades. However, in the early 1960s, Bulova developed its Accutron tuning-fork electronic watch, and the digital quartz electronic watch from Pulsar followed shortly after.[1]

Despite the influx of electronic watches (and now smart watches), leading makers of mechanical watches have survived—and even prospered—in this age. Today, automatic watches are sold anywhere from under $100 to tens or even hundreds of thousands of dollars.

Why would someone pay a premium for a watch that is not particularly accurate, is heavy, is often bulky, and has to be kept wound when not in use? I'm sure the answer is different for every collector, but I'd guess that they, like me, enjoy the elegance, prestige, sophistication, sense of history, and fine mechanical machinery that can't be achieved with its electronic counterparts—though the iWatch comes close in some respects. And like any collectible, one automatic watch is never enough—which brings us to the Watch Winder.

---

1. The transition from mechanical to electronic watches has been described as a prime example of Thomas Kuhn's concept of a paradigm shift, which he describes in his 1962 book *The Structure of Scientific Revolution*.

Required Tools

Drill and drill bits

Tapered reamer set

Small vise-grip pliers

Center punch

Weld-On 4 and Weld-On 16 acrylic bonding fluid

Assorted sandpaper, including grades from 220 to 600 and grade 1500 for final polish

Jewelers rouge or other liquid plastic polish

(Optional) Circular saw

(Optional) Thread-locking fluid

(Optional) Wire-wrap tool

(Optional) Rotary tool (For example, you could get a Dremel tool with an abrasive cutoff wheel.)

Parts List

If you want to build a Watch Winder like the one pictured, you will need several pieces of acrylic and some other hardware, which I detail in this section.

Acrylic

The following acrylic parts can easily be cut from a standard sheet of acrylic. Without the disks, which I recommend you purchase separately, everything can be cut from two 12×12-inch acrylic sheets (one 3/8-inch thick, one 1/4-inch thick). If you prefer, you can find vendors that will laser-cut acrylic to your dimensions. (ZLazr, among many others, is equipped to do that.) It will cost a little more than doing it yourself but will make it both cutting and finishing easier.

Four pieces with dimensions 1/4 × 2 × 1 1/2 inches (long sides of the watch basket; can be 3/8 inches)

Four pieces with dimensions 1/4 × 1 × 2 inches (short sides of the watch basket; can be 3/8 inches)

Two pieces with dimensions 3/8 × 3 × 2 inches (bearing holders of bearing box)

Two pieces with dimensions 3/8 × 2 × 1 1/2 inches (mounting side of bearing box)

One piece with dimensions 1/4 × 1 × 2 inches (motor mount)

Two round pieces, 3/8 inches thick and 5 inches in diameter (watch basket ends)

Two pieces with dimensions 1 1/2 × 5 × 1/4 inches (side supports for stand)

One piece with dimensions 3/8 × 3 × 5 1/2 inches (base for stand)

One piece with dimensions 3 × 1 1/2 × 3/8 inches (lightbar)

One piece with dimensions 2 1/2 × 2 1/2 × 3/8 inches (shield mounting)

Two 3.5 mm standoffs with M/F M3-05 threads (motor mounts)

Three 1.5 mm standoffs with M/F M3-05 threads (shield mount)

There are several online vendors you could purchase the acrylic for this project from; just search for *acrylic sheet* on Google to find one near you. In the United States, *http://www.zlazr.com/* seems to be good. At the time of this writing, I talked with the owner personally, and he said he can handle the kind of cutting required for this project with no problem.

Other Hardware and Circuit Components

One Arduino Nano or clone

One Hall effect switch, such as Melexis US5881LUA (Dimensions for side supports should be 1 1/2 × 5 × 1/4. See "Building the Stand" on page 115.)

One driveshaft, 8 inches long and 1/4 inches in diameter with 28 threads per inch (I suggest brass because it's easy to work.)

Two ball bearings (R4A-2RS)

Six jam nuts, 1/4-inch-28

Two decorative bolts, 1/4-inch-28, 1 inch long (I used chromed Allen bolts.)

Ten ZTX649 transistors

One SN754410 quad H-bridge

Ten 470-ohm resistors

One 10-kilohm, 1/8 W resistor

One 0.1 μF ceramic capacitor (C1)

One 10 μF tantalum capacitor (C2)

One custom shield as described in "The Shield" on page 108, or perf board (You can also have the shield custom fabricated from ExpressPCB; see "Making Your Own PCBs" on page 13.)

One gear head motor (I used a 6V, 20 RPM motor called the Amico 20 RPM 6VDC 0.45 A.)

One LM7805 voltage regulator

Fourteen LEDs, in assorted colors (I purchased both clear and frosted versions. The higher-output units tended to be clear.)

Assorted hookup wire and wire-wrap wire

Ten stakes for LED wire wrap or soldering (Try Pololu item #966 or Electronic Goldmine item #G19870.)

One length brass round that's 3/8 or 1/2 inches in diameter and approximately 3/4 inches long (I used one with a 3/8-inch diameter. Brass stock is readily available in 6- and 12-inch lengths, which can be cut to size with a hacksaw.)

One 6-inch length of piano wire that's 0.39 inches in diameter

One niobium, or neodymium, magnet, approximately 3/8 inches round and 1/8 inches thick

One flat-head 4-40 screw

Six M3×3/8-inch screws

Seven M3-05×1/2-inch screws

(Optional) One Amico H7EC-BCM counter

(Optional) Eight 270-ohm resistors (Use these when you build the breadboard prototype if you choose to follow my exact instructions in "The Breadboard" on page 98.)

Downloads

Before you start building, go to *https://www.nostarch.com/arduinoplayground/*, download the resource files for this book, and look for the following files for Chapter 4:

Templates *MotorMountAndBearingBox.pdf*, *BaseAndLightbar.pdf*, *WatchBasket.pdf*

Mechanical drawing *MotorAssembly.pdf*

Shield *WatchWinder.pcb*

Sketch *WatchWinder.ino*

Basic Watch Winder Requirements

Some initial research suggested that a watch winder should rotate a watch between 600 and 1,200 revolutions per day to keep it in top shape. But that is not completely correct. I subsequently discovered that the range was actually much wider, and according to at least two websites of leading automatic watches, watches cannot be overwound because they have a built-in protection system. I also learned that watches should be rotated both clockwise and counterclockwise to keep lubricant in the right places and to avoid possible uneven wear over a very long period of time. There is a wealth of information about this subject on the web, both on sites for individual watch manufacturers as well as on sites for watch winders.

Apparently the total number of turns is the important part, not necessarily the sequencing of the turns or getting exactly the same number of turns in each direction. (There is a possible downside to winding, if a watch is wound too much over an extended period.) That doesn't sound so daunting, right? I thought so, too.

Using an Arduino to Control Winder Revolutions

A purely utilitarian watch winder just has to serve its function, rotating the watches so the pendulums swing. But it's more interesting to have a winder with extra features. As mentioned in "Why a Watch Winder?" on page 90, some winders are dressed up with fancy exotic wood boxes to display the watches.

However, this is an Arduino project, and extra technical features and LEDs should reflect the flexibility and versatility of the platform. In a developmental model, the original sketch instructed the electronics to turn a motor first in one direction and then the other, using delays to ensure that the requisite 600 to 1,200 revolutions occurred each day.

But it turns out that some watches need more than the minimum number of revolutions, and some can get away with less. The easiest way to change the number of revolutions is by adjusting the various delays in the sketch as needed. You could even add hardware to the circuit to allow you to adjust the number of turns per day with a potentiometer, as I describe in "Design Notes" on page 124.

To drive the motor itself, I used an H-bridge IC. It accepts control logic from the Arduino and lets you reverse the polarity to the motor from a single power supply to allow the motor to rotate in both directions.

NOTE *For more information on H-bridges, see "Using an H-Bridge" on page 48.*

Using a Hall Effect Sensor to Monitor Rotations

Then, there was the matter of how to meter the number of turns the device made to assist the timing and give some more information to the sketch. The number of turns per unit time is a function of the motor, and while the timing I provided for the motor specified could conceivably work, it might not be consistent for all motors.

For example, I sampled three motors of the same model at the same voltage, and each ran at a slightly different speed. Further, if you elect to substitute another motor with a different rotational speed, the rotation count would be different. And, in beta testing, one user experienced difficulty running a 6V motor on 5V. (See "Motor Voltage" on page 126.) Because the number of turns per unit time is a function of the motor, these inconsistencies could present a problem if timing alone determined the total number of rotations; some mechanism to monitor the number of revolutions is needed.

To assure consistent timing, I decided to meter the number of turns the device made. Thus, I attached a small magnet to the rotating shaft that turns the watches and mounted a Hall effect device, or a sensor that detects a magnetic field, in line with the magnet. A small reed switch could be substituted for the Hall effect sensor if you wanted.

When the watch and the magnet rotate, the Hall effect switch turns on only when in close proximity to the magnet, causing the switch to turn on and off once per rotation. Each time the Hall effect switch changes state, the Arduino increments an internal counter. Combined with the sketch, this ensures the proper number of turns per day is made in all cases, regardless of the speed of the motor. Unlike the reed switch, the Hall effect switch does not require any buffering or debounce, as discussed in "The Sketch" on page 102, because a Schmitt Trigger is included in the device's circuit. If you elect to use a reed switch, you may have to add the debounce into the sketch.

When using a Hall effect switch with a permanent magnet, you just have to be careful how you move the magnet around. Some mechanical watches are damaged by close proximity to a strong magnetic field because the hairspring becomes magnetized, resulting in a change in physical characteristics that cause timing to be off. While the magnet specified is small and unlikely to cause a problem, I strongly recommend you keep any magnet at least an inch away from any watch—mechanical or electronic.

The Schematic

Figure 4-3 shows the schematic diagram of the circuit used for the Watch Winder. Notice that the output from the Hall effect device has a pull-up resistor tied to the positive supply. This holds the input to Arduino pin A0 high until the Hall effect switch, or reed switch, encounters a strong enough magnetic field, which closes the switch and brings the pin low. The Hall effect device uses what is essentially an open collector on its output, so without the pull-up resistor, the collector would be left floating and could give a false trigger.

The two capacitors prevent the LM7805 regulator from oscillating on its own and drawing excessive power. Although I looked at both the input and the output of the regulator with an oscilloscope and saw no oscillation, I decided to add the capacitors as a preventative measure. I selected them based on previous projects, and they work well.

I was trying to develop a spectacular look for the Watch Winder, as befits some of the timepieces it holds, so I used higher-power LEDs, as described in the "Parts List" on page 93. These LEDs have a light output of as much as 100,000 to 200,000 or more millicandela (MCD). But that raised yet another problem. The Arduino Nano's processor chip, an ATmega328, can source or sink only 40 mA per output pin. Further, the entire chip is rated at only 200 to 300 mA for its entire current drain. Because the 100,000+ MCD LEDs draw around 30 to 60 mA each, something had to be done.

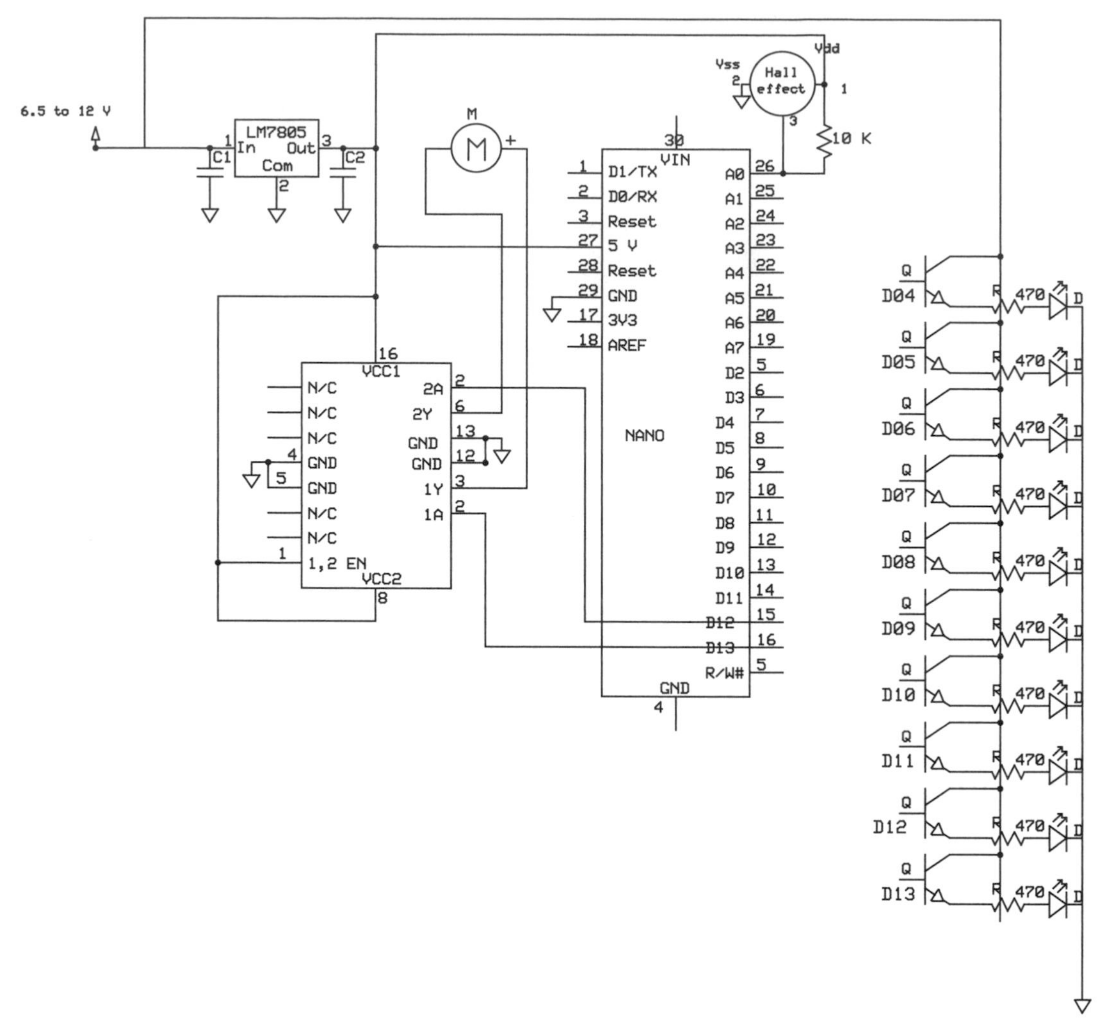

Figure 4-3: The Watch Winder circuit. The transistors connect to the digital outputs of the Nano, while A0 is tied high through the 10-kilohm resistor.

One 1 A transistor per LED is included in the schematic to pick up the load. The collectors of the NPN transistors—the positive side—go to VIN rather than the 5V that powers the Nano and H-bridge, so the LEDs take no toll on the voltage regulator, even though the emitters follow the base and send 5V to the LEDs.

The Breadboard

Just like other projects we've discussed, the Watch Winder started out as a breadboard, shown in Figure 4-4. This allowed me to sound out the technology and do the preliminary tuning of the sketch.

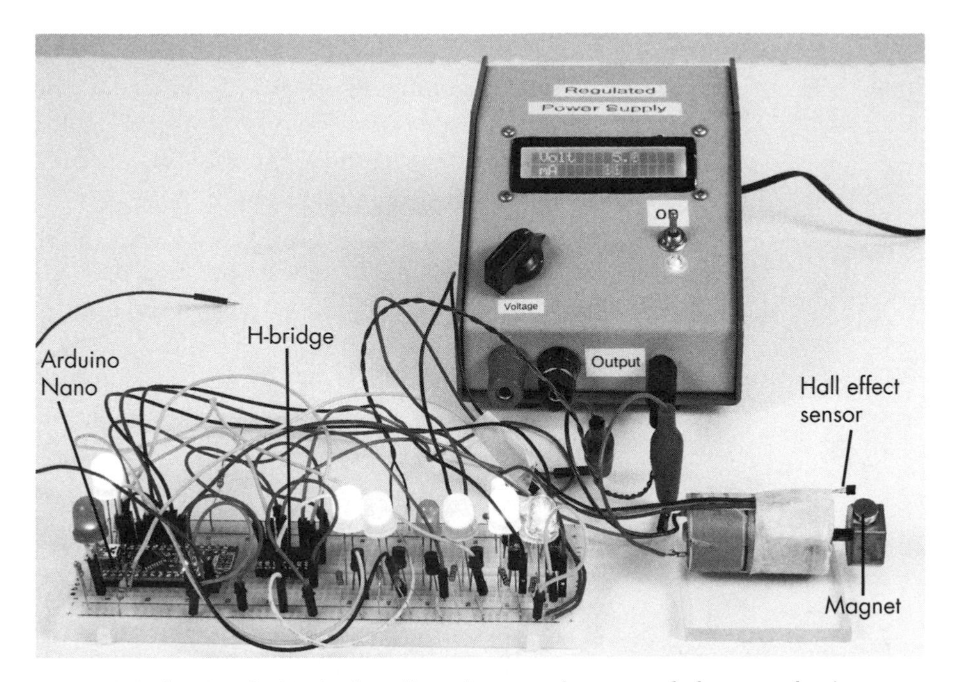

Figure 4-4: The Watch Winder breadboard was used as a proof-of-concept for the project. Here, I powered it with the Regulated Power Supply from Chapter 3.

I suggest building a breadboard for this project first so you can see where everything goes and why. With a breadboard, you also get to play with the sketch and LEDs without having to unsolder and resolder with each change. I used a 6.5-inch long breadboard to hold everything. I did take a couple of shortcuts on the breadboard, which are noted in the instructions; you can also just build straight from the schematic, instead.

To wire up the breadboard, take the following steps:

1. Connect the red stripe on the right side of the breadboard to the corresponding red stripe on the left. These are your positive rails.
2. Connect the blue stripe on the right side of the breadboard to the corresponding blue stripe on the left. These are your negative rails (ground connections).
3. Insert the Arduino Nano at one end.
4. Connect the 5V pin of the Nano to the red positive rail.
5. Connect the GND pin of the Nano to the blue negative rail.
6. Insert the LM7805 regulator, and connect the output pin to the red positive rail. (Figure 4-5 shows the regulator pinout.)

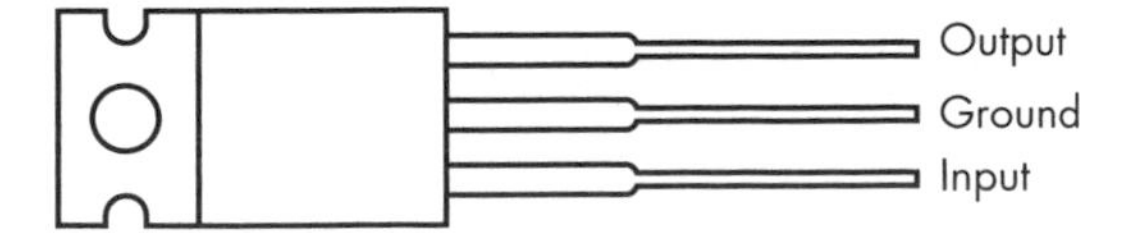

Figure 4-5: LM7805 regulator pinout in a TO-220 package

7. Connect the ground terminal of the regulator to the blue negative rail.
8. The input terminal of the regulator will connect to a blank row in the breadboard, which will connect to the +7.5V to 9V supply.
9. Connect capacitor C1 from the input of the regulator to ground.
10. Connect capacitor C2 from the output of the regulator to ground.
11. Insert the SN754410 H-bridge several rows away from the Nano, straddling the gutter in the middle of the breadboard. (Figure 4-6 shows the H-bridge pinout.)

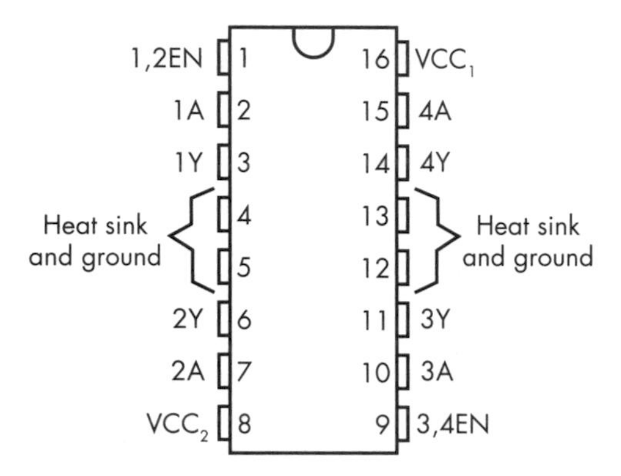

Figure 4-6: SN754410 H-bridge pinout in a DIP form factor

12. Connect pins 4, 5, 12, and 13 of the H-bridge to ground.
13. Connect pins 8, 9, and 16 of the H-bridge to the red positive rail.
14. Attach pins 14 and 11 of the H-bridge to the motor with leads at least 10 to 12 inches long. The connections to the motor will have to be soldered unless you use alligator clips or clip leads.
15. Attach approximately 8-inch wires to all three leads of the Hall effect sensor. Connect the leads attached to the positive and negative leads of the Hall effect sensor to the red positive rail and blue negative rail, respectively.
16. Connect the wire attached to the third pin of the Hall effect sensor to pin A0 on the Nano. The Hall effect sensor will be taped (I used masking tape) to the motor body (this works because the leads are insulated) in such a position that the active part of the device will be close to the magnet attached to the shaft as it goes around.
17. Connect a 10-kilohm resistor from pin A0 on the Nano to the red positive rail.
18. Connect pin 10 of the H-bridge to pin D13 on the Nano.
19. Connect pin 15 of the H-bridge to pin D12 on the Nano.

20. Insert five ZTX649 transistors, each using three rows, on one side of the breadboard. (Figure 4-7 shows the transistor pinout.)
21. Connect the collector of each transistor to the red positive rail.

These transistor connections differ from the final schematic, where the collectors will be tied to the 9V input voltage.

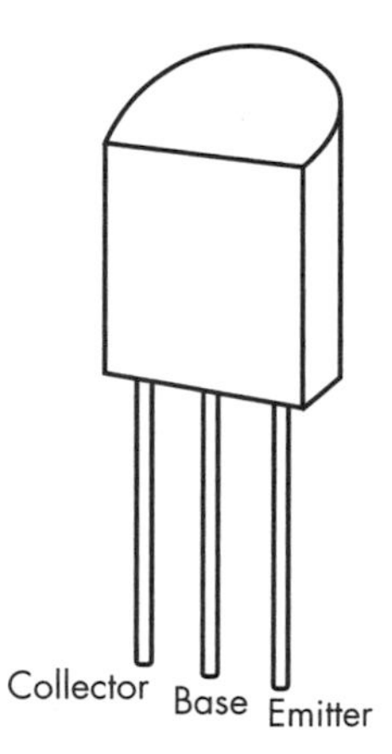

Figure 4-7: ZTX649 transistor pinout in a TO-92 package

22. Take five LEDs, and connect the positive terminal of an LED to the emitter of each transistor.
23. Connect the negative terminal of each LED to ground through a 270-ohm resistor.
24. Connect three additional transistors, LEDs, and 270-ohm resistors on the opposite side of the breadboard in a similar manner to the five groups connected in steps 20–23.

NOTE *Steps 20–25 are different than the schematic, as the LEDs are driven directly from the Nano, for ease of experimentation. In the schematic, they are connected using a transistor. Using the 470-ohm resistor instead of the 270-ohm resistor limits the current to the Nano.*

25. Connect the positive terminal of another LED to pin D12 of the Nano.
26. Connect the negative terminal of the pin D12 LED to an empty row on the breadboard.
27. Connect one end of a 470-ohm resistor to the negative terminal of the pin D12 LED, and connect the other end of the resistor to ground.
28. Connect the positive terminal of another new LED to pin 13 of the Nano.
29. Connect the negative terminal of the pin 13 LED to an empty row on the breadboard.
30. Connect one end of a 470-ohm resistor to the negative terminal of the pin 13 LED, and connect the other end of the resistor to ground.
31. Connect pins D4, D5, D6, D7, D8, D9, D10, and D11 on the Nano to the bases of the transistors feeding the LEDs.

Because the breadboard is for illustration only, the order that the connections are made in doesn't matter unless you want to reprogram the Arduino while the circuit is on the breadboard.

The magnet was mounted to a plug on the motor shaft using double-sided adhesive foam tape. For the plug, you can use almost anything—a cork, a rubber stopper, and so on—as long as it puts the magnet in a position so it will be about 3/8 inches from the sensor as the magnet rotates.

After you complete the breadboard, upload the *WatchWinder.ino* sketch to the Arduino. Just follow the instructions in "Uploading Sketches to Your Arduino" on page 5.

The Sketch

The Watch Winder employs functions and arrays to show different flashing sequences on the LEDs without rewriting the sequence each time. There are also some excruciatingly long delays: about 829 rotations in 24 hours translates to a motor at 20 RPM being on for approximately 32.5 minutes out of 1,400 minutes in the day. This means that if the sketch were to handle an entire day of turning, it would be idle for 1,367.5 minutes a day.

But you can divvy up the rotations so that the sketch can be repeated and need only some fraction of the 24 hours to complete. For example, if an hour is selected as the length of time it takes for the sketch loop to complete, the motor has to do some 24 turns. It could do 12 each way or some other combination.

In the following sketch, I also made an effort to make the lights and motor movements as visually interesting as possible, leaving very little time when nothing is happening—but that's an artistic choice.

```
/*This gives about 829 revs/day*/

const int HallPin = A0; //Identify those things that will not change
const int CWpin = 12;
const int CCWpin = 13;

const int LED11 = 11;
const int LED10 = 10;
const int LED9 = 9;
const int LED8 = 8;
const int LED7 = 7;
const int LED6 = 6;
const int LED5 = 5;
const int LED4 = 4;

int autoDelay = 1000;
int timer = 500;
int timer2 = 3000;
int repeats = 10;

int previous;
int HallValue = 1; //Response from the Hall effect sensor
int time = 0;
int state;
int count = 0;
int q = 0;
int i;
int j;
```

```
int ledPins[] = {
  11, 4, 7, 6, 8, 10, 5, 9,
};
int pinCount = 8;

void blinkIt() {
  //Initiate rapid blink sequence
  for(int thisPin = 0; thisPin < pinCount; thisPin++) {
    //Turn the pin on:
    digitalWrite(ledPins[thisPin], HIGH);
    delay(timer2);
    //Turn the pin off:
    digitalWrite(ledPins[thisPin], LOW);
    delay(timer2);
  }

  //Loop from the highest pin to the lowest:
  for(int thisPin = pinCount - 1; thisPin >= 0; thisPin--) {
    //Turn the pin on:
    digitalWrite(ledPins[thisPin], HIGH);
    delay(timer2);
    //Turn the pin off:
    digitalWrite(ledPins[thisPin], LOW);
    delay(timer2);
  }
}

void flashIt() {
  //Initiate rapid blink sequence
  for(int thisPin = 0; thisPin < pinCount; thisPin++) {
    //Turn the pin on:
    digitalWrite(ledPins[thisPin], HIGH);
    delay(timer2);
    //Turn the pin off:
    digitalWrite(ledPins[thisPin], LOW);
  }

  //Loop from the highest pin to the lowest:
  for(int thisPin = pinCount - 1; thisPin >= 0; thisPin--) {
    //Turn the pin on:
    digitalWrite(ledPins[thisPin], HIGH);
    delay(timer2);
    //Turn the pin off:
    digitalWrite(ledPins[thisPin], LOW);
  }
}

void allatOncefast() {
  {
    digitalWrite(LED4, HIGH);
    digitalWrite(LED5, HIGH);
    digitalWrite(LED6, HIGH);
    digitalWrite(LED7, HIGH);
    digitalWrite(LED8, HIGH);
    digitalWrite(LED9, HIGH);
```

```
    digitalWrite(LED10, HIGH);
    digitalWrite(LED11, HIGH);

    delay(500);

    digitalWrite(LED4, LOW);
    digitalWrite(LED5, LOW);
    digitalWrite(LED6, LOW);
    digitalWrite(LED7, LOW);
    digitalWrite(LED8, LOW);
    digitalWrite(LED9, LOW);
    digitalWrite(LED10, LOW);
    digitalWrite(LED11, LOW);

    delay(500);
  }
}

void allatOnce() {
  {
    digitalWrite(LED4, HIGH);
    digitalWrite(LED5, HIGH);
    digitalWrite(LED6, HIGH);
    digitalWrite(LED7, HIGH);
    digitalWrite(LED8, HIGH);
    digitalWrite(LED9, HIGH);
    digitalWrite(LED10, HIGH);
    digitalWrite(LED11, HIGH);

    delay(4000);

    digitalWrite(LED4, LOW);
    digitalWrite(LED5, LOW);
    digitalWrite(LED6, LOW);
    digitalWrite(LED7, LOW);
    digitalWrite(LED8, LOW);
    digitalWrite(LED9, LOW);
    digitalWrite(LED10, LOW);
    digitalWrite(LED11, LOW);

    delay(2000);
  }
}

void setup() {
  pinMode(HallPin, INPUT);  //Identifies inputs and outputs
  pinMode(CWpin, OUTPUT);
  pinMode(CCWpin, OUTPUT);

  Serial.begin(9600);

  for(int thisPin = 0; thisPin < pinCount; thisPin++)  {
    pinMode(ledPins [thisPin], OUTPUT);
  }
}
```

```
void loop() {
  int HallValue = (digitalRead(HallPin)); //Sets value of initial Hall effect

  if(HallValue == HIGH && previous == LOW) {
    if(state == HIGH)
      state = LOW;
    else
      state = HIGH;

    //Increments counter each time the Hall effect sensor passes the magnet
❶   count++;
  }

  /* The "Serial.print" line was used in development. I left it in so that
     you can experiment and look at some of the values on a serial
     monitor. You might even want to change the parameters of what you
     are looking at in the monitor.
  */
  Serial.print("HallValue      ");
  Serial.println(HallValue);
  Serial.print("count                    ");
  Serial.println(count);
  Serial.print("CCW                          ");
  Serial.println(" ");

  if(count == 1) {
    digitalWrite(CCWpin, HIGH);
    digitalWrite(CWpin, LOW);
  }

  if(count == 3) {
    digitalWrite(CWpin, HIGH);
    digitalWrite(CCWpin, HIGH);
  }

  if(count == 3) {
    for(i = 0; i < repeats; i++) {
      allatOncefast();
    }
    count = count + 1;
  }

  if(count == 3) {
    digitalWrite(CWpin, LOW);
    digitalWrite(CCWpin, HIGH);
  }

  if(count == 4) {
    digitalWrite(CWpin, HIGH);
    digitalWrite(CCWpin, HIGH);
  }
```

```
if(count == 4) {
  for(q = 0; q < repeats; q++) {
    blinkIt();
  }
  count = count + 1;
}

if(count == 5) {
  for(j = 0; j < repeats; j++) {
    allatOnce();
  }
  delay(50);
  count = count + 1;
}

if(count == 6) {
  digitalWrite(CCWpin, LOW);
  digitalWrite(CWpin, HIGH);
}

if(count == 7) {
  digitalWrite(CWpin, LOW);
  digitalWrite(CCWpin, LOW);
}

if(count == 7) {
  for(i = 0; i < repeats; i++) {
    flashIt();
  }
  count = count + 1;
}

if(count == 8) {
  digitalWrite(CCWpin, HIGH);
  digitalWrite(CWpin, LOW);
}

if(count == 10) {
  for(i = 0; i < repeats; i++) {
    allatOncefast();
  }
  count = count + 1;
}

if(count == 11) {
  digitalWrite(CCWpin, LOW);
  digitalWrite(CWpin, LOW);
}

if(count == 11) {
  for(i = 0; i < repeats; i++) {
    blinkIt();
  }
```

```
  delay(2000);
  count = count + 1;
}

if(count == 12) {
  digitalWrite(CCWpin, LOW);
  digitalWrite(CWpin, HIGH);
}

if(count == 13) {
  digitalWrite(CCWpin, HIGH);
  digitalWrite(CWpin, HIGH);
}

if(count == 13) {
  for(i = 0; i < repeats; i++) {
    flashIt();
  }
  count = count + 1;
}

if(count == 14) {
  for(i = 0; i < repeats; i++) {
    allatOnce();
  }
}

if(count == 14) {
  digitalWrite(CWpin, LOW);
  digitalWrite(CCWpin, HIGH);
  delay(autoDelay);
}

if(count == 17) {
  digitalWrite(CWpin, HIGH);
  digitalWrite(CCWpin, HIGH);
}

if(count == 17) {
  for(i = 0; i < 20; i++) {
    blinkIt();
  }
}

{
  for(i = 0; i < repeats; i++) {
    allatOncefast();
  }
  count = count + 1;
}
```

```
  if(count == 18) {
    digitalWrite(CCWpin, HIGH);
    digitalWrite(CWpin, LOW);
    delay(2000);
    digitalWrite(CCWpin, LOW);
    digitalWrite(CWpin, HIGH);
    delay(2000);
  }

  if(count > 20) {
❷   count = 0;
  }
  previous = HallValue;
}
```

First, the sketch creates several constants, integers, and arrays, which assist with timing turns by reading from the Hall effect sensor and counting the turns. Next, come a few function definitions: `blinkIt()` and `flashIt()` blink the LEDs in different patterns, while `allatOnceFast()` and `allatOnce()` blink the LEDs all at the same time with different delays.

As usual, the `setup()` function tells the Arduino which pins are inputs and outputs. At the start of the `loop()` function, the Hall effect sensor is read, and the sketch increments the counter at ❶ as needed, printing a few useful debugging values to the serial monitor along the way. This sketch uses the `count` value to turn different sequences on or off and limit the repetitions. However, because `count` is reset at the end of the sequence at ❷, it cannot be used as a totalizer.

Finally, for various counts, the sketch uses `if` statements to hard-code different patterns for turning the watches and flashing the LEDs; I show a few here, but I encourage you to set up your own. The sketch is written with many functions you can use as-is or repeat in a `for` loop to give multiple iterations.

The Shield

As in some of the other Arduino projects, the shield is not terribly complex, but it looks a little busy. For simplicity, this shield is a single-sided board. The circuit uses an LM7805 voltage regulator to handle excess current that could result from using a different motor. The on-board regulator built into the Nano is intended only for currents less than 300 mA.

NOTE *I have used the regulator in this project at up to 500 mA, but the regulator tends to get pretty warm, and I don't feel comfortable using it at that level.*

You may be able to leave the regulator out; the collectors of the transistors feeding the high-output LEDs are configured as emitter-followers and wired directly to the positive 9V supply, so they are not contributing to the load on the regulated 5V.

Figure 4-8 shows the foil pattern for the shield (left), and the silk-screen layer (right).

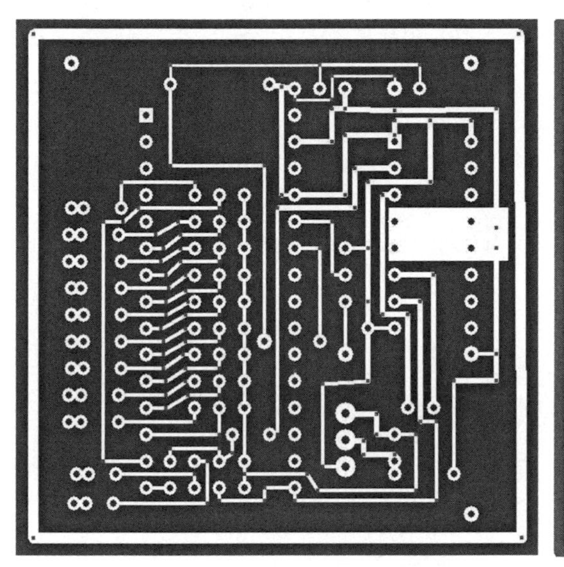

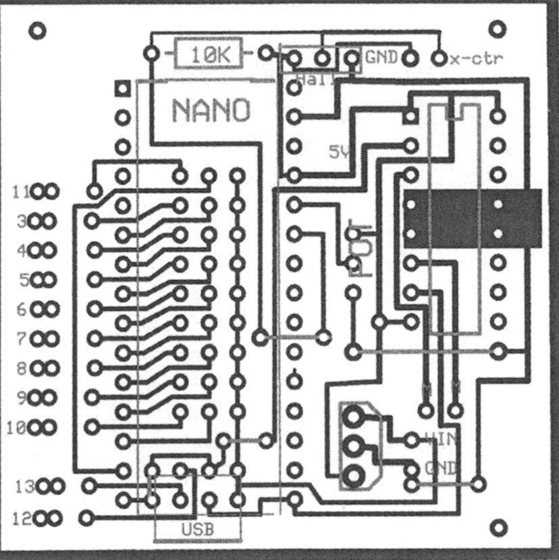

Figure 4-8: The foil pattern of the shield (left) and the silk-screen layer (right). The Watch Winder shield's silk-screen layer shows the approximate placement of the Nano, H-bridge, contacts for the external counter, Hall effect sensor, potentiometer, LED connections, jumpers, input voltage (VIN), and ground (GND). The PCB Express file is available to download from https://www.nostarch.com/arduinoplayground/*.*

Notice the contacts for the Hall effect switch at the top of the board, labeled *Hall*. I soldered wires that connected the Hall effect device directly to the PCB, though you could use a connector if you prefer.

In the center of the board, I left connections for a potentiometer (POT) for external adjustment of the period, an option I describe in "Total Rotation Adjustment" on page 124. The numbers of the digital outputs are labeled at the left-hand side.

If you choose to assemble the Watch Winder shield, note that the Nano is meant to be plugged into female headers soldered onto the shield and that the transistors are underneath the Nano board. Push the transistors down far enough before you solder them so they will not be in the way of the Nano when it's plugged in. I had to place the transistors fairly close to fit all the connections into the PCB layout. The ZTX649 transistors I selected fit well enough within the 0.100-inch spacing allowed by the footprint of the Nano.

You will also have to add a few jumper wires to complete the connections on the shield. They are marked on the silk-screen pattern. In Figure 4-8, those appear as five black lines. Don't forget to include them when wiring up the board. I also left out the capacitors from the LM7805 regulator's input and output to ground; in the finished board, they are soldered on externally. If you want those capacitors, simply solder a 10.0 μF tantalum and a 0.1 μF ceramic capacitor directly to the pins of the regulator, as shown in the schematic in Figure 4-3.

Overview of the Motor Assembly

When you've had enough fun watching the LEDs blink on the breadboard, watching the motor start and stop, and playing with the sketch, it's time to address the mechanical side of this project, which offers a few special challenges. This winder won't be functional until the motor has something to turn; see Figure 4-9 for a detailed diagram of the motor, motor mount, transmission, bearing box, and driveshaft, which comprise the turning assembly.

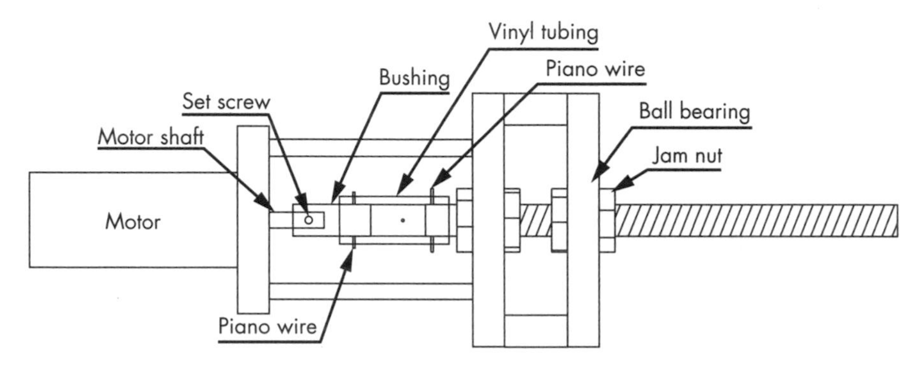

Figure 4-9: The construction entails making a small box that retains the bearings through which the driveshaft is mounted and held in place by jam nuts. The motor, mounted on standoffs, is connected to the driveshaft, and the watch basket will be attached to the other end of the driveshaft.

The driveshaft will need to be mounted through the bearings, and the two can be held together with jam nuts. I chose a fairly standard R4A-2RS bearing, which is a relatively common part and has a 3/4-inch outer diameter, a 1/4-inch inner diameter, and a 9/16-inch thickness. I suggest ball bearings because the prebuilt winder I bought used the brass bushing of the motor as the only bearing, and that's what failed. Because the inside diameter of the bearing was 1/4 inches, I decided to use a standard threaded rod at a 1/4-inch × 20 tpi (turns per inch) or 1/4-inch × 28 tpi rather than attempt to press-fit a 1/4-inch bar into the bearing.

NOTE *I used jam nuts to fasten the rod to the bearings because they were thinner and less obtrusive looking than regular nuts, but you could also use standard nuts.*

Construction

Construction of the winder provided a number of challenges, particularly working with the acrylic material—which was unfamiliar to me before this project. Though there were some rough spots to get over, I learned by trial and error some ways to get the job done. Figure 4-10 shows the completed Watch Winder on its side.

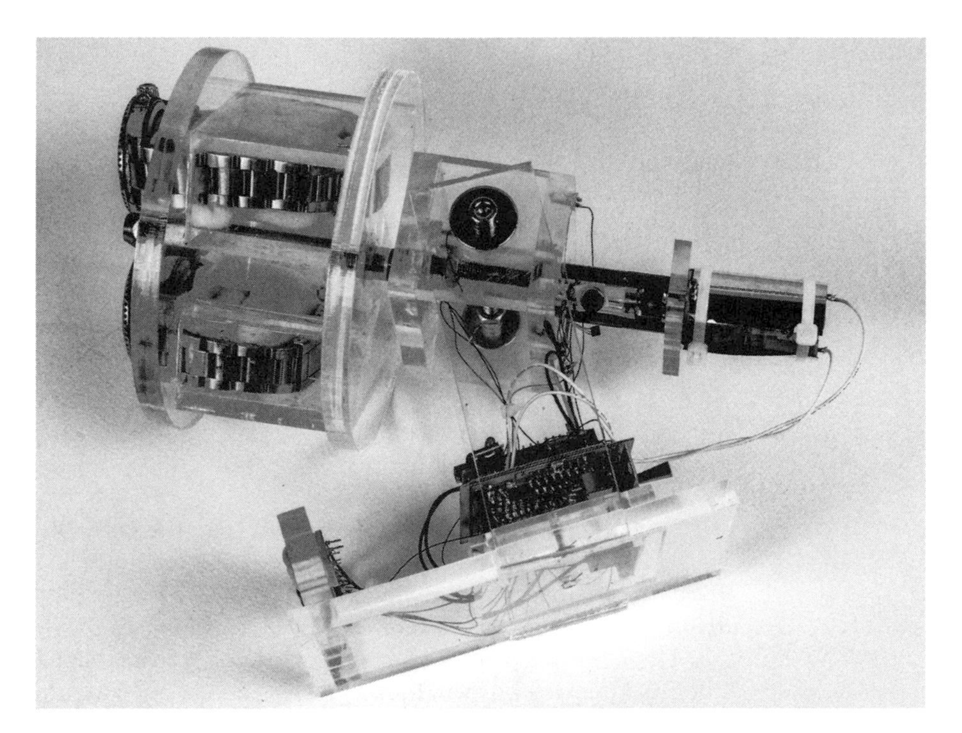

Figure 4-10: The completed Watch Winder on its side, showing the base fabrication, bearing box, motor mount, and watch basket

The toughest part was cutting the acrylic and drilling the holes for the ball bearings. There are several ways to cut acrylic, and none of them is particularly easy. If you use a supplier that will laser-cut the acrylic parts for you, this will be a lot easier. Several companies offer that service for a little more than the price of the raw materials. I mentioned one of them, ZLazr, earlier.

If you have access to a circular saw, that's about one of the easiest DIY approaches. Otherwise, just about any saw will do. I've used a hacksaw, which works better than most if you take your time. (If you go too fast, the acrylic will heat up and start to melt, causing the saw to bind.) I even know some people who have had success scoring and snapping the acrylic sheet. Just use whichever approach works best for you.

See "Acrylic" on page 93 for a list of acrylic shapes needed for the bearing box, the watch basket, and the stand. Cut these pieces now, if you've not done so already, and take the following steps to build the pieces.

Preparing the Motor Mount and Bearing Box Acrylic

First, print the motor mount template from this chapter's folder (see Figure 4-11), cut it out, and align it on the acrylic for the motor mount using the centerline.

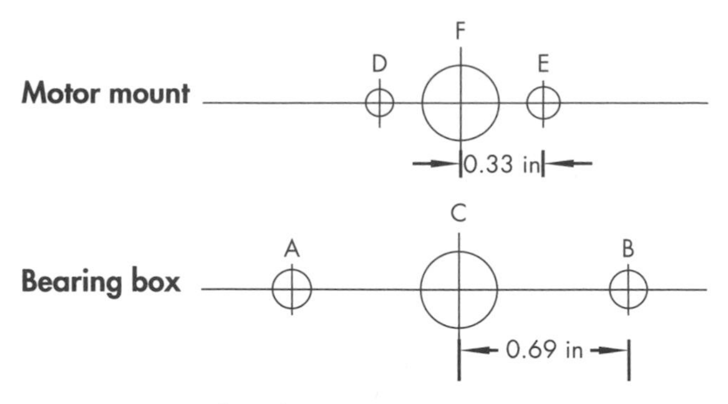

Figure 4-11: Templates for motor mount and bearing box. You can download the templates at https://www.nostarch.com/arduinoplayground/.

Tape the template to the acrylic, lining up the centerline on the center of the acrylic piece, and mark the drill centers for holes A, B, D, E, and F. To mark the holes, just punch the centers with a center punch or nail. Now, drill them; use a 1/8-inch bit for A, B, D, and E and a 3/8-inch bit for F.

Then, set this piece aside, and gather the acrylic for the bearing box. The final bearing box will look like Figure 4-12 once we put it all together.

Figure 4-12: Completed bearing box, before final trim and polish. One bearing is temporarily in place.

Use the bearing box template to mark the drill centers for holes A, B, and C on one of the bearing mount pieces. C will be the bearing hole, and A and B will be for the motor mount's standoffs. Center punch and drill one of the bearing plate's two smaller holes with a 2.5 mm (or #39) drill and tap the hole with a M3-05 tap. These holes will accept the standoffs for mounting the motor. Use the same template to mark only the bearing hole

(C) on the acrylic for the opposite side of the bearing box. Then, take the two pieces of bearing box acrylic that won't hold bearings, and mark the center by drawing lines from corner to corner. Center punch them, and drill 1/4-inch holes for mounting to the stand.

WARNING *When drilling any size hole in the acrylic, securely clamp the piece, as shown in Figure 4-13. Do not hold it manually. If the drill or hole saw binds, the acrylic will want to spin. In any case, keep the drill at a slow speed and advance it very gradually into the work.*

Figure 4-13: Drilling a large hole in the acrylic. Note that the acrylic piece is securely clamped down.

The best solution I found to make the holes for the bearing and other large holes in the acrylic is to drill a relatively small hole—perhaps 1/4 or 3/8 inches—and ream them out with a tapered reamer to the finished dimension (see Figure 4-14). This is, by far, the safest and easiest approach and the one I strongly recommend.

When you ream out the bearing hole, make sure to ream from both sides. This will result in the center of the hole being a slightly smaller diameter than the outsides. Ream until the bearing is a tight fit and then, if necessary, you can use an anaerobic bonding agent to fill in around the edges.

The final prep stage for the acrylic is to sand and finish it. How you cut the acrylic originally will determine how much finishing it will take to get the edges ready. If you had the pieces laser cut, little finishing will be required. For all finishing, I used ascending grades of sandpaper, starting with 220 grit and going up to 1500—that is, 220, 320, 400, 600, and then 1500. Automatic sanders—orbital, belt, vibratory, and so forth—often are too rough, and without special care, they will melt the acrylic. If you use one, try it on a scrap piece first. The sanding process worked well even though some extra sanding was required on roughly cut sections. Additional sanding was required on the opposing piece to make everything fit together as a rectangle—or as a cube in the case of the bearing box.

Figure 4-14: Using a tapered reamer to enlarge the bearing hole to the finished dimension

Use a liquid polish or jewelers rouge to achieve the final polish. Make sure to remove all the wax from the polish from the surface before bonding. Try not to round the edges of the sections so the thinner bonding agent (Weld-On 4) will work well. You want to assure that you have sufficient bonding surface in contact to make a secure bond.

Bonding the Acrylic for the Bearing Box

Now, it's time to use a bonding agent to connect the pieces of your bearing box. Fortunately, bonding the acrylic actually turned out to be somewhat easier than I anticipated.

Where the edges are smooth but not too badly rounded, Weld-On 4 thin bonding fluid should work quite satisfactorily. It partially dissolves the acrylic and forms an actual weld. The most difficult part is keeping the fluid from running where it shouldn't go. If you have larger gaps, or have rounded the edges, try Weld-On 16, which has a higher viscosity and a clear acrylic filler, to fill gaps and voids where necessary.

In both cases, you should follow the instructions on the product, but here's how the acrylic weld works in general: just clamp the dry pieces of acrylic together, and then, using a needle applicator included with the bonding agent, apply a thin layer of acrylic cement to each joint. Capillary action will draw the cement into the joint. For joints where the surfaces are a little more uneven, you can apply the thicker Weld-On 16 to one surface and then attach it to the other surface. Clamping is required for only a few minutes, but allow several hours for final curing.

The system doesn't need a lot of strength, but it should not fall apart when touched. For the parsimonious, a paint stripper like Klean Strip also works well to bond the acrylic. (The chemical behind the bonding agent in Weld-On is methyl chloride, which is the key ingredient in the paint stripper.) Klean Strip is less than one-fourth the price of Weld-On.

NOTE *There are several tutorials on bonding acrylic on the web, too. If in doubt, look one up, and experiment with a few scrap pieces of acrylic first before trying it out on the pieces you worked so hard on.*

After bonding the bearing box, as shown in Figure 4-9, and bonding the side pieces to the bottom, check the alignment. Run the threaded rod through the bearings, and put the jam nuts in place without overtightening them. Make sure the bearings don't bind. If they are not well centered, this can happen, but usually, they will align themselves as you tighten the jam nuts a little. If your alignment is off, you may need to adjust the holes a little with the reamer and touch up with some acrylic cement, but I never ran into that problem. For now, remove the driveshaft from the bearing box.

Building the Stand

The stand is the least complex part of the project. It comprises the two side supports, 1 1/2 × 5 × 1/4 inches, and the base, 3/8 × 3 × 5 1/2 inches. I included the lightbar, 3 × 1 1/4 × 3/8 inches, and the shield mounting, 2 1/2 × 2 1/2 × 3/8 inches, with the stand (see Figure 4-15).

First, drill 1/4-inch holes in each side support a 1/2 inch from the top, centered left to right. Then, bond the two side supports to the base 1 1/2 inches in from the edge of the base that will be the back. Next, drill the holes for the LEDs in the lightbar. I used five LEDs (red, blue, white, yellow, and green) that were 10 mm in size. You can use whatever color combination you choose.

Finally, drill and mount the shield. To find the center of the piece, mark from corner to corner. Then, in the center, drill a #43 hole and tap for a 4-40 screw. Drill a corresponding hole, centered and 2 inches from the rear edge, in the base. Next, use the shield itself, or the drawing from the ExpressPCB print, as a template to drill a hole for the three mounting screws. In designing the board, I failed to leave room for a fourth screw. However, three are more than sufficient, as there is no mechanical force on the board. I used a

2.5 mm (#39) drill and tap for a M3-05 screw that the standoffs will fit into. Figure 4-15 shows the dimensions of the parts and the partially assembled base, including supports, the lightbar, and the shield mount.

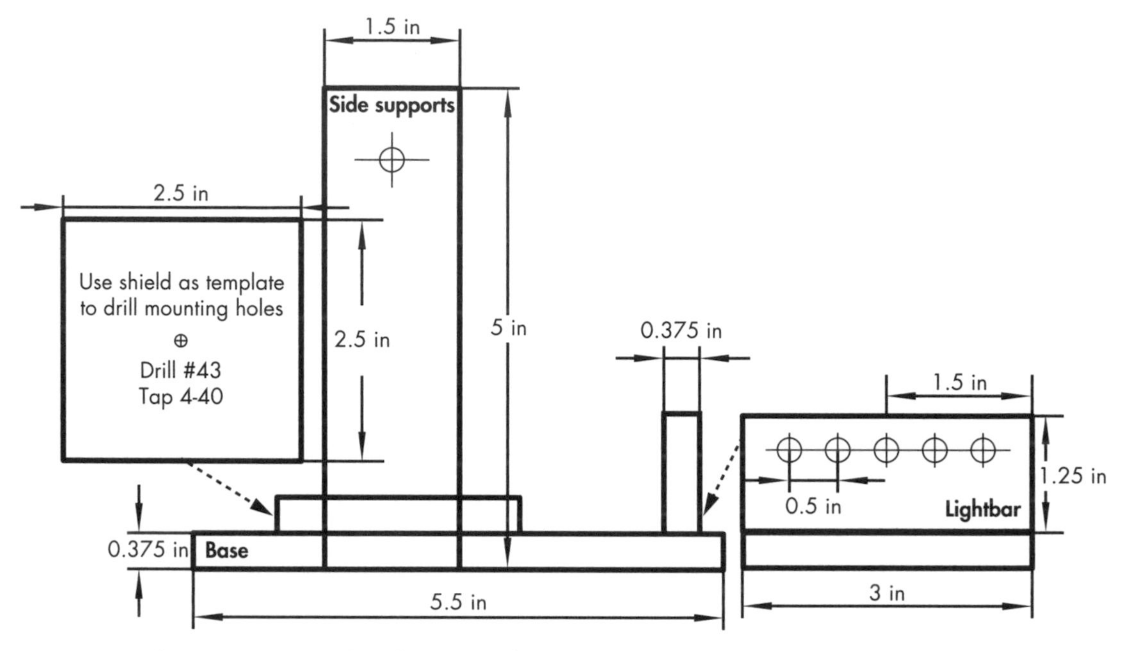

Figure 4-15: The components and configuration of the base and lightbar for the Watch Winder

Preparing the Motor and the Driveshaft

Despite a concerted effort to mark and drill the holes accurately, you may still end up with misalignment between the motor and driveshaft, so this build aims to keep the coupling flexible. My solution might not be on the hit parade of industrial engineers, but I used a length of vinyl tubing to couple the motor to the driveshaft. This coupling has been working for more than a year with no sign of deterioration or problems.

A 1-inch length of heavy-wall vinyl tubing with an outside diameter of 7/16 inches and an inside diameter of 3/16 inches should do the job.

It won't fit the motor shaft or the 1/4-inch threaded shaft without a bit of work, though. (I drove the sales clerk at Lowe's batty buying six of each size they had in stock.) We simply need to reduce the diameter of the threaded shaft and craft a small bushing for the motor shaft.

Trimming the Threaded Shaft

First, trim the diameter of the threaded shaft. Clamp a hand-held electric drill in a vise or parallel clamp, using a folded towel to keep from damaging the drill, and place the shaft where the drill bit would normally go (see Figure 4-16). Then, turn on the drill, and use a sharp file to trim the shaft.

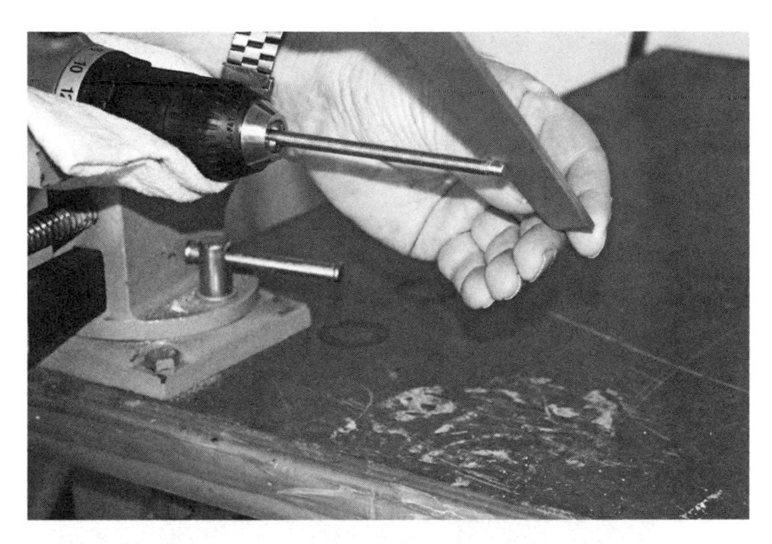

Figure 4-16: Reducing the diameter of the threaded shaft

It should take only a minute or two to reduce the shaft diameter to a little over 3/16 inches for a tight fit on the vinyl tubing.

Creating the Motor Bushing

Next, take a small piece of round stock approximately 3/4 inches long and 3/8 to 1/2 inches in diameter. (I used a 3/8-inch diameter, as it required less work.) Drill a 11/64-inch (#21) hole in the end approximately 3/8 inches deep. The bushing hole needs to be as close to the center as possible, so you might want to mark it with a center punch first (see Figure 4-17).

Figure 4-17: Clamp the shaft of the piece used as the bushing in a vise or pair of vise-grip pliers, and center punch a mark before drilling the hole. Get as close to the center as possible.

Then, place the short piece of stock you cut in the drill as you did the threaded shaft with the center hole toward the drill. File the end without the hole in it. Reduce the diameter by about 1/4 inches to approximately equal the filed end of the driveshaft.

Next, drill a 0.041-inch hole through the bushing, about 3/8 inches from the edge of the bushing.

While you're at it, drill a corresponding hole in the driveshaft. To center punch and drill the driveshaft and the bushing holes, file the end flat so the center punch can find a purchase (see Figure 4-18). These are the holes that will accept the piano wire through the vinyl tubing.

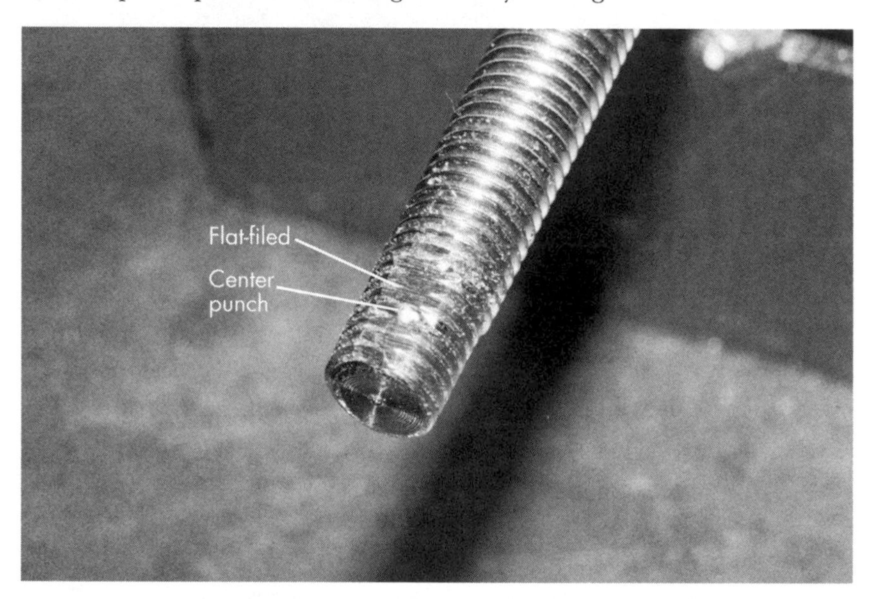

Figure 4-18: The easiest way to drill the holes in the threaded rod and bushing is to file a small flat on the shaft and center punch.

Finally, drill a 2.5 mm (#39) hole and tap an M3-05 hole in the bushing, perpendicular to the hole drilled for the motor shaft. This will accept a set screw for the motor shaft. If you prefer, you can drill a #43 hole and tap for a 4-40 set screw. This set screw holds the bushing in place on the motor shaft, and the two 0.041-inch holes drilled in the driveshaft and motor bushing, respectively, will pin the vinyl tubing in place with piano wire.

Cutting Piano Wire Pins and Completing the Motor Assembly

Cut two pieces of 0.039-inch wide piano wire, 1/2 to 5/8 inches long. This can be a bit of a job. I used an abrasive cutoff wheel on a Dremel tool. Using the corner of a small grinding wheel attached to the drill should work to put a groove in the wire. Once you have a groove in the wire, you can snap it by hand. You can also use any sharpening stone to score the wire, and it should easily snap.

Once that's done, the rest of the assembly should go easily. First, mount the bushing to the motor; put a drop of thread-lock liquid on the set screw before tightening. Then, fit one end of the vinyl tubing over the bushing, and run the piano wire through the 0.041-inch hole in the bushing. Hold one end of the wire tightly in a pair of pliers (small vise-grip pliers work well), and force it through the vinyl, into the hole in the motor bushing, and into the vinyl on the other side. If this proves difficult, try heating the piano wire with a small flame, and then it should go through easily. Figure 4-19 shows the tubing over the motor bushing.

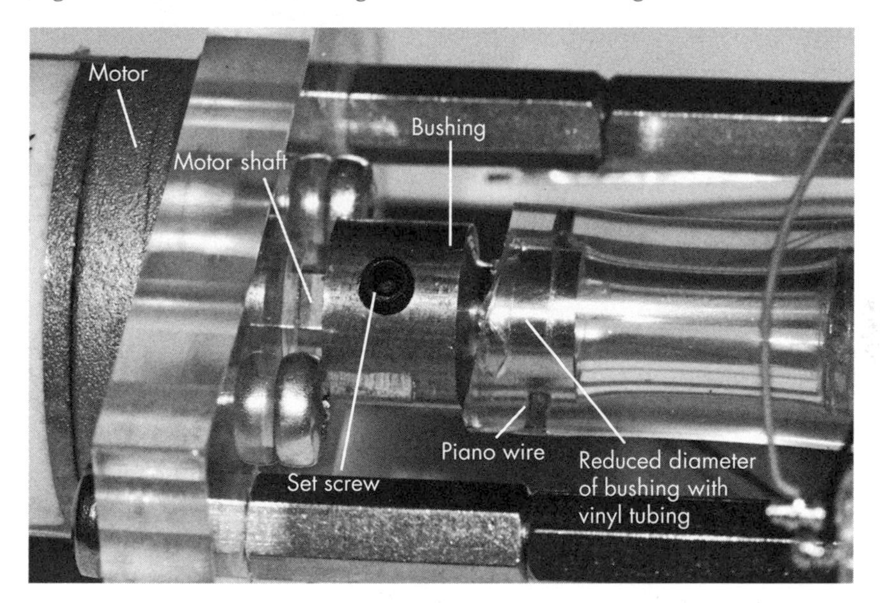

Figure 4-19: The motor shaft with the bushing comprises half of the transmission to the winder. The other half is the reduced driveshaft that goes through the bearing and holds the watch basket.

Attach the motor mount to the motor using M3×3/8-inch screws. Next, screw the motor standoffs into the bearing cage; see Figure 4-20 for how to place them. Take the motor assembly—that is, the motor, mounting plate, bushing, and vinyl—and fit it to the bearing cage standoffs using M3-05×1/2-inch screws.

To make the holes in the vinyl for the driveshaft, just install the driveshaft into the bearing box without the jam nuts, push the vinyl tubing onto it, and install the piano-wire pin as I described in the previous section. If you are concerned about the piano-wire pins coming out, you can wrap a wire tie over them or cover them with a piece of tape. (I never had a problem.) For now, remove the piano-wire pin from the driveshaft and remove the driveshaft from the bearing box until you're ready for the final assembly.

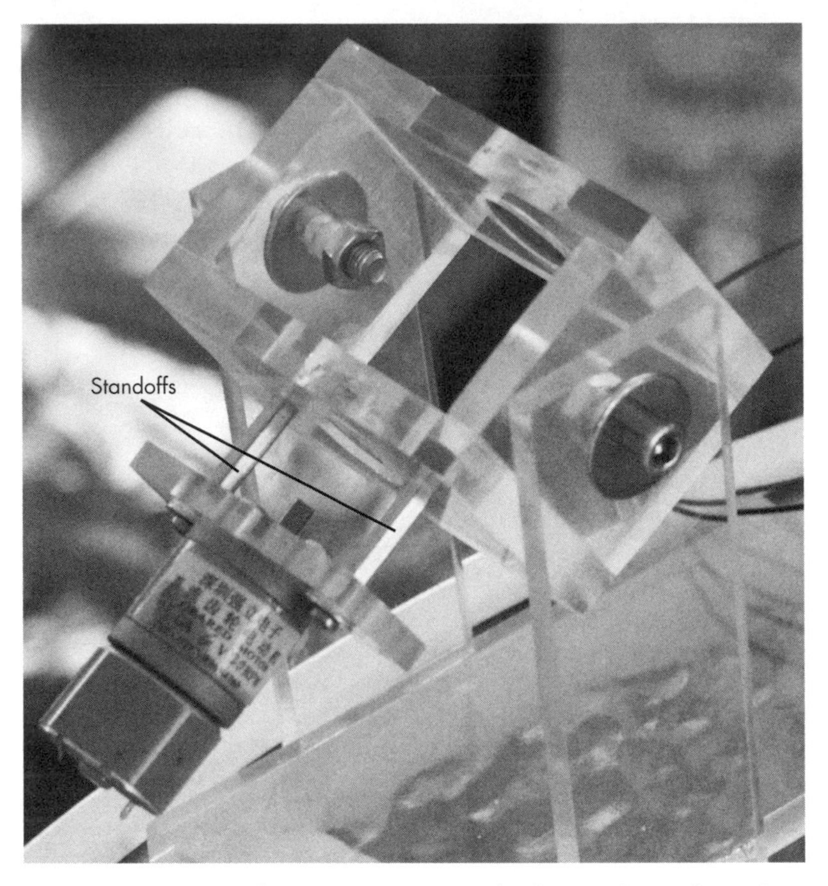

Figure 4-20: Motor and motor mount connected to bearing box with standoffs. The standoffs are threaded into the bearing box while the motor mount is fastened with 3 mm × 1/2-inch screws and washers. The bearing box is mounted to stand with 1-inch-long, 1/4-inch × 28 screws and nuts.

Making the Watch Basket

There are many ways to construct the part of the project that holds the watches. I chose to make my watch basket from acrylic. The construction is relatively straightforward, though it does require some patience. First, take two 5×3/8-inch acrylic disks and carefully mark the center on each. Drill 1/4-inch holes in the center of each. Then, on what will be the top disk, mark the rectangles shown in Figure 4-21 (left).

If you've not done so already, cut the rectangular acrylic pieces I described in the "Parts List" on page 93 for the watch boxes and assemble them following the same instructions given under "Bonding the Acrylic for the Bearing Box" on page 114. (In most of the samples I've made, I used 1/4-inch acrylic; however, 3/8-inch acrylic works fine.) Then, carefully mark out the openings on the disk (see Figure 4-21). Now, bond the watch baskets to one disk, as shown in Figure 4-22. Then, cut the openings to match up with the inside of the watch baskets.

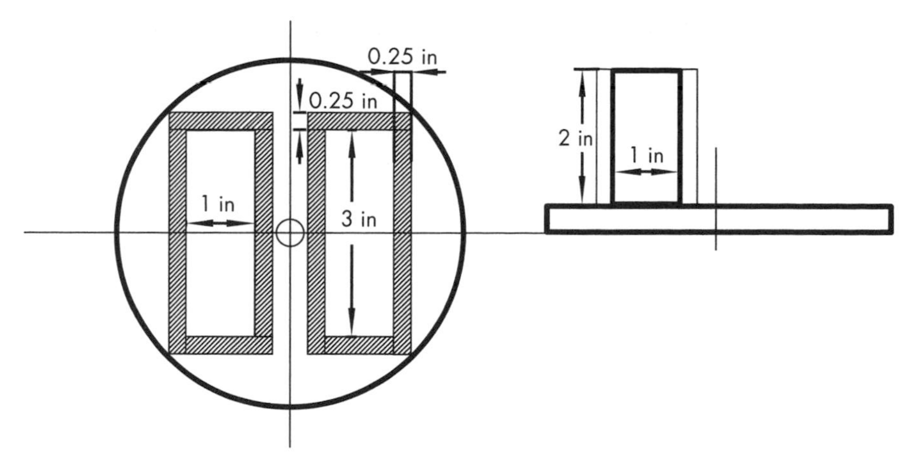

Figure 4-21: Cut out dimensions for the watch holder basket. This pattern can be downloaded as a PDF file from http://www.nostarch.com/arduinoplayground/.

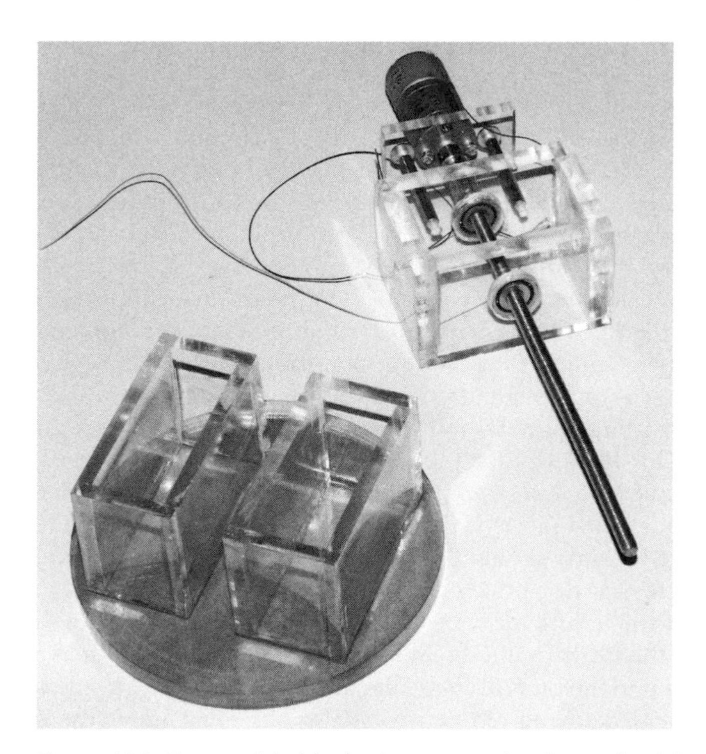

Figure 4-22: The watch holder baskets mounted to the acrylic disk. Above them is the fully assembled motor, bearing box, and driveshaft.

The simplest way to cut the rectangles out is to drill a hole at the corners, being careful not to drill into the watch basket, and use a keyhole or coping saw (or if you're careful, a saber saw) to cut the openings. They have to be cut only on one disk, which will be the top. You can clean up the edges of the cuts a little with a file or sandpaper, but don't spend too much time as it will not be noticeable with the watches and cushions in place.

Finally, mount the basket on the driveshaft. First, thread two nuts onto the driveshaft and lock them against each other—that is, tighten one nut against the other. Then, add a washer, followed by the lower disk of the basket, the top assembly with the watch baskets, a washer, and a nut at the top. If you desire, go to the hardware store to search for a decorative nut to cap off the project.

NOTE *Bonding the bottom disk isn't necessary if you use locking nuts as I described and tighten the basket securely. I didn't bond the boxes to the lower disk, and it has not been a problem.*

At this point, you can mount the finished assembly to the stand using the two decorative 1/4 inch × 28 bolts and nuts. You can also bond the lightbar to the base using acrylic cement and drill the hole for mounting the shield mounting plate.

Adding the LEDs

You're just about done. Locate and mount the LEDs on the acrylic anywhere you like, and wire them up to the shield. You can see where I placed mine in Figure 4-1. (The way the acrylic conducts the light produces some neat effects.) You may want to drill some blind holes to mount the LEDs in. Simply drill a hole the diameter of the LED but not completely through the acrylic. If you're careful, you can probably do a neat job in running the wires so they can barely be seen.

Because the LEDs are critical to the ultimate appearance of the winder, their placement and mounting is an important component of the finished product. Drilling the holes for mounting the LEDs can be a little tricky because if you bought a variety, they may have slightly different diameters. As a starting point, try 3/16-inch holes for 5 mm LEDs and 25/64-inch holes for 10 mm LEDs. I found the best way to get the right size was to drill a sample hole in a scrap piece of stock and try it before venturing to drill into a finished piece. If the hole is a bit small, a simple touch up with the tapered reamer should fix that. If a hole ends up a little large, try filling it in with some acrylic cement, such as Weld-On 16.

If you have some wire-wrap wire and a wire-wrap tool, you can wire-wrap the LEDs to the shield instead of soldering. This makes a neat connection to the back of the LEDs and is relatively inconspicuous because of the small diameter of the wire. It also lets you connect the wire close to the LED, as in shown Figure 4-23. If you soldered it, you might risk overheating the junction of the LED.

Wire-wrap or solder the leads from the LEDs to the shield. I soldered a header to the shield so that it was easy to wire-wrap or solder the leads from the LEDs to the shield directly. You do have to solder leads from the motor to the other end of the shield, but it should not be a problem.

The wires from the Hall effect sensor can be soldered to the shield and to the leads of the sensor. Measure the wires so they fit neatly where you plan to mount the Hall effect sensor. You can use a connector, but let's keep it simple. The sensor itself and the magnet are mechanically mounted using double-sided foam tape.

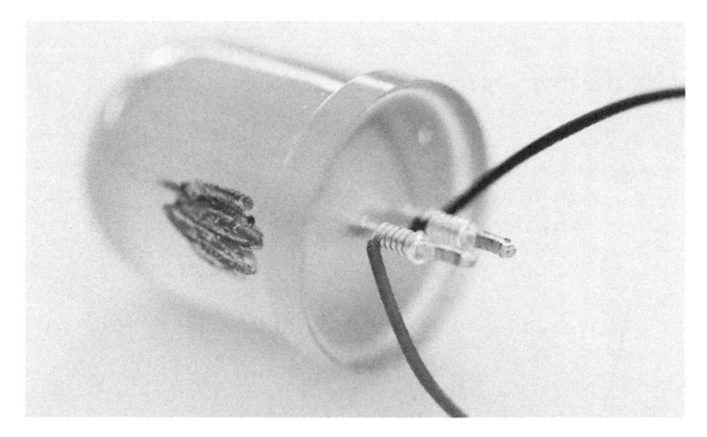

Figure 4-23: Wire-wrap wire on an LED. These wires are very fine (30 gauge) with thin insulation, so they are unobtrusive. Small wire ties can neaten up the wiring. I suggest marking the LED's positive terminal ahead of time, as I've done here, and wiping it off later.

Leaving the Components on Display

Now, what to do with the shield and Nano? The theme of this project has been transparency, so I suggest letting everything hang out: mount the bare board on standoffs right out in the open with the switch and power jack at the back (see Figure 4-24).

Figure 4-24: The Nano and shield mounted on the completed Watch Winder. Only three standoffs were used on the shield.

I placed the electronics directly under the bearing box, between the uprights holding the entire assembly to the base. Then, I mounted the board on a separate piece of acrylic screwed to the base with a flat-head 4-40 screw.

Keeping the Watches in the Basket

To hold the watches in the watch basket, I simply cut a block of fine foam sponge, and it worked well. If you want something a little dressier than a sponge, you can sew small pillows that you put the watch band around. I don't have any sewing ability, so I stuck with the sponge.

NOTE *The open frame works well overall, but it could collect dust. If you have a fastidious streak, you could build an acrylic box to cover the entire winder from 3/16-inch-thick acrylic sheet. Or you could just buy a can of dust spray, as I did. Some have also suggested to me that the entire winder could be mounted on a piece of hardwood, such as walnut or some other decorative hardwood, to add a finishing touch.*

Design Notes

Now that you've seen how I built the Watch Winder, I'll walk you through some key design decisions I made that you might want to do differently.

Total Rotation Adjustment

It's possible to vary the total number of turns the Watch Winder makes without changing the sketch, though I chose not to do this.

You can use a potentiometer to create a variable voltage and input it to one of the analog inputs of the Arduino. Then, you can substitute that value for one of the delays in the sketch to vary the number of revolutions per day. Here's how to install the potentiometer and the tweaks you'd need to make to the sketch.

Hardware Changes

Connect the upper and lower terminals of a potentiometer to the positive and negative rails of the system. Solder pads have been included on the shield for this purpose. You needn't use a full-size potentiometer; a small trimmer (10 turn is best) will do nicely, and it saves a lot of space. Connect the potentiometer's center pin (in the shield) to an analog pin of the Nano. Because the Hall effect switch uses pin A0, I suggest using analog input A1. Solder points have been provided in the shield.

Software Changes

On the sketch, there are several things you must do. First, tell the Nano that input A1 is in play. Go to the top of the sketch to add the following:

```
const int revSet = A1;
```

Then, a little further, identify the value the potentiometer is set at as follows:

```
int revNumber = 0;
```

revSet is the arbitrary name I gave to the input A1, and revNumber is the arbitrary name I gave to the number you will substitute in the sketch for one of the delays.

The potentiometer will give values from 0 to 5 volts. Because the analog input, A1, is connected to a 10-bit ADC, it will generate 1,024 digital values between 0 and 1,023. In other applications, it's been necessary to map these 1,024 values to some other set of values. However, in this particular situation, it's easiest to use the values as is.

In the sketch file, move to the line after `void loop() {` and assign the value of A1 to revNumber as follows:

```
revNumber = analogRead(revSet);
```

Go back to where we define some of the delays in the sketch. Change

```
int timer = 500;
```

to

```
int timer = 0;
```

Finally, go back to where you entered `revNumber = analogRead(revSet);` and after that, enter the following:

```
timer == revNumber;
```

Now, each time the sketch calls for the timer value, the revNumber value should be used automatically, which will give you a wide variation in delay. The resulting variation runs from 200 to 1,200 revolutions per day.

How Many LEDs to Use and Where to Put Them

I originally imagined the Watch Winder having only two LEDs to indicate the direction of rotation. The first version used LEDs attached to the motor-direction pins, D12 and D13, of the Nano (see Figure 4-2). One pin was on for the duration of rotation in one direction, and vice versa. Red and green LEDs were used to indicate which direction the winder was going in, like running lights on a boat.

But that's still pretty boring, and there were all those pins sitting there not doing anything. Furthermore, if you invited a friend over to see your winder, it would sit there, doing nothing most of the time—and so would your visitor. So I decided to spice up the project with several more decorative LEDs. I also decided to add more variability by having the sketch provide some animation, calling for the watches to turn a varying number of times—sometimes shorter but more often. I even added a "ping-pong effect."

Because the half of my brain dealing with artistic matters apparently never developed, I'll leave the placement of the LEDs to you. On the unit in Figure 4-1, there are four in the bearing box and five in a lightbar, but you could place them anywhere.

I arbitrarily chose nine LED channels, and in some cases, I used two LEDs per channel. I used two LEDs for each direction of the motor—the two channels, D12 and D13, that serve double duty driving the LEDs as well as driving the motor. D2 and D13 power two LEDs each. D4 through D10 power the other LEDs, the two behind the watch basket—D9 and D10, each with two LEDs—and the five out in front in the lightbar—D4–D8. D11 is reserved for future developments you may want to include.

Motor Voltage

One beta tester of the Watch Winder experienced difficulty running a 6V motor from the 5V supply. The complaint was insufficient torque. The solution, should you run into this problem, is to run the second supply (VCC_2) of the H-bridge directly from the 9V supply. To do this, you are either going to have to cut the traces to pin 8 or remove that pin and solder it separately to the 9V supply. Because the motor runs so intermittently, there is little risk of burning it out. It may not be an elegant solution, but it works. Incidentally, out of about 20 different motors sampled, that was the only one that experienced that problem. And as addressed earlier, the higher speed of the motor at the higher voltage will have no, or little, effect on the number of rotations.

How Many Rotations Does the Watch Winder Make?

If you really have to know how many rotations the Watch Winder makes, here's a solution. The internal counter serves to sequence the sketch but does not accurately reflect the total number of revolutions the motor makes. While we could have counted the rotations internally, it would have required a separate readout or being hooked up to the serial monitor. But if you need to keep count, you can add a small external counter. Because you will need it only on rare occasions when changing the revolution count, you can plug it in—a provision is made in the shield—when you need it and save it for other projects when you don't. The external counter in Figure 4-25 is self-powered and costs under $8. See the "Parts List" on page 93 for details.

The external counter is not required; however, it could be a nice accessory to include for this and other projects. It is not included in the design because it is used only on occasion and can be plugged in. I used a two-pin female header on the shield—the connections are labeled *GND* and *X-ctr* on the screen layer of the shield—and included a two-pin plug on wires from the counter. The count connection goes from ground to pin 4 of the counter and from the Hall effect sensor to pin 1. You can add a reset button from ground to pin 3 of the counter. The counter comes from the manufacturer with little information, so Figure 4-25 shows a view from the back with the pushbutton reset on the counter on your right.

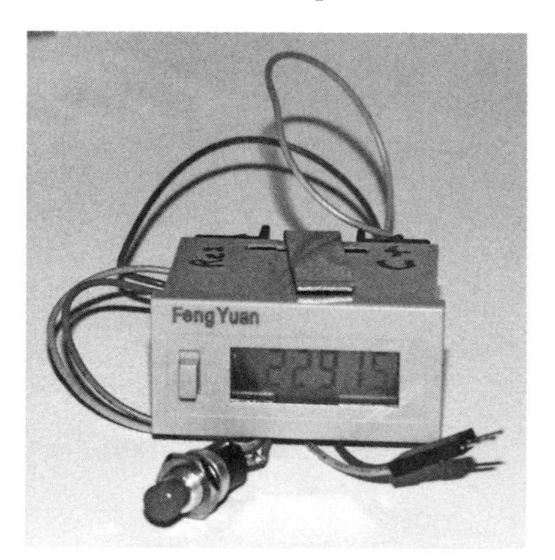

Figure 4-25: An external hardware counter with a reset button added

Even if this Watch Winder didn't keep my watches wound, I think it would still be a great sculpture. And when you are done with this project, perhaps your next Arduino build will be just that: a kinetic, blinking, moving piece of art.

5

THE GARAGE SENTRY PARKING ASSISTANT

This project is a reliable electronic device to gauge the distance you need to pull your car into your garage. If you park in a garage, you're probably familiar with the problem: how far do you pull your car into the garage to make sure there's room in front for whatever is there and enough space behind so the garage door will close? Some people suspend a tennis ball on a string from the ceiling and stop at the point when the ball meets the windshield. That works fine, but the ball is a pain to set up and adjust, and it often gets in the way if you want to use the garage for something other than parking the car.

Arduino offers a better solution. This Garage Sentry project is the electronic version of the classic tennis-ball-on-a-string device, only better. The Garage Sentry accurately detects when your car reaches exactly the right position in the garage and sets off an alarm that blinks so you know when to hit the brakes.

In addition, at the end of the chapter, I'll show you how to modify the basic Garage Sentry into a deluxe version that alerts you when you're getting close to the perfect stopping point.

INSPIRATION BEHIND THE GARAGE SENTRY

This project evolved out of playing with an *ultrasonic transceiver module*, a device that emits sound waves and then detects them after they travel to an object, reflect off that object, and travel back to the module. The output of the module allows a microcontroller to measure the time it takes to travel to and from the object and, knowing the speed of sound, determine the distance. To test the ultrasonic transceiver's sensitivity and limits, I used the battery-operated breadboard version in my garage, which had enough space to move objects around for different distances. It turns out cars are great reflectors for ultrasonic energy. From this experimentation, I was inspired to turn my test apparatus into a Garage Sentry.

Required Tools

This project doesn't require many tools or materials, but you will need the following tools for both the standard and deluxe versions:

Drill with a 3/8-inch or 1/2-inch chuck (powered by battery or with 110/220V from the wall)

Drill bits for potentiometer (9/32 inches), power input (1/4 inches), and LED (3/8 inches)

Soldering iron and solder

Tapered reamer set

Philips head and slotted screwdrivers

Pliers (I recommend needle nose.)

(Optional) 1/4-inch tap

Parts List

You'll need the following parts to build the basic Garage Sentry:

One Arduino Nano (or clone)

One HC-SR04 ultrasonic sensor

Two high-intensity LEDs (>12,000 MCD)

Two 270-ohm, 1/4 W (or more) resistors (to limit current to the LEDs)

One 20-ohm, 1/8 W potentiometer

Two NPN-signal transistors rated for a collector current of at least 1.5 A (I used ZTX649 transistors.)

One enclosure (I recommend a blue Hammond 1591 ATBU, clear 1591 ATCL, or something similar.)

(Optional) One 0.80-inch aluminum strip for mounting bracket

(Optional) Two 1/4-inch × 20-inch × 3/4-inch bolts with nuts

One section (approximately 1×1 inch) perforated board (can include copper-foil rings on one side)

One 3.5 mm jack

Two 2-56×3/8-inch screws and nuts

Two additional 2-56 nuts to use as spacers

One 9V, 100 mA plug-in wall adapter power supply (Anything from 7.5V to 12V DC at 100 mA or upward should work well.)

One piece of double-sided foam tape about 3-inches long

One LM78L05 (TO-92 package) regulator (for the breadboard build only)

28- or 30-gauge hookup wire

(Optional) Wire-wrap tool and wire

Because the basic version doesn't require a lot of additional components, I suggest building the circuit on a standard perforated circuit board instead of a shield. To power your circuit, you can use a 9V, 100 mA wall adapter plugged into a 3.5 mm jack (see Figure 5-1). You shouldn't need an on/off switch.

Figure 5-1: I used a Magnavox AC adapter, but any similar power supply with a DC output from 7.5V to 12V should work. These are readily available online and cost from under $1.00 to about $3.00.

Be sure to use two bright LEDs that are clearly visible, even when a car's headlights are on. Bright LEDs range from 10,000 MCD (millicandela) to more than 200,000 MCD. The brighter, the better; just remember that brighter LEDs require more power, so the current-limiting resistor will need a higher power rating for the brighter lamps. The 270-ohm current-limiting resistors result in a current drain of about 30–40 mA each with the 12,000 MCD LEDs I used at 5V. (Power equals volts times amps, or $P = VI$, so at 40 mA and 5V, you'd have 0.20 W.) It's best to use a 1/2 W or greater resistor even though you can easily get by with a smaller value—as I did with 1/4 W—because the LEDs are on only intermittently.

Deluxe Parts

In addition to the components for the basic Garage Sentry, you'll need the following extra components if you want to build the deluxe version:

Two high-intensity green LEDs
Two high-intensity amber LEDs
Two additional 270-ohm, 1/4 W resistors
Two additional NPN-signal transistors
One Hammond 1591 BTCL enclosure (to replace the 1591 ATCL)
One PCB (shield)

Downloads

Sketches *GarageSentry.ino* and *GarageSentryDeluxe.ino*
Drilling template *Transducer.pdf*
Mechanical drawing *Handle.pdf*
Shield file for Deluxe Garage Sentry *GarageSentryDeluxe.pcb*

The Schematic

Figure 5-2 shows the schematic for the Garage Sentry. R1 and R3 are the 270-ohm resistors for the LEDs and should be 1/4 W or larger. If a higher wattage resistor is not available, you could place several resistors in parallel to gain the required wattage. First, find the right resistor value with the formula:

$$\frac{1}{R_{\text{total}}} = \frac{1}{R_1} + \frac{1}{R_2} + \frac{1}{R_3} + \ldots + \frac{1}{R_n}$$

You can also use an automatic calculator, such as the one at *http://www.1728.org/resistrs.htm*, which is a lot easier than doing the math yourself.

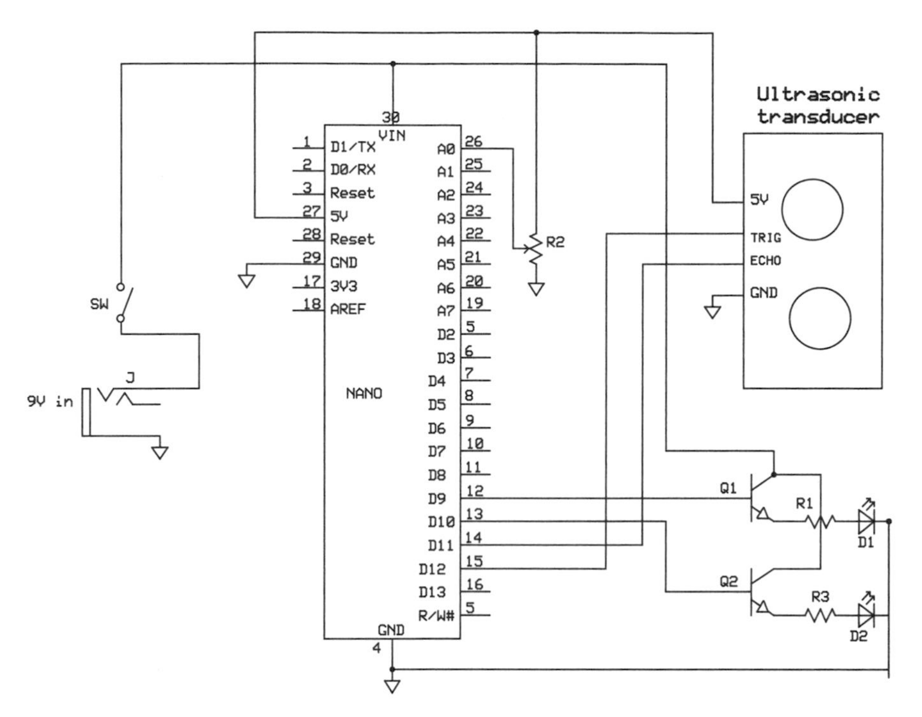

Figure 5-2: Schematic diagram of the Garage Sentry

To avoid extra calculations, select resistors of the same value. This way, the same amount of current flows through each one. For example, two 1/8 W resistors in parallel will give you a 1/4 W value.

If you do use resistors of different values, you will have to calculate the current flowing through each and the total dissipation.

This schematic also leaves you with room to customize your alarm. While this version of the project uses LEDs to create a visual alarm, with a slight modification, you can easily create an audible alarm as well. Simply replace either the red or blue LED with an audible device, such as a Sonotone Sonalert, and the alarm will sound. To replace an LED, you would need to connect the Sonalert across that LED's connections; just make sure to get the polarity correct. Alternatively, you could keep both LEDs and add an audible device for a third warning.

NOTE *In this project, the Nano takes advantage of its on-board voltage regulator, which is why there's no external regulator in the schematic.*

Basics of Calculating Distance

This project measures the time it takes for a sound to originate, bounce off an object, and be received back at the point of origin, and it uses that time to calculate the distance between the object and the sensor.

The basic distance calculation is not much different from determining the distance of a storm by counting the seconds between a lightning flash and a thunderclap. Each second represents a distance of 1,125 feet, or about 0.2 miles. Given that sound travels at 1,125 feet per second in air at sea level, if there's a five-second delay between a lightning flash and the thunderclap, you can determine that the storm is roughly a mile away. In the case of the Garage Sentry, once you know how long it takes for the sound to make a round trip and know the speed of sound, you can calculate the distance according to the time-speed-distance formula:

$$\text{Distance} = \text{Speed} \times \text{Time}$$

How the Garage Sentry Works

This project takes advantage of *ultrasonic sound*, which, unlike thunder, is above the hearing range of most individuals. If your hearing is good, you can detect sound ranging from about 30 Hz to close to 20 kHz, although hearing attenuates quickly above 10 kHz or 15 kHz.

NOTE *For reference, middle C on the piano is 261.6 Hz. Young children (and most dogs) can often hear high frequencies, but hearing, especially in the upper registers, deteriorates quickly with age.*

The ultrasonic transceiver module used in this project sends out pulses at a frequency of about 25 kHz and listens for an echo with a microphone. If there is something for the signal to bounce off, the system receives the return echo and tells the microcontroller a signal has been received and to calculate the distance. For the Garage Sentry, the unit is placed in the front of the garage, and the signal is sent out to bounce off the front—or rear if you are backing in—of your vehicle. To calculate your car's distance from the ultrasonic transceiver, the Arduino measures the time it takes for the signal's round trip from the transceiver to the target and back. For example, if the Arduino measures a time of 10 milliseconds (0.010 seconds), you might calculate the distance as:

$$\text{Distance} = 1125\frac{\text{ft}}{\text{s}} \times 0.010\ \text{s} = 11.25\ \text{ft}$$

Ah, but not so fast. Remember the signal is traveling to the car and then back to the microphone. To get the correct distance to the vehicle, we will have to divide by two. If the controller measures 10 milliseconds, then the distance to your car would be:

$$\text{Distance} = \frac{1{,}125\frac{\text{ft}}{\text{s}} \times 0.010\ \text{s}}{2} = 5.625\ \text{ft}$$

The HC-SR04 ultrasonic module sends out a signal at the instruction of the Arduino (see Figure 5-3). Then, the sketch instructs the transmitter to shut down, and the microphone listens for an echo.

Figure 5-3: The ultrasonic sensor module. The back of the module (bottom) has connection terminals at the bottom.

If there is an object for the signal to bounce off, the microphone picks up the reflected signal. The Arduino marks the exact time the signal is sent out and the time it is received and then calculates the delay.

The HC-SR04 module is more than a speaker and microphone, though. The module includes transducers—a loudspeaker and mic—and a lot of electronics, including at least three integrated circuits, a crystal, and several passive components. These components simplify its interface to the Arduino: the 25 kHz tone is actually generated by the module and turned on and off with the microcontroller. Some of the components also enhance the receiver's, or the microphone's, sensitivity, which gives it a better range.

The range of the HC-SR04 ultrasonic transducer is approximately 10 to 12 feet. The returning signal is always a lot weaker than the transmitted signal because some of the sound wave's energy dissipates in the air (see the dotted lines in Figure 5-4).

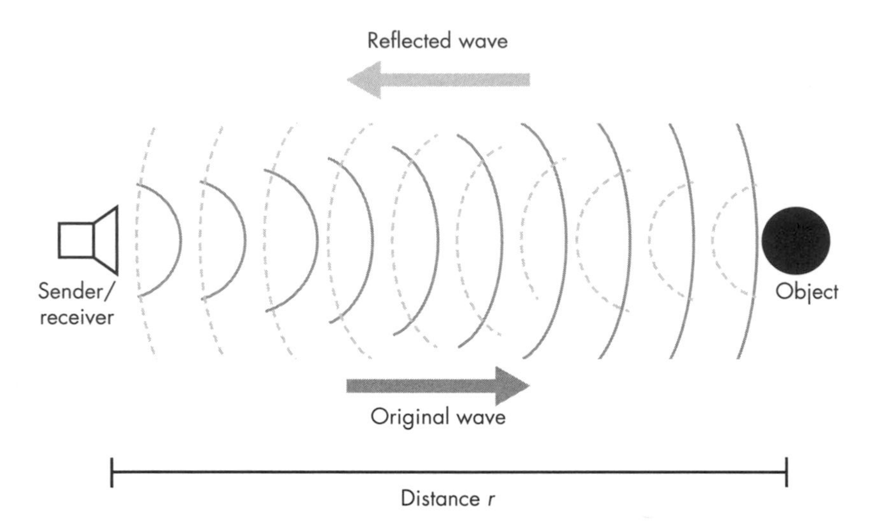

Figure 5-4: In this project, sound is transmitted from a sender, bounces off an object, and is received.

The arithmetic to calculate the distance between the sender and the object is not difficult. You take the number of microseconds it takes for the signal to return, divide by the 73.746 microseconds it takes sound to travel an inch, and then divide by two because the signal is going out and coming back. The full arithmetic for this appears later in "Determining Distance" on page 141.

The sketch provides a response in inches or centimeters depending on your preference. We'll use inches for setting up the distance for the alarm, but converting to centimeters simply requires a remapping of the analog input and setting the numbers a bit differently. The sketch also does the basic arithmetic for determining the centimeter measurement for you.

With the high-level overview out of the way, let's dig in to how you'll wire the Garage Sentry.

The Breadboard

The entire Garage Sentry fits on a small breadboard, so you can set it up, program it, power it with a battery, and walk around to test it out. As you play with it, I'm sure other applications of ultrasonic technology will come to mind.

The breadboard I assembled, shown in Figure 5-5, is powered by a 9V battery. Usually, you could wire the battery directly to the VIN of the Nano and use the Nano's built-in voltage regulator. But you'll power the Nano with a USB cable when you program and test it for the first time, so on the breadboard, you'll set up the positive and negative rails for 5V for both the Nano and the ultrasonic module. To avoid risking damage to the Nano or the module and avoid overcomplicating the build, I included a single-chip external voltage regulator (LM78L05) so the entire breadboard runs on 5V. Take a look at Figure 5-6 to see how it's wired up.

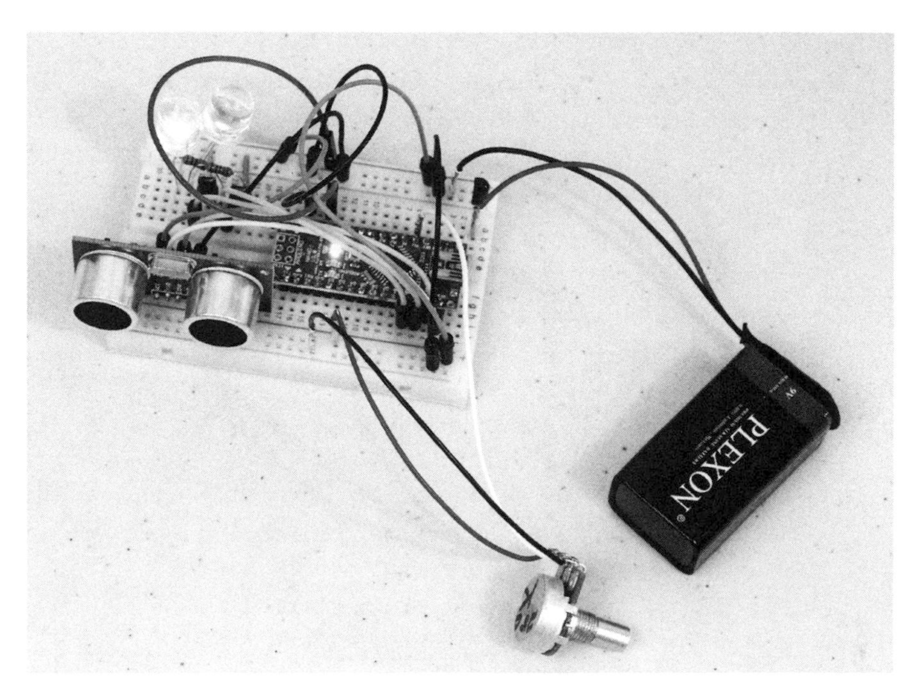

Figure 5-5: Here's the breadboard wired up. I used a 9V battery so I could experiment in different environments. Both LEDs look illuminated because of the length of the exposure of the camera.

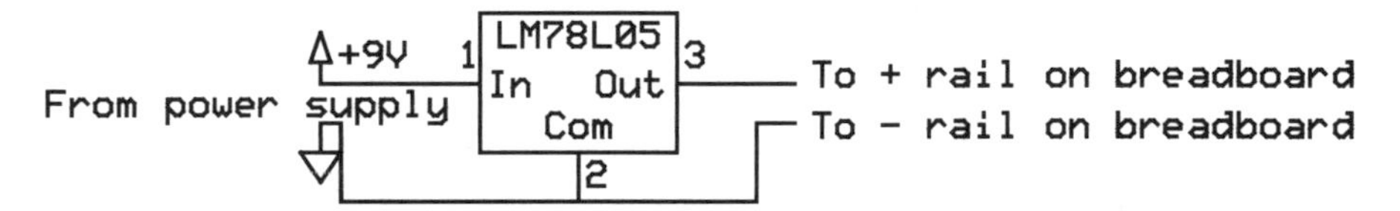

Figure 5-6: This is how the LM78L05 TO-92 regulator is wired up on the breadboard. Bypass/filter capacitors are not required.

Here's a blow-by-blow list of the steps to wire the breadboard:

1. First, put the ultrasonic module at the lower end of the breadboard facing out, and plug the Nano in to the breadboard, leaving four rows of connections above it.
2. Make sure the positive and negative rails (red and blue stripes) on the left and right are connected properly—red to red, blue to blue. If you connect red to blue, it will cause a major problem.
3. Connect the red positive rail to the 5V power supply (pin 27 of the Nano, labeled *5V*). This is necessary if you are operating from the USB connector.
4. Connect pin 4 of the Nano (labeled *GND*) to the breadboard's blue negative rail.
5. Connect VCC of the HC-SR04 transducer to the red positive rail.
6. Connect GND of the HC-SR04 transducer to the blue negative rail.

7. Connect TRIG of the HC-SR04 transducer to pin 15 (D12) of the Nano.
8. Connect ECHO of the HC-SR04 transducer to pin 14 (D11) of the Nano.
9. Insert two ZTX649 transistors into the breadboard. Select an area where all three pins of each transistor can have their own row.
10. Connect pin 12 (D9) of the Nano to the base of transistor Q1.
11. Connect pin 13 (D10) of the Nano to the base of transistor Q2.
12. Connect the collectors of both transistors to the red positive rail.
13. Connect the emitter of transistor Q1 to one end of a 270-ohm resistor.
14. Connect the other end of the 270-ohm resistor connected to the emitter of transistor Q1 to a blank row on the breadboard.
15. Connect the emitter of transistor Q2 to one end of another 270-ohm resistor. Connect the other end of the 270-ohm resistor connected to the emitter of transistor Q2 to another blank row on the breadboard.
16. Connect the + (long end) of LED (D1) to the 270-ohm resistor and the other end to the blue negative rail.
17. Connect the + (long end) of LED (D2) to the second 270-ohm resistor and the other end to the blue negative rail.
18. Connect one end of the 20-ohm potentiometer to the red positive rail.
19. Connect the opposite end of the potentiometer to the blue negative rail.
20. Connect the wiper (center) of the potentiometer to analog pin A0 (26) of the Nano.

You should be good to go! If you use the AC connection, simply connect it to the VCC connection of the Nano.

To add a battery connection, include the 78L05 with its center pin to ground (negative rail), the input to the positive side of the battery, and the output to the positive rail (see Figure 5-7). Connect the negative terminal of the battery to the negative rail.

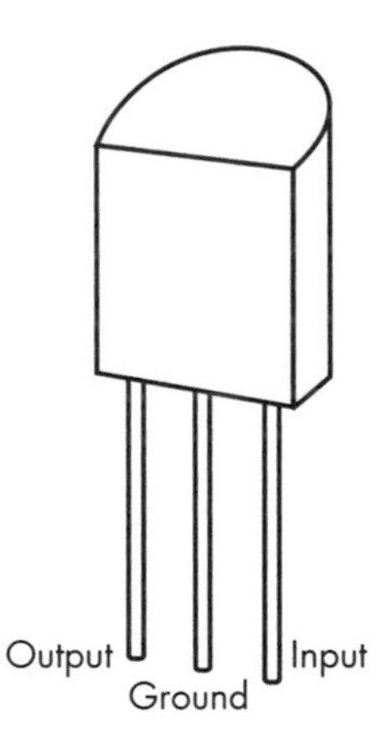

Figure 5-7: Pinout of the 78L05 regulator

The Sketch

Once the breadboard is complete, the sketch can be loaded onto the Nano. Download the *GarageSentry.ino* file from *https://www.nostarch.com/arduinoplayground/*. To load the file onto the Nano, follow the instructions outlined in "Uploading Sketches to Your Arduino" on page 5. Remember to select the correct board type. Once it's loaded, the unit is ready for experimentation.

The sketch for the Garage Sentry serves several functions. It tells the ultrasonic sensor to generate a wave and detects how long it takes the echo to return. It then calculates the distance based on that time and, if necessary, alerts you to stop by turning on the LEDs. Here's the sketch in full; I'll walk you through it next.

```
/* Garage Sentry
*/

int ledPin = 10;
int ledPin1 = 9;
int count;
int analogPin = A0;
int val;
int y;

void setup() {
  Serial.begin(9600);
  pinMode(ledPin, OUTPUT);
  pinMode(ledPin1, OUTPUT);
  pinMode(analogPin, INPUT);
}

void loop() {
  val = analogRead(analogPin);
  long duration, inches, cm;
  //Give a short LOW pulse beforehand to ensure a clean HIGH pulse:
❶ pinMode(12, OUTPUT);  //Attach pin 12 to TRIG
  digitalWrite(12, LOW);
  delayMicroseconds(2);
  digitalWrite(12, HIGH);
  delayMicroseconds(5);
  digitalWrite(12, LOW);

  pinMode(11, INPUT);  //Pin 11 to receive ECHO
  duration = pulseIn(11, HIGH);

  //Convert the time into a distance
  inches = microsecondsToInches(duration);
  cm = microsecondsToCentimeters(duration);
  val = map(val, 0, 1023, 0, 100);
  if(inches == 0)
    digitalWrite(ledPin, LOW);

  if(count == 0 && inches > 0 && inches < val) {
❷   for(y = 0; y < 200; y++)
    {
      digitalWrite(ledPin, HIGH);
      digitalWrite(ledPin1, LOW);
      delay(100);
      digitalWrite(ledPin, LOW);
      digitalWrite(ledPin1, HIGH);
      delay(100);
    }
```

```
    count = count + 1;
  }

  digitalWrite(ledPin1, LOW);
  if(inches > 10) {
    //delay(1000);
    count = 0;
  }
  Serial.print(inches);
  Serial.print("   inches ");
  Serial.print(count);
  Serial.print(" count   ");
  Serial.println();
  Serial.print(" Val       ");
  Serial.println (val);
  delay(100);
}
long microsecondsToInches(long microseconds) {
  return microseconds / 74 / 2;
}
long microsecondsToCentimeters(long microseconds) {
  return microseconds / 29 / 2;
}
```

First, we define several variables, establish parameters, and load libraries (if any). In this case, define `ledPin` and `ledPin1`, which will serve as the alarm. Other definitions (`int`) include `cm` and `count` (a variable that will be used internally), `analogPin` (as A0), `val` (to hold the limit information), and `y` (used in the loop).

Inside the setup() Function

Next is the `setup()` function. Here, you set up Arduino features that you might want to use; this sketch includes the serial monitor, which you probably will not need in the final product but is often useful in debugging code, particularly if you want to change the code. This sketch sets the rate of the monitor at 9600 baud, which is standard in many applications. It also defines the mode of the pins you'll use as either input or output. You could set the `pinMode` values at almost any point in the code, including before or inside the setup; they're also often defined within the main loop, particularly if the definitions are expected to change.

Inside the loop() Function

The `loop()` function is where everything really happens. The loop continually executes unless it's delayed or halted by a command. So even when it appears that nothing is happening, the controller is continually cycling through the code. In this application, one of the first tasks the controller performs in the loop is to set the variable `val` to store the input from the potentiometer connected to the analog pin (`analogPin`).

In order to initiate the ultrasonic module's transmit/receive function, the sketch first calls for a low signal to be sent to the transmitter (`TRIG`) to purge the module to assure that the following high signal will be clean. You can see this in the lines starting at ❶.

Next, there's a delay to let things settle before the sketch writes a high to the transmit pin, which orders the transmitter to transmit an ultrasonic signal. This is followed by another delay, and then the sketch drives digital pin 12 low to turn off the transmitter and activates the receiver by calling the `pulseIn()` method.

Determining Distance

If there's no echo—that is, if `inches == 0` or inches approaches infinity—the controller continues to run the code until it reaches the end and then starts again at the beginning. If it detects an echo, the number of microseconds between turning the transmitter on and receiving signal (`duration`) is then converted to both inches and centimeters. This gives us a measurement of how far the transceiver is from the object. Note that throughout this explanation, I will refer to inches, but you could follow along in centimeters, too.

The `microsecondsToInches()` and `microsecondsToCentimeters()` commands convert the time measurement to inches and centimeters, respectively, according to the arithmetic discussed in "How the Garage Sentry Works" on page 134. The data type `long` is used, as opposed to `int`, because it provides 4 bytes of data storage instead of just 2, and the number of microseconds could exceed the 2-byte limit of 32,767 bits. So far, so good.

In a regular formula, the distance arithmetic looks like this:

$$\text{in} = \frac{\text{time}}{2} \div \frac{74\,\mu\text{s}}{\text{in}}$$

$$\text{cm} = \frac{\text{time}}{2} \div \frac{29\,\mu\text{s}}{\text{cm}}$$

In either case, we first divide by 2 because the signal travels from the transducer to the target and back, as previously discussed. In the inches function, we then divide the halved number of microseconds by 74, and in the centimeters function, we divide by 29. (It takes 74 microseconds for the signal to travel 1 inch, and 29 microseconds for it to travel 1 centimeter; I arrived at those numbers by following the arithmetic in "Time-to-Distance Conversion Factors" on page 142.)

> **TIME-TO-DISTANCE CONVERSION FACTORS**
>
> You could simply trust my math and copy the time-to-distance conversion code, but you can apply this arithmetic to any project using a similar ultrasonic module or other sensor, so I encourage you to work through the math yourself.
>
> As I describe in "How the Garage Sentry Works" on page 134, the speed of sound is roughly 1,125 feet per second. Multiply that by 12 inches per foot to get 13,500 inches per second.
>
> To get the number of seconds per inch, you simply divide this value by 13,500 inches:
>
> $$\frac{1\text{ s}}{13{,}500\text{ in}} = 0.000074\frac{\text{s}}{\text{in}}$$
>
> It takes about 74 microseconds, or 0.000074 seconds, for sound to travel an inch. To determine the distance in centimeters, go through the same exercise, but use 343 meters per second for the speed of sound, multiply it by 100 centimeters per meter, and take the reciprocal.

Triggering the Alarm

The sketch is not done yet. Now we have to look at the number of inches (or centimeters) measured and compare it to the predetermined value—`val`, in this case—to see whether the alarm should be activated. To establish the variable `val` as a numeric value, take a potentiometer (R2) straddling the power supply on either end and tie the wiper to pin A0 (see Figure 5-2). Because A0 is the input to a 10-bit analog-to-digital converter, it converts that voltage (between 0V and 5V) to a numeric digital value between 0 and 1,023. Reading that value with an `analogRead()` command results in a value between 0 and 1,023 depending on the position of the potentiometer.

That value is then used to establish the trigger point for the alarm. But allowing all 1,024 values would essentially allow the distance to be set from 0 to 1,023 inches. Because the control rotates only 270 degrees, to adjust between, say, 40 and 42 inches would represent a very minuscule rotation—beyond the granularity of most potentiometers.

To scale this for the potentiometer, the sketch maps the value so the entire rotation of the potentiometer represents a distance of only about 100 inches with the following line of code:

```
val = map(val, 0, 1023, 0, 100);
```

Mapping the potentiometer value changes the maximum distance from 1,023 inches down to 100 inches while leaving the minimum distance of 0 inches unchanged. You can map any set of values so the Garage Sentry's target distance can be from *X* to *Y*, with full rotation of the potentiometer, so when you set up your Garage Sentry, you may want to test it and this range until it's right for your garage.

A conditional control structure sets the limit for the alarm. This structure makes sure that the LED is turned off when the measured distance is 0, regardless of whether the sketch is using inches or centimeters. First, the value `inches` is compared to `val` in the following expression:

```
count == 0 && inches > 0 && inches < val
```

If this statement is true, the alarm is set off and the `for` loop at ❷ is activated (see page 139), which alternately blinks the LEDs 200 times before timing out and turning the LEDs off.

The `for` loop just counts from 0 to 200, but that can be easily changed. After each count, it turns on an LED, delays briefly and turns off the same LED, delays slightly and turns on a second LED, delays slightly and turns off the LED, and then goes to the next count. At the end of the 200 count, the system turns off the LEDs and the program continues to the next line where it is reset. That is, the program starts again at the beginning.

Construction

Figure 5-8 shows the enclosure for the Garage Sentry.

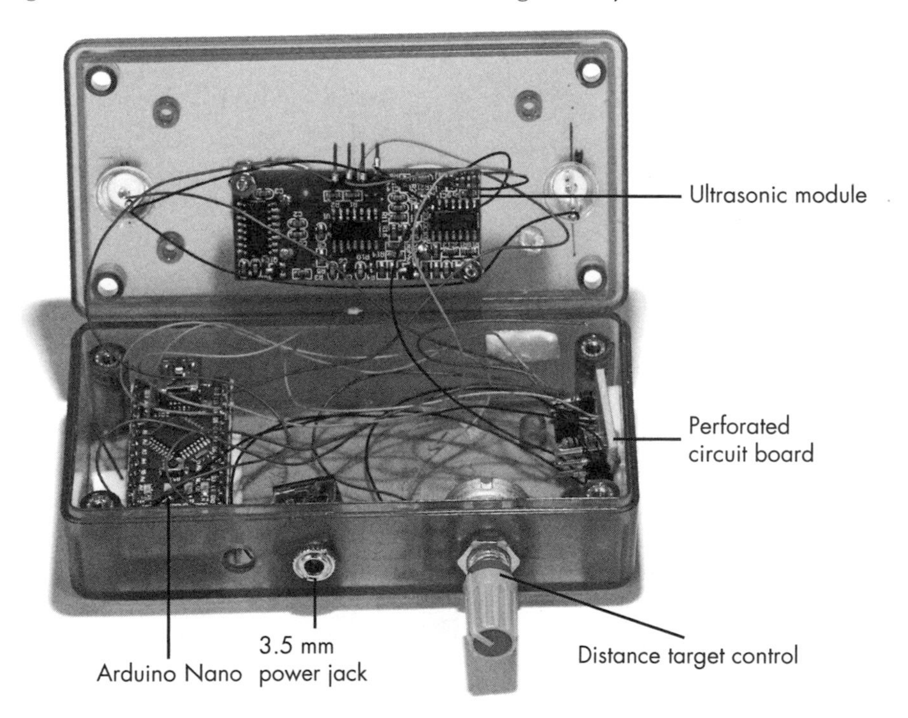

Figure 5-8: The basic Garage Sentry uses wire wrap for the final connections.

The trickiest part of the Garage Sentry is mounting the ultrasonic module on the enclosure. Because the module can send out sound waves only in a straight line, you need to be able to adjust its direction so that the ultrasonic sensor can hit its target and receive the echo. But the module includes only two mounting holes, diagonally opposed from each other, so there's no easy way to fasten it to a flexible mounting. We'll tackle that first.

Drilling Holes for the Electronics

To solve this problem, I mounted the transceiver directly to the enclosure and just aimed the enclosure as required. To mount the module, drill 5/8-inch holes in the enclosure and use standoffs to hold the board securely. See the template in Figure 5-9 for drilling measurements. A PDF of the template is available in this book's online resources at *https://www.nostarch.com/arduinoplayground/*, in case you want to print it and lay it over your enclosure as a guide. The enclosure I recommend is made of polycarbonate plastic and is less likely to crack than styrene or acrylic; however, it tends to catch the drill, so be careful.

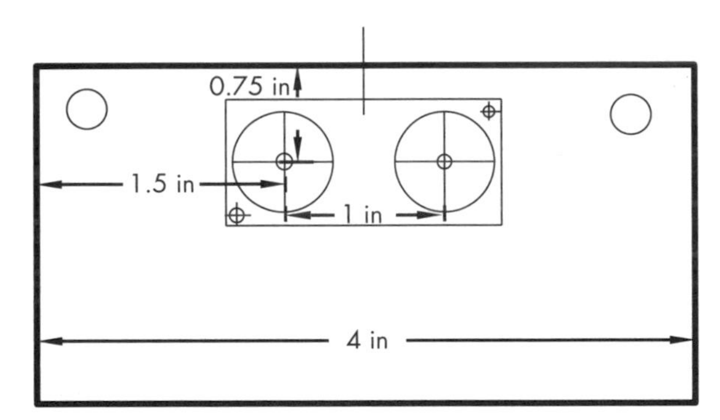

Figure 5-9: Template for drilling transducer holes

There are several ways to drill the 5/8-inch holes. If you are good at drilling, you could simply use a 5/8-inch drill bit and bore the holes directly. But I discovered that the holes can be bored safely and easily by first drilling a hole about 1/4 to 3/8 inches in diameter and then enlarging it with a tapered reamer, available from Amazon for under $15. The larger reamer in the Amazon set will ream a hole up to 7/8 inches in diameter, and it is handy to have around for other projects. Use a 1/8-inch drill to drill the holes for the standoffs, as shown in the drawing, which you can use as a template.

If you ream out the hole, make sure to ream from both sides. Enlarge the hole to a size that holds the transducer elements tightly—but not too tightly. While this is not the most precise way to bore a hole and would probably be frowned upon by professional machinists, it works well enough here.

WARNING *Regardless of the size of the hole, do not hold your work piece with your hand when drilling. Always clamp it securely. If the drill binds, the work will want to spin or climb up the drill. Drill at a slow speed and go gently.*

Next, drill the holes for the potentiometer, power jack, and two LEDs. Select a drill size based on the particular power jack and potentiometer you have. I used a 9/32-inch drill for the potentiometer, a 1/4-inch drill for the 3.5 mm jack, and a bit of approximately 25/64 inches for the LEDs. The size of the 10 mm LEDs tends to vary a bit from manufacturer to manufacturer, so I would recommend that you select a smaller drill bit, say 3/8 inches, and ream until the LED fits tightly. Because the LED is tapered, ream from the rear of the enclosure so that the LED will fit better.

The location of both the potentiometer and power jack is not important, but make sure that neither crowds the transducer or Nano. You want them to be on the bottom of the enclosure so that they are accessible after the enclosure is mounted (see Figure 5-12).

Mounting Options

Before you stuff the Arduino, ultrasonic sensor, and perforated board circuit into your enclosure, figure out how you want to mount the Garage Sentry. There are several ways to mount the enclosure onto whatever surface you need.

Velcro Strips

If you have a good flat surface to mount the assembly to, you could simply affix the enclosure with adhesive Velcro (see Figure 5-10). Two sentries have been in place in my garage that way for several months, with no sign of slippage or deterioration.

Figure 5-10: Adhesive Velcro mounting strips used to mount the Garage Sentry enclosure

A U Bracket That Can Be Aimed

If you don't have a good surface and need to aim the module at an angle, mounting it on a U bracket that lets the sensors swing up and down or left and right will work. In this section, I'll describe how to build the U bracket mount shown in Figure 5-11.

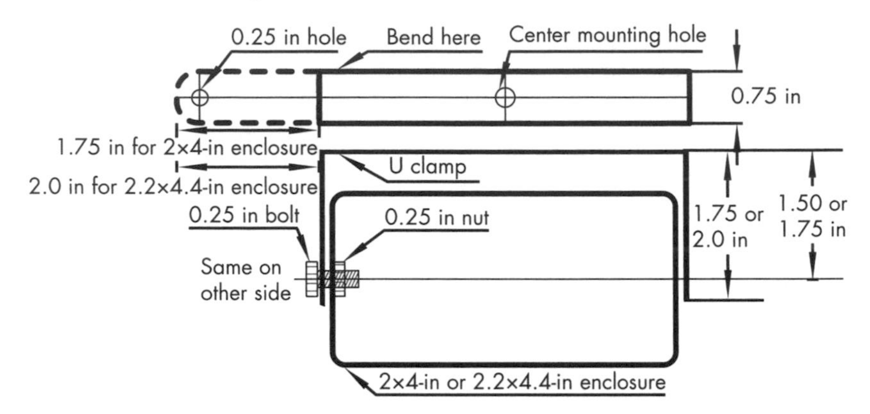

Figure 5-11: This drawing illustrates the size and shape of the optional U bracket handle and how it connects to the enclosure. Where you see two measurements for a single dimension, the smaller applies to the basic Garage Sentry, while the larger applies to the deluxe version.

To make the U bracket for the basic Garage Sentry with the 1591 ATCL 2×4-inch enclosure, take a strip of 3/4-inch × 0.080-inch × 5 1/2-inch long aluminum (available at Ace Hardware, Home Depot, or Lowe's), and drill 1/4-inch holes 5/16 inches from the ends of the aluminum strip. Drill corresponding holes with a No. 7 or 15/64 drill in the side of the enclosure centered on the ends, and thread the holes with a 1/4-inch-20 tap. Bend the aluminum strip 1.5 inches from each end for the standard version and 2 inches for the deluxe version (see Figure 5-11). Using a vise is the easiest way to bend the metal, but if that's not convenient, you can sandwich it between a bench and piece of metal, clamp it down, and bend it by hand.

To make the U bracket for the Deluxe Garage Sentry with the 1591 BTCL 2.2×4.4-inch enclosure, use a 6 3/8-inch long strip of the same material, and drill the 1/4-inch holes 1/2 inches from the ends. Then, bend the aluminum at right angles at 3/4 inches from either end for the standard version and 1 inch from each end for the deluxe version.

To fasten the U bracket to either enclosure, you can start by drilling a hole in the center of each end of the enclosure. It's simplest to drill a No. 7—15/64 is close enough—hole at either end of the enclosure. The easiest way to center the holes is to draw a line along each diagonal on both ends. Where the lines intersect is the center. Thread the holes with a 1/4-inch-20 tap, and you'll be able to fasten the enclosure to the U bracket directly. The threads in the thin ABS plastic will not be very strong, so be careful not to overtighten the bolts.

When you're finished attaching the bracket to your enclosure, it should look like Figure 5-12.

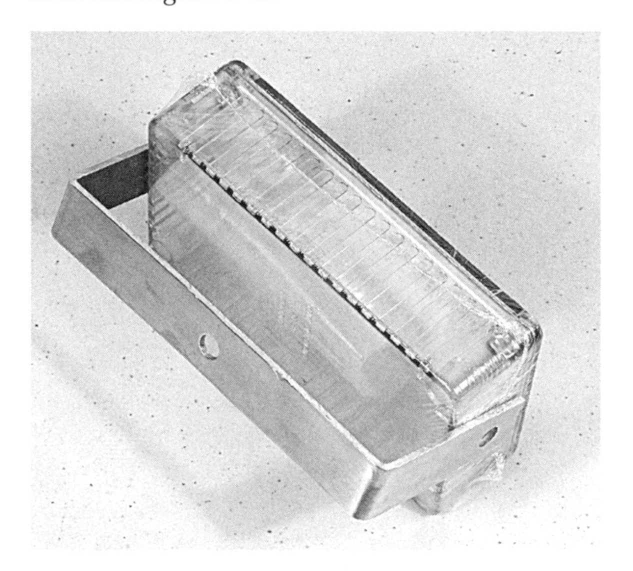

Figure 5-12: The enclosure for the basic Garage Sentry can be mounted with the bracket so it can be tilted or rotated to point the transducers in the correct direction.

Soldering the Transistors and Current-Limiting Resistors

After testing your circuit on a breadboard and deciding how to mount the Garage Sentry, solder the driver transistors and current-limiting resistors to a small section of perforated phenolic or FR-4 predrilled board. Use the schematic in Figure 5-2 or the instructions in "The Breadboard" on page 136 as a guide to wiring and soldering the components in the perforated board.

Make the connections in the schematic, but otherwise, there is no right or wrong way to assemble the perf board. I do recommend using perforated board with copper pads for each hole to simplify soldering. Solder all the hookup wires for the power, potentiometer, Nano, ultrasonic module, and LEDs before attempting to mount the board on the inside of the enclosure.

When you're done soldering, mount the perforated board anywhere in the enclosure where you can find room. I used double-sided foam adhesive, and it worked well. Mount the Nano, LEDs, and ultrasonic module next.

Wiring the Pieces Together

Finally, use 30-gauge hookup wire to connect the Nano, ultrasonic sensor, perforated board circuit, and LEDs according to the schematic in Figure 5-2. Optionally, you can use wire-wrap wire and a wire-wrap tool to wire up the sections, but it is not necessary and can be expensive if you don't already have the tool and wire. Wiring the components and fitting them in the enclosure may be a little messy, but it saves building a shield.

The Deluxe Garage Sentry

That's it! Or is it?

I have been using the standard model in my garage for several months; it does what it's supposed to do and does it well. But it seems like something's missing. The alarm goes off when you reach the desired spot in the garage, but why not have it give you a little warning before you get there so you can slow down as you approach the stopping point?

The idea is to have the system warn you at some pre-established distance from the stopping point so you don't have to stop suddenly. It isn't much extra effort to add two more LEDs to go off at different distances. Figure 5-13 shows the Deluxe Garage Sentry.

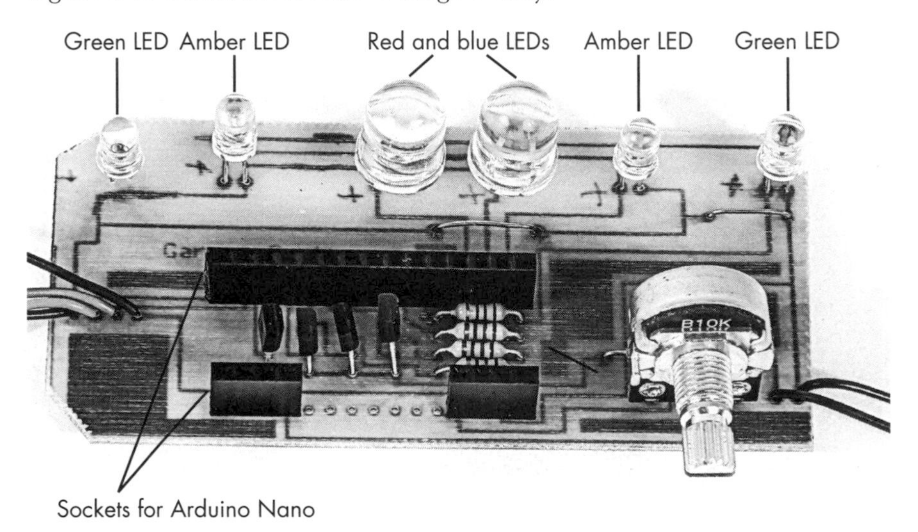

Figure 5-13: The Deluxe Garage Sentry sets off three stages of alarms.

Now, let's discuss how to assemble the Deluxe Garage Sentry.

The Deluxe Schematic

Hand-wiring everything in the standard version is tedious. So for the deluxe version, I developed a shield (PCB) that holds the LEDs, potentiometer, Nano, transistors, and current-limiting resistors. Adding the LEDs and extra transistors required some changes in the circuitry. Figure 5-14 shows the revised schematic.

Note that this circuit drives the transistors as emitter followers. As such, the base shows a high resistance, and therefore no resistor is required between the Arduino and the resistors Q1 through Q4. If you were driving using a common emitter, however, you would need a resistor, as current would flow through the base-emitter junction, short out the driver, and burn out the transistor.

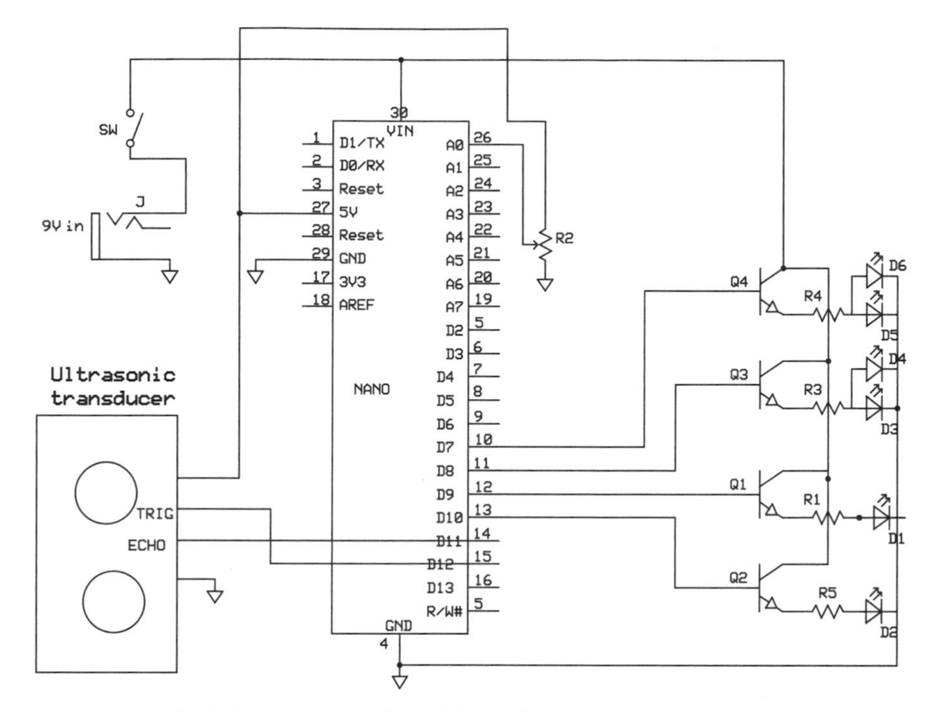

Figure 5-14: The deluxe schematic has additional LEDs, driver transistors, and current-limiting resistors on the right-hand side.

NOTE *To improve the Garage Sentry further, you could also conceivably double or otherwise increase the range using special transducers and electronics, but in this application, the 10-foot operating range is more than enough.*

The Deluxe Sketch

Before you build the Deluxe Garage Sentry, download *GarageSentryDeluxe.ino* from this book's resource files at *https://www.nostarch.com/arduinoplayground/*, and upload it onto your Arduino Nano according to the instructions provided in "Uploading Sketches to Your Arduino" on page 5. The sketch is basically the same as the basic Garage Sentry sketch, but it's updated to include the new LEDs.

```
/* Deluxe Garage Sentry: goes with the shield PCB
*/

int ledPin = 8;
int ledPin1 = 7;
int ledPin2 = 10;
int ledPin3 = 9;
int count;
int analogPin = A0;
int val;
int y;
```

```
void setup() {
  //Initialize serial communication:
  Serial.begin(9600);
  pinMode(ledPin, OUTPUT);
  pinMode(ledPin1, OUTPUT);
  pinMode(ledPin2,OUTPUT);
  pinMode(ledPin3,OUTPUT);
  pinMode(analogPin, INPUT);
}

void loop() {
  val = analogRead(analogPin);

  long duration, inches, cm;

  pinMode(12, OUTPUT);
  digitalWrite(12, LOW);
  delayMicroseconds(2);
  digitalWrite(12, HIGH);
  delayMicroseconds(5);
  digitalWrite(12, LOW);

  pinMode(11, INPUT);   //Attached to ECHO
  duration = pulseIn(11, HIGH);

  //Convert the time into a distance
  inches = microsecondsToInches(duration);
  cm = microsecondsToCentimeters(duration);

  val = map(val, 0, 1023, 0, 100);
  //Map the value of the potentiometer to 0 to 100

  if(inches == 0)
    digitalWrite(ledPin, LOW);

❶ if(count == 0 && inches > 0 && inches < val + 15)
    digitalWrite(ledPin2, HIGH);
  else digitalWrite(ledPin2, LOW);

❷ if(count == 0 && inches > 0 && inches < val + 7.5)
    digitalWrite(ledPin3, HIGH);
  else digitalWrite(ledPin3, LOW);

  if(count == 0 && inches > 0 && inches < val) {
❸   for(y = 0; y < 200; y++) { //Repeating blink sequence
      digitalWrite(ledPin, HIGH);
      digitalWrite(ledPin1, LOW);
      delay(100);
      digitalWrite(ledPin, LOW);
      digitalWrite(ledPin1, HIGH);
      delay(100);
    }
    count = count + 1; //Turn off instruction
  }
```

```
  digitalWrite(ledPin1, LOW);

  if(inches > 10) { //Reset if inches > 10
    delay(1000);
    count = 0;
  }

  Serial.print(inches);
  Serial.print("   inches ");
  Serial.print(count);
  Serial.print(" count   ");
  Serial.println();
  Serial.print(" val       ");
  Serial.println(val);
  delay(100);
}

long microsecondsToInches(long microseconds) {
  return microseconds / 74 / 2;
}

long microsecondsToCentimeters(long microseconds) {
  return microseconds / 29 / 2;
}
```

The most notable difference between the deluxe sketch and the standard sketch is that in the two `if-else` statements at ❶ and ❷, the green and amber LEDs are activated at different distances, based on the stopping point. For example, if the stopping point is set to 36 inches, or `Val` in the sketch, then the green LED turns on at `VAL + 15`, or 52 inches, and the amber LED turns on at `VAL + 7.5`, or 43.5 inches. This way the green LEDs will turn on when the car is 15 inches from the final stopping point, and the amber LEDs will turn on when the car is 7.5 inches from the stopping point. These numbers were selected arbitrarily, and you can change them.

The red and blue LEDs start flashing when the car has reached the stopping point. You can see how the LEDs are flashed at ❸.

The Deluxe Shield

Figure 5-15 shows the shield for the Deluxe Garage Sentry. If you want to build this shield, download this book's resource files, look for the file *GarageSentryDeluxe.pcb*, follow the etching instructions in "Making Your Own PCBs" on page 13, and solder your components to the board. You can also take the file and send it out to one of the service bureaus to have the board made for you.

The potentiometer is soldered directly to the shield, though you'll still have to solder wires to the power jack and to the ultrasonic module. As you can see in Figure 5-15, the connections for the ultrasonic sensor are located on the left-hand side.

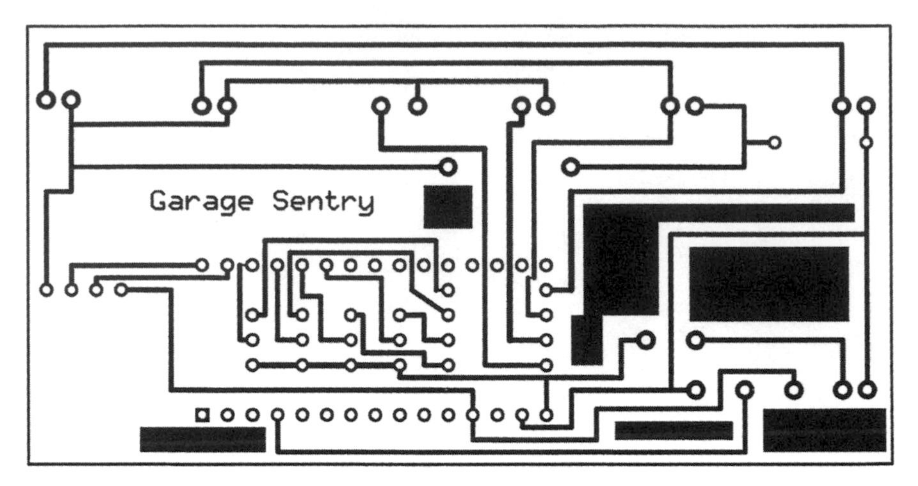

Figure 5-15: This is the shield for the Deluxe Garage Sentry, which simplifies the individual wires that had to be soldered to complete the earlier version.

The green LEDs are on the outermost edges, the amber LEDs are next, and the flashing red and blue LEDs are in the middle. The connections for the potentiometer are on the lower right-hand side, next to the first two connections, which are ground and VIN (from left to right). The potentiometer helps hold the shield in place in the enclosure.

This version uses high-power LEDs that draw a fair amount of current. Because the sentry uses the 5V voltage regulator on the Nano, the transistors driving the LEDs are wired directly to the 9V input voltage, allowing the unit to function without a separate voltage regulator. The LEDs are configured with the driver transistors as emitter followers, so the voltage to the LEDs will "follow" the voltage on the base of the transistor—that is, 5V—and not present LEDs with 9V.

There are several jumpers required on this shield, including the jumper for the power, which connects to the collectors of the transistors and connects the raw input to the (VIN) Nano. There are also jumpers to connect the ground to the LEDs.

I mounted the transistors and current-limiting resistors under the Nano to save some space. Also, note that the connections for the ultrasonic module in the standard version call for a right-angle female header, but doing that in the deluxe version means the length of the male part gets in the way of the LEDs, so I simply soldered the connections to the PC board and to the ultrasonic module to keep the wires out of the way.

A Bigger Box

Both the green and amber LEDs operate as pairs, so only a single driver transistor is required for each pair. But with all this new circuitry, the deluxe board does not easily fit in the same enclosure as the standard version.

You'll need to find a larger enclosure for the deluxe version, which provides you with some other benefits. With a larger, clear polycarbonate enclosure, like Hammond 1591 BTCL, the LEDs can stay inside the enclosure and still be visible so you won't have to drill holes to mate with the LEDs in the PC board. You'll have to drill only four holes: the two large holes for the ultrasonic sensor, a hole for the power jack, and one for the potentiomcter. These holes make it possible to mount the ultrasonic sensor on top of the Nano board with double-sided foam tape, which is in turn mounted to the shield (see Figure 5-16). In other words, you create a sandwich with the Nano in the middle, the shield on the bottom, and the ultrasonic module on the top. This design eliminates the need for the mounting screws used in the first version.

Figure 5-16: Compared to the basic Garage Sentry, there is virtually no hand wiring in the deluxe version. The driver transistors and current-limiting resistors for the LEDs are located under the Nano board.

Simply use the same template you used in the standard version, and drill (and/or ream) the two 5/8-inch holes for the two ultrasonic elements. The shield itself can be fastened to the bottom of the enclosure with small flathead screws and nuts or with more double-sided foam tape. When you affix the potentiometer to the enclosure, it should hold the board in place, as its leads are soldered to the shield. Figure 5-16 shows the deluxe version; note how much neater it is than the basic version from Figure 5-8.

Before drilling the hole for the potentiometer, carefully measure the height of the potentiometer hole and drill the hole slightly oversized so that the shaft and screw can be inserted at an angle into the enclosure. You'll also note in Figure 5-16 that the corners of the printed circuit board have been clipped off so as not to get in the way of the studs used for the top screws of the enclosure.

I suggest mounting the power connection on the same surface of the enclosure as the potentiometer—that is, the bottom. Here, the power connection and potentiometer will be accessible after mounting, leaving the top free to fit snugly against a shelf. The unit can just as easily be mounted upside down with the adjustment and power jack on the top.

Figure 5-17 shows the completed Deluxe Garage Sentry mounted on my garage workbench with the car in place.

Figure 5-17: Completed Deluxe Garage Sentry mounted on my garage workbench with the car in place

The completed sentry unit works flawlessly. Depending on your particular garage and where you place the unit, you might want to adjust the sketch to make the green and amber lights turn on at different distances. The unit pictured has been working perfectly for almost six months now, and I don't know how I'd be able to pull my car in the garage without it.

6

THE BATTERY SAVER

This project was actually born back in the 1970s, when I built a very similar device for the first time. Its purpose is to disconnect a vehicle's battery when an inadvertent drain would discharge the battery to the point where the vehicle would not start. Before the advent of computerized automobiles, it was common for drivers to park and leave the lights turned on only to come back to find a dead battery. Then, they had to find a way to call a service station (cell phones weren't available) and get a boost. Worse, if the battery died and the car was left sitting long enough, the battery would become useless and have to be replaced. After remaining in a discharged state between 12 and 18 hours, most lead-acid batteries would go totally dead and could not be rejuvenated.

But that was then. Now, many vehicles—particularly those equipped with automatic lighting systems—protect against such inconveniences with automatic (often delayed) shutoff for electrical systems. However, several types of vehicles still don't have automatic shutdown systems or alarms to warn you that the lights are on. This project, shown in Figure 6-1, is particularly useful for those vehicles.

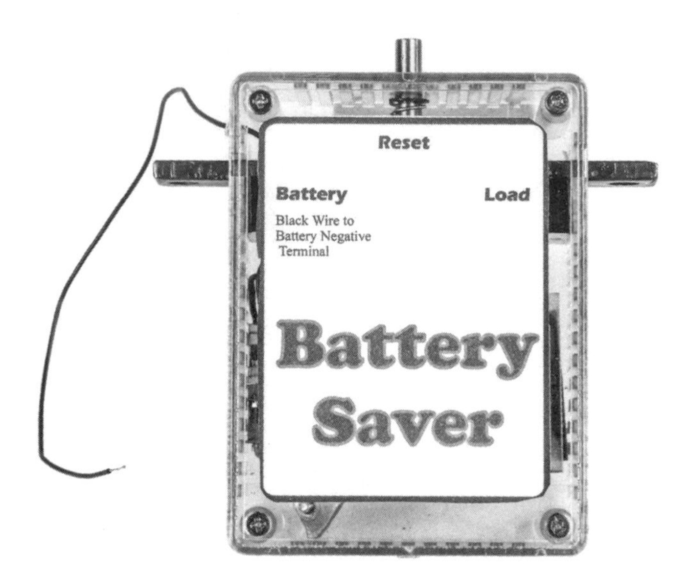

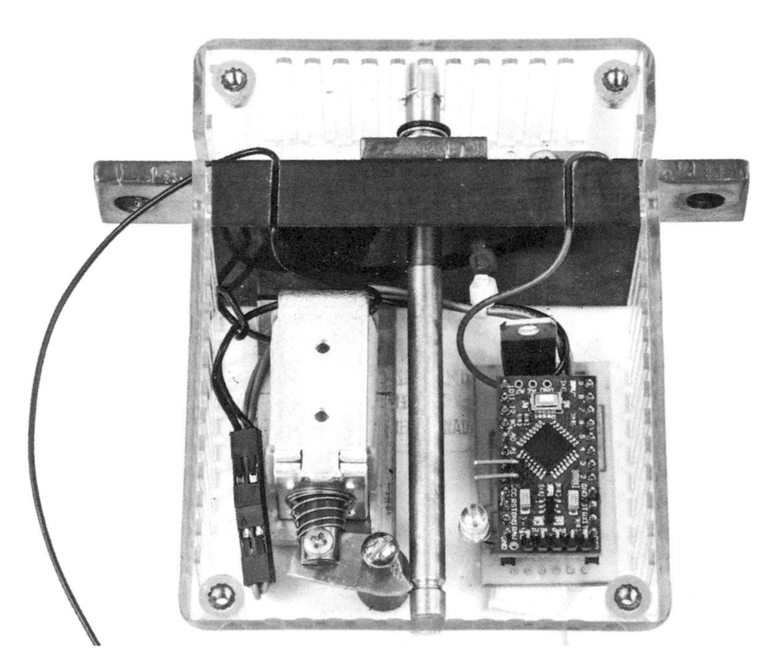

Figure 6-1: The finished Battery Saver, boxed and ready to go (top), as well as a look inside (bottom)

Boats, Tractors, and Other Vehicles

Many road-worthy vehicles could benefit from the Battery Saver, but this project is targeted at other systems where an accidental discharge of a battery could be annoying—and expensive. Boats are particularly vulnerable. Even my small runabout has experienced problems. On more than one occasion, the running lights were left on during the day, and I did not notice until a couple of days later. The battery was totally dead, and it had to be replaced.

Leaving the running lights on isn't the only way to accidentally drain a boat battery. Most inboard boats have a blower system designed to safely expel potentially explosive gases and fuel vapors from the bilge. Conventional wisdom (and the Coast Guard) says to keep the bilge blowers on before starting the engines, the entire time the boat is in service, and for at least 10 minutes after shutdown. It's very easy to forget the blowers are on and then find you need to replace the battery or batteries the next time you use the boat.

NOTE *Before building the Battery Saver for a safety system like the blowers on a boat, read "Notes of Caution" on page 158. There are some considerations to keep in mind that may require some minor wiring changes to the electrical system.*

Boats aren't the only vulnerable vehicles, though. Riding mowers and tractors are also at risk for three reasons:

- They are usually used only intermittently.
- Parking and/or headlight switches are often placed where it's easy to accidentally bump them when getting on or off the vehicle. The switches can also be damaged by flying debris or rough terrain.
- They are usually used in daylight hours, so it can be hard to tell when the lights are on.

The Battery Saver can also protect powered tools, like tow-behind sprayers. It's easy to leave these tools turned on when finished, and while most sprayer pumps don't take too much current, leaving one on for, say, a day or two, will likely drain the battery.

Required Tools

Drill and drill bits

Center punch

4-40 tap

Hacksaw

400-grit sandpaper

Countersink (Almost any 82° countersink will do. The material being formed is relatively soft, so no special materials are required.)

Needle files (See Figure 6-20. While only one is required, they usually come in sets.)

Triangle file

NOTES OF CAUTION

If you create the Battery Saver as described in this chapter, it should work well. However, it is always possible for Murphy's Law to cause a problem, so here are a few points to consider before you build.

Should the Battery Saver fail, it can be reset, but in the meantime, it will remove all power to the vehicle's electrical system. *Do not* use the Battery Saver on any system where an electrical failure could cause catastrophic problems (such as on an aircraft) or result in bodily harm or property damage. That said, I have used the following unit in a car, on three agricultural tractors, and on two boats for well over a year (and earlier versions for several years) with no failures.

The Battery Saver includes a very high-current switch. It is possible that, under certain conditions, a resistance could develop such that when current is applied, the switch becomes extremely hot—perhaps hot enough to be a fire hazard. (For what it's worth, this has never happened even on prototype versions.) All efforts have been made to eliminate problems like this—for example, the block that holds the Battery Saver's copper contacts is made of a fire- and melt-resistant phenolic material—but overheating remains a hazard. Always check the heat of the Battery Saver before you touch it to make certain that it is not hot enough to burn you. Even the reset plunger could become warm enough to burn if the vehicle or Battery Saver malfunctions.

This last warning is specific to mariners. If you keep your boat in the water, chances are it has a built-in bilge pump and automatic float switch to keep it afloat when you're not aboard. If you have multiple batteries, make sure the bilge pump is wired to a battery that isn't in the circuit with the Battery Saver, as in Figure 6-2. Otherwise, find the wire that provides current to the bilge pump/switch, and simply bypass the Battery Saver.

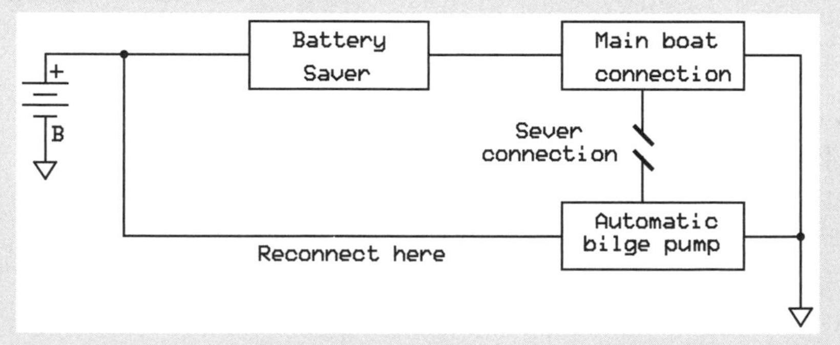

Figure 6-2: Wiring the Battery Saver on a boat

Parts List

This parts list might look somewhat like a scavenger hunt, but everything can be found easily from a variety of sources. Getting the particular size or quantity, however, may be challenging, so read the list carefully to make sure you have everything before you start. And before shopping, look ahead to Figure 6-13 on page 170 to see some of the more unusual parts, like the copper contacts.

- One Deek-Robot Pro Mini Arduino clone microcontroller board (There are several available, and some have different pinouts—particularly for pins A4 and A5. Figure 6-3 shows the pinout for the one I used. Other units with different pinouts should work, but the connections on the shield must be changed.)
- One LM7805 voltage regulator
- 12V solenoid (This project uses an Electronic Goldmine G19852.)
- One 10-kilohm resistor
- One 5.6-kilohm resistor
- One 4.7-kilohm resistor
- One 470-ohm resistor
- One 4.7V Zener diode
- One 1N4002 diode or equivalent
- Two ZTX649 transistors
- One 1/2-inch phenolic sheet
- One 6×3/4×3/16–inch copper bar, which is part of a high-current switch (This is actually a piece of copper bus bar that is generally used in large electrical installations. See Figure 6-4 for an example. To find this, you may need to do a little digging with an online search for *copper bus bar.* I bought a piece that was 3 feet long.)

A7 A6 GND
10 11 12 13 A0 A1 A2 A3 VCC RST GND RAW
A4 A5
9 8 7 6 5 4 3 2 GND RST RXI TX0
DTR TX0 RXI VCC GND GND

Figure 6-3: Pinout of the Deek-Robot Pro Mini Arduino clone

Figure 6-4: A copper bus bar, drilled out for the Battery Saver

One ABS plastic enclosure, like the Hammond 1591 STCL

Two 1/4-inch brass rounds (You'll use one of these as the pylon.)

A small piece of sheet metal to use as the release

One e-clip

Four 1/2-inch, 4-40 flathead screws

One 3/8-inch, 4-40 roundhead screw

3 oz of Permatex Silicone RTV sealant

One matching pair of inline connectors (You can buy simple, cheap, inline connectors online, or you can make your own. I used Pololu's 1×2 connector housing and male and female crimp pins, shown in Figure 6-5. See "Connectors Used in This Book" on page 18 for crimping techniques. You can also use telephone-style chicklets if you can find them.)

Figure 6-5: A pair of mating connectors used to connect the solenoid

28-gauge hookup wire

One solder lug (If you can't find one, you can work around it by taking the power wire from the Arduino and wrapping it under the head of the screw, as shown in Figure 6-7.)

One battery cable to fit your storage battery on one side and a lug to attach to the Battery Saver on the other

Downloads

Sketch *BatterySaver.ino*

Templates *ReleaseLever.pdf, BatterySaverEnclosure.pdf*

Shield *BatterySaver.pcb*

The Schematic

While the circuit is relatively simple, as shown in Figure 6-6, there are a couple of key design elements to be aware of.

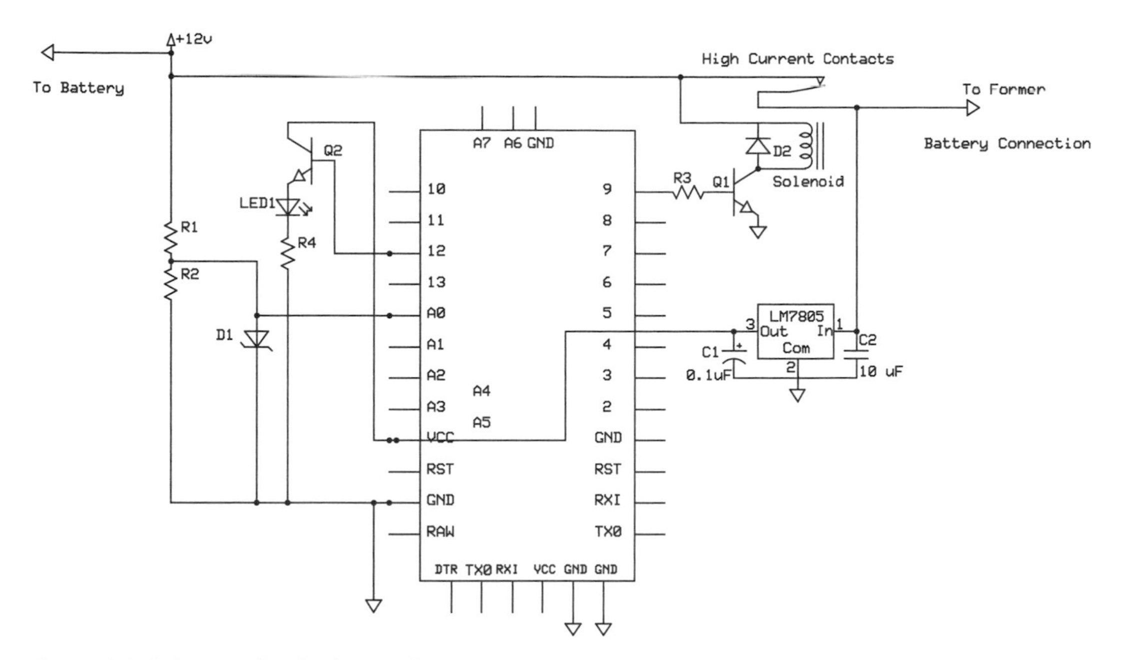

Figure 6-6: Schematic for the Battery Saver

Notice that the circuit uses an LM7805 voltage regulator. In theory, the small regulator included in the Pro Mini board would be more than satisfactory because the load is relatively light. But under the hood of a car and around high-current systems and high-voltage electronics, there is a lot of stray electromagnetic energy bouncing around. While it's improbable that this energy would cause a problem for the Battery Saver, it is well within the realm of possibility. The more robust 1.5 A LM7805 regulator offers the Pro Mini another level of protection. In addition, capacitors C1 and C2 have been included to bypass any AC sneaking into the circuit and to prevent unwanted oscillation. Similarly, the 4.7V Zener diode (D1) protects the input of the Arduino from a voltage spike—which is very likely—on the 12V supply. It limits the voltage to A0 to only 4.7V.

The configuration of resistors R1 and R2 may look familiar from other projects. They comprise a voltage divider to lower the 12V supply to a level below the 5V maximum the Arduino input can tolerate. To be on the safe side, I selected a value of 10 kilohms for R1 and 5.6 kilohms for R2. Both are standard values. This design should allow the battery voltage to jump to 14V before reaching the point where the Zener diode kicks in.

Diode D2 (1N4002 or equivalent) provides yet another level of defense: it protects against an inverse current that could be created when the magnetic field in the solenoid collapses. This is a standard protection device in inductors with an iron core, which can store magnetic energy and release it rapidly into the coil. The reverse voltage can reach relatively high levels and create significant currents, which could, in this case, damage the driver transistor and other components in the system.

Transistors Q1 and Q2 are both ZTX649 silicon NPN transistors. I chose these transistors because they have sufficient drive capability and are inexpensive and readily available. I used the same model as in other projects. The high side of Q2 is brought to the red positive rail (VCC), as the LM7805 regulator supplies plenty of current. (You could just as easily connect the high side, or collector, of Q2 to the 12V supply, because Q2 is wired as an emitter follower and the voltage at the emitter will follow only the voltage at the base.) R4, a 470-ohm resistor, limits the current to LED 1.

LED1 indicates when the unit is operating. The circuit containing Q2 and LED1 provides a blinking LED with a very short on cycle, which is consistent with this volume's goal of blinking LEDs as often as possible.

NOTE *The Deek-Robot Pro Mini also has a red LED that indicates when it is turned on. This additional, although very small, constant current drain could contribute to the discharge of the battery. In the units that I installed, I unsoldered one end of the Pro Mini's LED, which is kind of a delicate operation.*

Transistor Q1, wired as a common emitter circuit, provides the drive for the solenoid. Resistor R3 (4.7 kilohms) drives the transistor and also protects it—a direct connection would allow too much current to flow in the base-emitter junction, resulting in potential damage. The high side of the solenoid is connected to positive 12V to allow maximum voltage and current to flow through the solenoid coil without taxing the voltage regulator.

How the Battery Saver Prevents Draining

The Battery Saver comprises a very high-current switch that can be electrically turned off and a sensor circuit to detect when the battery is in jeopardy of dying. The high-current switch is required because it interrupts the main power from the battery, which includes the feed to a starter motor that can draw up to several hundred amps.

The high-current switch—which could also be considered a relay—is made of the three pieces of copper bus bar: a release bar, a solenoid, and a release lever. When the pieces of copper bus bar are connected, the battery is connected to its circuit as normal. When the solenoid is pulled in, the power is disconnected.

The sensor circuit uses the power of the Arduino microcontroller. The Arduino continually monitors the battery, and if it senses that the power is diminishing, it calls for the high-current switch (relay) to be thrown. There are many ways to determine when a battery is reaching exhaustion, and this project simply looks at the voltage left in the battery. Table 6-1 shows the voltage versus the remaining charge in a standard 12V, lead-acid storage battery.

Table 6-1: Battery Charge State Versus Voltage

Battery charge level	Battery voltage
100%	12.7V
90%	12.5V
80%	12.4V
70%	12.3V
60%	12.2V
50%	12.1V
40%	11.9V
30%	11.8V
20%	11.6V
<10%	11.3V

The battery charge level quickly deteriorates with only a minimal reduction in voltage. In order to have at least 40 to 50 percent of the charge remaining, the drain on the battery must be stopped somewhere between 11.9V and 12.2V (shaded in Table 6-1), a voltage I will refer to from now on as the *trigger point*. In practical applications, empirical evidence shows that there is still plenty of juice left, even when battery voltage drops to 12.0V, 11.90V, or even below. A battery in good condition under ideal circumstances could perform with perhaps as little as 30 percent of its capacity, but that would depend on the load, ambient temperature, and other factors, so I'm taking a very conservative view of about 12+V. Newer design batteries tend to do better.

While the state of charge is a good indication of remaining capacity in a battery, other factors—like internal resistance, battery discharge rate, and so on—could impact the usable state of charge remaining in a battery. A more accurate measurement of the capacity remaining in a fully charged battery might be current used, which can be calculated if the total energy in the battery is known. For example, if a battery has a capacity of 1,100 ampere-hours (A·h), you could calculate the point where 550 A·h remain and disconnect the battery then. However, because this project targets systems with batteries that range widely in capacity, I decided that measuring the battery voltage would be more than adequate.

Arduino to the Rescue

The Arduino microcontroller steps up to the task of measuring the battery voltage and throwing the disconnect switch at the appropriate voltage, but if that's all it did, you could be in trouble. When turning on a vehicle, the starting motor uses a lot of current. Depending on the state of the battery (internal resistance and so on) and ambient conditions (temperature, for example) during the cranking process, the battery voltage can conceivably fall well below the critical shut-off voltage. The Battery Saver depends on the Arduino to give the vehicle enough time to start the motor.

Further, should the system shut down due to an inadvertent drain, you'll need to ensure that when the Battery Saver is reset, it doesn't immediately sense a critical battery voltage and shut down again. All of these functions are handled by the microcontroller, under orders from the sketch. While there are probably several different ways of handling it, the sketch implements *timing sequences*—rules to follow for when to take certain actions—to allow for the voltage drop during engine cranking. Similar timing rules give the user time to restart the engine after the Battery Saver is reset.

The Breadboard

Even though this project doesn't require a lot of extra components and peripheral equipment, I still believe it's useful to go through the exercise of wiring a breadboard. A working prototype gives a definitive proof of concept, and it allows you to play with the hardware and software prior to committing to the finished version. When working with the breadboard, I tested the circuit with a solenoid similar to the one used in the finished product (see Figure 6-7).

Figure 6-7: The breadboard I used to test the Battery Saver concept and put the software together. Testing your circuit with a solenoid is not required.

Figure 6-8 shows the breadboard operating the high-current switch; "Construction" on page 170 describes how to build that switch. To test the circuit without making the high-current switch first, or to use a solenoid by itself (see Figure 6-7), you can connect a lamp, LED, relay, or some other component in place of the switch. Remember: if you use an LED or other polarized device, make certain to get the polarity correct.

Figure 6-8: The breadboard circuit operating the solenoid after the high-current switch has been assembled. Note the use of clip leads to hook up the Battery Saver to the breadboard. Diodes D1 and D2 and capacitors C1 and C2 were not included in the breadboard primarily because they are needed only when the project is in use.

The basic breadboard is fairly straightforward. Just follow these instructions to assemble it:

1. Connect the breadboard's red positive rails together, and connect the blue negative rails together. Do not connect the positive rails to the negative rails as that will result in a direct short circuit.
2. Insert the LM7805 voltage regulator so that the three terminals span three different rows. (See "The Schematic" on page 160 for why an external voltage regulator was used rather than the Pro Mini's onboard regulator.)
3. Connect the input of the regulator to a positive 12V source. (See Figure 6-9 for the pinout of the LM7805.)

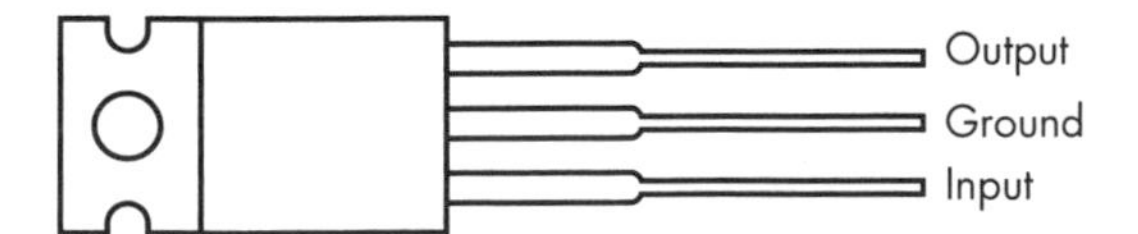

Figure 6-9: Pinout of the LM7805 regulator

4. Connect the center pin of the regulator to blue negative rail and the output pin to the red positive rail. When you power the circuit, the positive rail should have 5V coming from the output of the regulator.
5. Insert the Pro Mini microcontroller board in the breadboard. Its position is not critical; anywhere in the general vicinity of where it is in Figure 6-7 is fine.
6. Insert resistors R1 (10 kilohm) and R2 (5.6 kilohm) into the breadboard. It's easiest to insert them near the regulator. Then, connect one side of R1 directly to the regulator (input pin). The joining point of R1 and R2 should be located in an independent place on the board, and the other side of R2 goes directly to ground.
7. Connect a jumper from the point where R1 and R2 join to pin A0 of the Pro Mini.
8. Insert transistor Q1 (ZTX649) into the breadboard in an area with three open rows (see Figure 6-10 for the pinout).
9. Connect resistor R3 from the base of Q1 to pin D9 of the Pro Mini.
10. Connect the emitter of Q1 to ground.
11. Connect the collector of Q1 to the load (solenoid or other).
12. Connect the other side of the load to positive 12V.
13. Insert transistor Q2 into the breadboard. Connect its collector to one of the red positive rails.
14. Connect the base of Q2 to pin D12 of the Pro Mini.

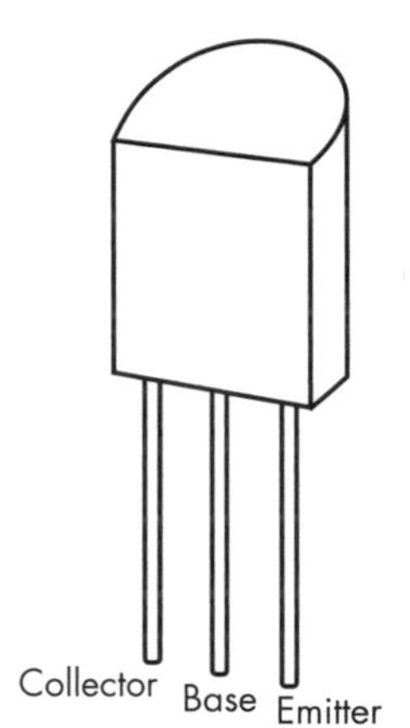

Figure 6-10: Pinout of the ZTX649 transistor

15. Connect the emitter of Q2 to the positive side of LED1.
16. Connect the negative side of LED1 to R4 (470 ohms).
17. Connect the other end of R4 directly to the blue negative rail.
18. Connect the negative side of the 12V supply to the blue negative rail.

Now, load the sketch onto the Pro Mini (see "Connecting and Programming an Arduino Pro Mini" on page 8), and take it for a test run. If you have a variable power supply with a voltage readout, setting the voltage will be easy; if you don't have a variable power supply, I suggest building the Regulated Power Supply in Chapter 3. If you use a variable supply without a readout, just use your multimeter to monitor the voltage and observe the trigger point.

Start the power supply at 13V (12.7V is normal for a charged 12V lead-acid storage battery), and the monitor LED should blink slowly. Gradually lower the voltage, and write down the voltage when the flashing goes from slow to rapid. That voltage is the trigger point, and it should be around 11.9V to 12V.

The Sketch

In developing the Battery Saver sketch, I went through several iterations to assure reliable functionality. One difficulty was avoiding false triggers when the sensed voltage jumped around because an engine or an accessory, such as a hydraulic tilt, was activated.

I used functions to adjust various sequences of operation independent of the main program. This isn't a complex sketch, but writing simple functions is a useful technique when you want to avoid repeating code. I could have avoided functions in the final sketch, but I let them remain because they work well, and writing the sketch this way provides a good lesson in the use of functions.

```
/*The Battery Saver sketch, which uses multiple functions
  to create timing sequences */

int led = 12;
int Battin = A0;
int Relay = 9;
int volts = 0;
int volts2 = 0;
int volts3 = 0;
int B = 387;  //Threshold trigger set point

void timer3() {  //Shut off timer function
  delay(200);
  volts = analogRead(Battin);  //Reset trigger point
  volts3 = map(volts, 0, 1023, 0, 500);
  if(volts3 > B) {
    digitalWrite(Relay, LOW);
  }
  else {
    digitalWrite(Relay, HIGH);
  }
}

void timer2() {  //Fast blink timer function -- low voltage
  if(volts2 < B) {
    for(int j = 1; j < 1800; j++) {
      digitalWrite(led, HIGH);
      delay(10);
      digitalWrite(led, LOW);
      delay(90);
    }
  }
}

void timer() {   //First timer function -- high voltage
  if(volts2 > B) {
    digitalWrite(led, HIGH);
    delay(200);
```

```
    digitalWrite(led, LOW);
    delay(1000);
  }
}
void setup() {
  Serial.begin(9600);
  pinMode(Relay, OUTPUT);
}
void loop() {
  delay(1000);
  volts = analogRead(Battin);
  volts2 = map(volts, 0, 1023, 0, 500);

  timer();
  if(volts2 < B) { //Set trigger point
    timer2();
  }
  if(volts2 < B) {
    timer3();
  }
}
```

In this sketch, `timer()`, `timer2()`, and `timer3()` are the three functions used. The `timer()` function is for normal operation when the battery voltage is above the trigger point. The trigger point is the voltage at which the Battery Saver goes into timeout mode prior to shutting down. The `timer2()` function sets off a rapid flashing sequence once the trip threshold is reached and provides the timing—the fast LED sequence—prior to shutoff. Once `timer2()` has finished, `timer3()` is invoked, provided the voltage remains below the threshold; this disconnects the battery by activating the solenoid. If the voltage increases above the trigger point at any time during the timeout period, the Battery Saver returns to normal operation after `timer3()` times out. Many of the variables in this sketch can also be changed to vary blinking and delay sequences and threshold; they're addressed in "Operating the Battery Saver" on page 180.

The Shield

As you might guess from the breadboard, the shield is straightforward, too. Compared with handwiring, however, I believe it's a lot easier and faster.

The PCB Layout

Figure 6-11 shows my finished PCB, and Figure 6-12 shows the layout image with silkscreen indicating the placement of the components. You can download a PCB layout for the shield at *https://www.nostarch.com/arduinoplayground/*.

Figure 6-11: The unpopulated Battery Saver board. The RLY connection is where you would connect the solenoid.

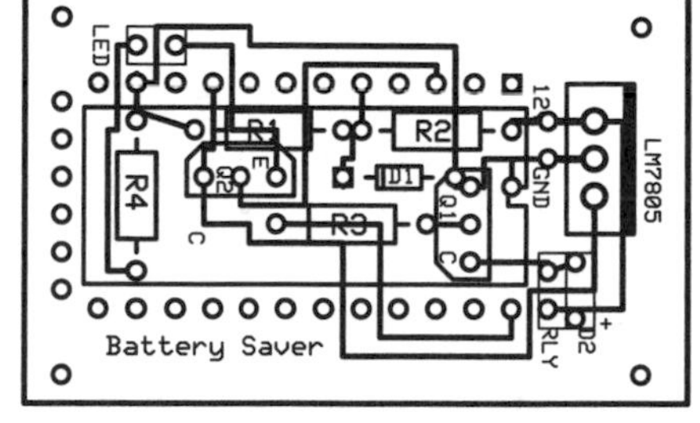

Figure 6-12: The foil pattern for the shield on the Battery Saver. The silkscreen image is in gray.

Most of the components are located under the Pro Mini to save space. Other than that, there is nothing special about populating the shield. As in other projects, it's not necessary to include headers for all the Pro Mini pins. Use all the pins that have connections with board traces and enough other pins to add mechanical stability. I always try to use at least one header for the very first pin, to aid in aligning the board while plugging it in.

The LED can be mounted directly to the board or can be mounted somewhere remotely with long wires. Just be sure to observe polarity. The Battery Saver may be located in an area where the operator can't see the LED, so placing it in a remote location is a practical solution.

Preparing the Shield and Pro Mini Controller

If you want to use a PCB, you can make the shield according to the layout provided with this book's resource files, whether you etch that yourself or send it off to be professionally manufactured. You could also design your own shield PCB if you're feeling ambitious or just solder everything to the prototyping board, but if you are using everything else from the parts list, just make sure your board has the same dimensions as the provided shield layout.

If you etch the PCB design I provide, drill the component holes next. I usually use a #66 drill. Solder a 2-inch wire to the plus 12V side of the shield, and solder the other side of the wire to a small solder lug. Solder a 15-inch wire to the ground terminal and then connect two wires to the solenoid connections. I used a small inline connector—made with Pololu #1950 crimp connector housings, #1931 male crimp pins, and #1215 female crimp pins—so I could remove the board easily if necessary. Almost any connector can be used.

Both transistors are located under the shield, so when you solder them, make sure to push them down enough that they clear the bottom of the Pro Mini. See Figures 6-11 and 6-12 for the transistor placement on the PCB.

I soldered the monitor LED directly to the board so it could be seen through the enclosure. However, you could also solder wires to the board and place the monitor LED in a location where it might be more visible outside your vehicle.

Construction

Building the rest of the Battery Saver involves a few mechanical challenges and requires the wits of a scavenger. Figure 6-13 shows all the parts of the Battery Saver laid out. There is nothing complex about any of the parts.

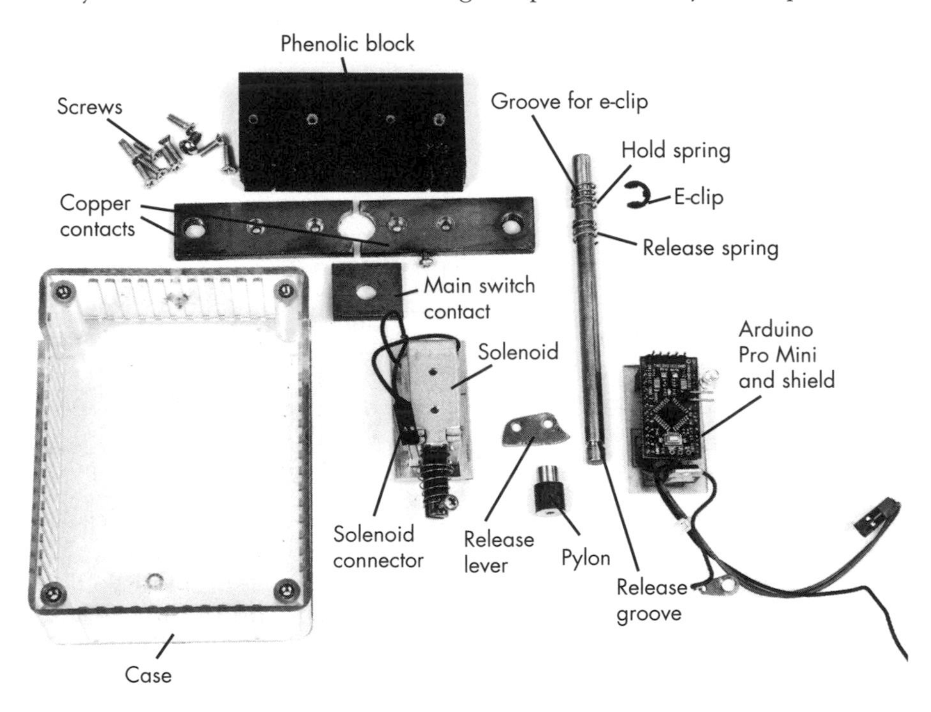

Figure 6-13: The components of the Battery Saver, completely disassembled. The cover of the enclosure is under the clear box. Note the short screws for fastening the spacer to the solenoid, so they don't damage the solenoid coil.

There are only a handful of components in the Battery Saver: the enclosure, the phenolic contact support, the shield and Pro Mini controller, the copper contact assembly, the solenoid and mounting, the release lever and pylon, the release rod, and the springs and e-clip. The assembly instructions that follow are a little involved, but if you get stuck, just use Figure 6-13 to keep things in perspective. When you're done, the Battery Saver should look like Figure 6-14.

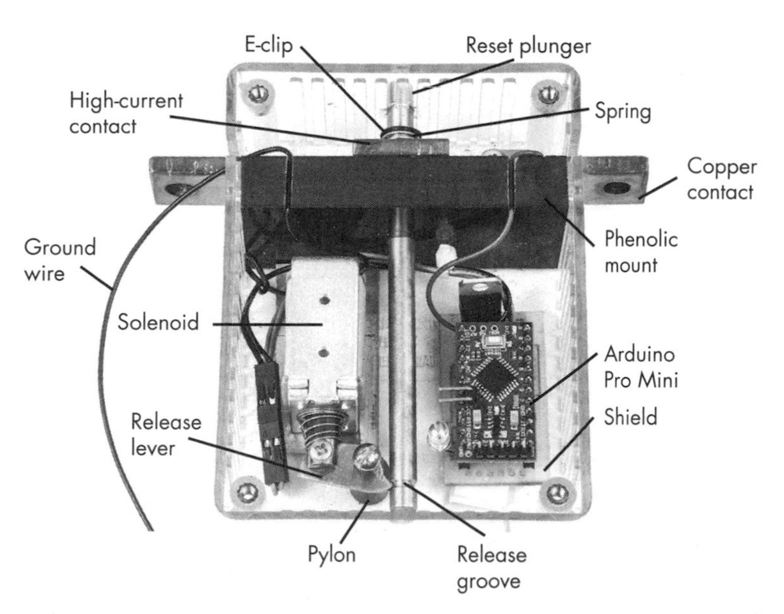

Figure 6-14: The completed Battery Saver with the cover off and the switch in the operating position. In this iteration, grooves were carved in the phenolic block for the positive and ground wires.

Preparing the Enclosure

Because the enclosure, a Hammond 1591 STCL, is an integral part of the design, I suggest starting there. Beyond a couple of holes and slots machined into the enclosure, there is very little else to do. Figure 6-15 shows the holes I cut in detail; a template is included in this project's resource files. There are only a few holes to cut, though, so you should be fine without the template if you follow these instructions carefully.

On the long sides of the enclosure, measure approximately 1 1/8 inches from the top—that is the side where the reset plunger will protrude through. This should locate you roughly at the fourth mounting rib. In the center of that rib, drill two #30—approximately 1/8-inch—holes 3/8 inches from either edge, and countersink both for a 4-40 screw. Do this on both sides of the enclosure.

On each side, at the top of the next rib (going toward the top of the enclosure), cut a 1-inch-deep groove for the copper contact assembly. The shaded part of the side view in Figure 6-15 shows where the groove should go on each side. The ABS plastic enclosure cuts easily with a hacksaw. Remove the cover and cut from the opening of the enclosure toward the back. For each groove, make two cuts so that when you remove the material between them, you will have a 3/16-inch-wide channel in both sides of the enclosure. A little over 1 inch from the edge at the opening will do, but it's not necessary to cut down to the back of the enclosure. You can even cut both sides at once, but just make sure the channels are directly opposite each other. Don't worry if your cuts are a little off; that can be corrected later with a file. After you make the cuts, break off the material in the center and clean up the channel with a small triangle or flat file.

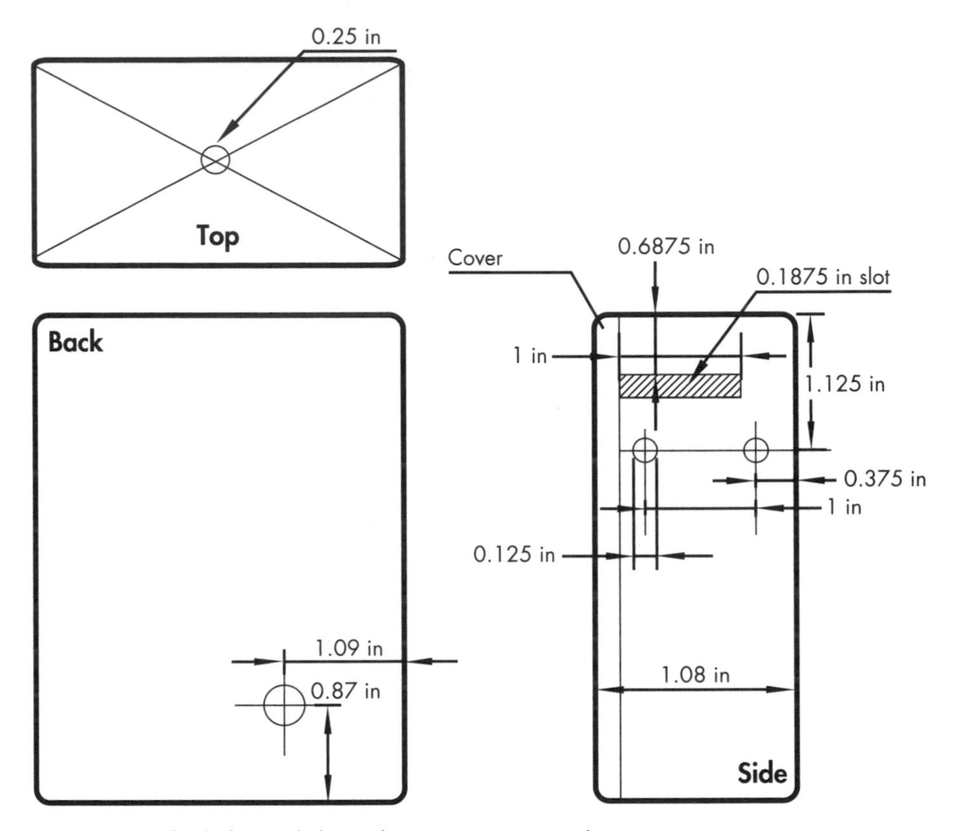

Figure 6-15: The holes and slot in the Battery Saver enclosure

Now, drill a 1/4-inch hole in the top of the enclosure, where the reset button will go, exactly centered on the top surface. Draw lines along that side's diagonals from corner to corner, as illustrated in Figure 6-15, and drill where the lines intersect. This hole is where the brass release rod for the reset plunger will go.

The Contact Support

The contact support is perhaps the easiest part to make. First, cut the phenolic sheet to a 3×1 3/8–inch block. The phenolic material cuts easily with a hacksaw. Insert the block in the enclosure so the two holes you drilled through the ribs are in the center of the block. Mark the holes on both sides, drill them with a #43 drill bit, and tap them for a 4-40 screw. While you are working on the phenolic contact support, you can cut two grooves, or channels, for the power and ground wires on either side (shown later in Figure 6-18). The location is not critical as these are used only to run the positive and negative wires for the shield. I used the small Dremel tool in

Figure 6-16 to cut the grooves. Use two 4-40×3/8-inch flathead screws on each side to screw the phenolic piece in place.

Figure 6-16: Dremel tool with small circular saw blade attached

With the phenolic contact support screwed in place, hold the drill as close to vertical as possible in the center of the largest face of the contact block. Drill a 1/4-inch hole that lines up with the 1/4-inch hole you previously drilled in the top of the enclosure. The easiest way to do this is to mount the contact support block and then use the hole in the top of the enclosure as a guide to drill the hole in the block. After drilling, put the contact support aside until you're ready for the copper contact assembly.

NOTE *The enclosure is not a perfect rectangle because some relief is included to allow it to come out of the mold easily. Factor this in as you line up to drill the center hole. When I drilled the hole, I held the enclosure in a vise to eliminate the effect of the relief.*

Preparing the Copper Contact Assembly

The copper contact assembly requires only a handful of holes. Cut a 4 3/4-inch section of the 3/16×3/4–inch copper bar, and drill the holes along the center of the 3/4-inch dimension, as outlined in Figure 6-17.

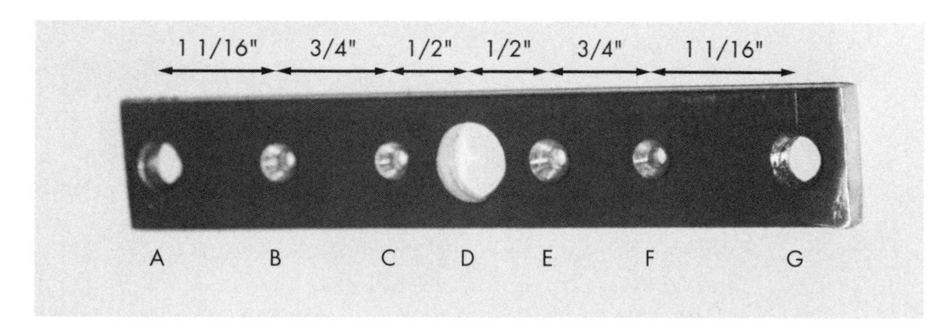

Figure 6-17: The base of the copper contact assembly, showing holes and spacing

For holes A and G, use a 9/32-inch bit. Holes B, C, E, and F should be drilled with a #30 bit; make sure to countersink them deeply enough for a 4-40 flathead screw. Finally, drill hole D with a 1/2-inch bit. Tap holes A and G for a 5/16×18×3/4–inch bolt, which will hold the battery cables.

CAUTION *While copper is not hard, it tends to grab a drill bit and climb up the bit. Always hold the copper piece in a vise, in a clamp, or with pliers when drilling. Never attempt to hold the copper with unprotected hands.*

When all seven holes are drilled, set the base of the copper contact on the phenolic contact support so that the 1/2-inch center hole in the copper is centered as closely as possible on the 1/4-inch center hole in the phenolic support and the bar is centered along the entire length of the contact support. Make sure the bar is equidistant from both sides of the support, hold the two pieces together firmly, and mark holes B, C, E, and F. Center punch the marks you just made in the contact support, drill them with a #43 drill, tap them for a 4-40 screw, and set the contact support aside again.

Now, mark the exact center of the copper contact, which should perfectly bisect the 1/2-inch hole. Cut the piece in two at this marker; a hacksaw should work well for this. Then, cut a 3/4×3/4–inch square of your leftover 3/16×3/4–inch copper bar, and mark the center by making diagonal lines from corner to corner. Center punch and drill a 1/4-inch hole where those lines intersect. This copper square will become the actual contact.

The final hole to make in the copper contact assembly is somewhere between holes E and F, on the outer edge of the bar; look for the little screw sticking out of the contact bar in Figure 6-18 (circled). Drill a #43 hole and tap for a 4-40 screw. This is where the switched positive voltage to the Pro Mini comes from. Figure 6-18 shows all the contact and mounting hardware drilled.

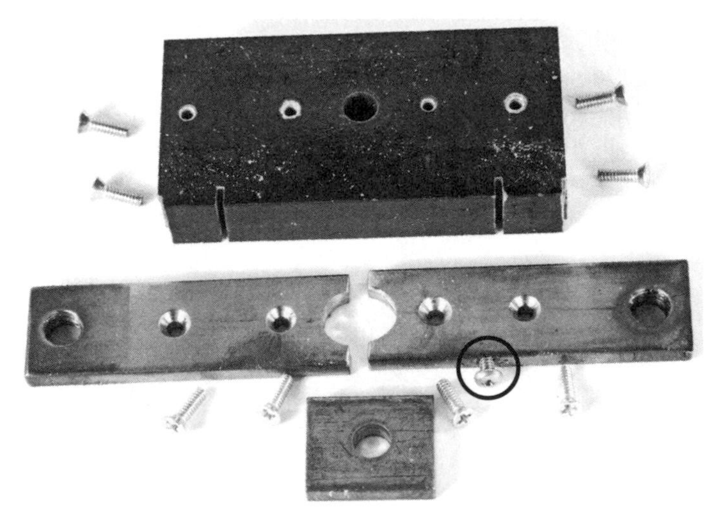

Figure 6-18: The contact and support hardware ready for assembly

Mounting Supplies for the Solenoid

Depending on your solenoid, the mounting process may vary slightly from the one described here. The frame of the solenoid I used had two holes tapped for a 4-40 screw in the bottom. However, there was precious little room between the frame and the coil, so I elected to look for an alternative mounting approach—a very aggressive double-sided adhesive. If you still want to mount the solenoid with screws, judge the screw length carefully. With either mounting approach, the solenoid was not high enough to line up with the release mechanism, so I had to add a platform.

Preparing the Release Rod, Springs, and E-Clip

As with the other mechanical components of the Battery Saver, the release rod to reset the contacts just needs a little TLC. Begin with a section of 1/4-inch brass rod, and cut it to 4 1/4 inches in length. Measure 1 1/8 inches down from one end, and make a groove for the spring retaining clip to fit into, as shown in Figure 6-19.

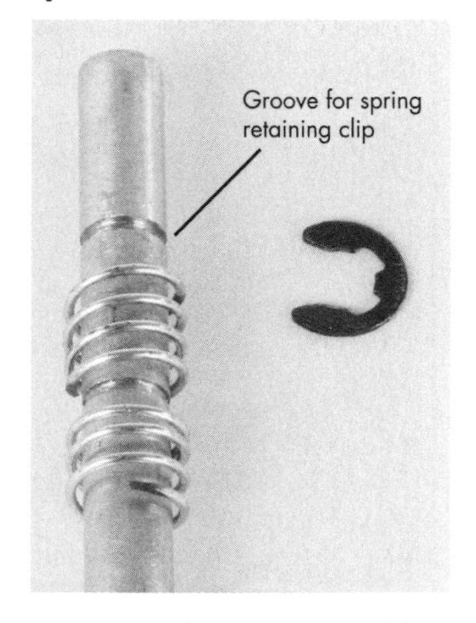

Figure 6-19: The upper section of the release bar. There are two grooves in the bar; the lower one was made in error. The springs used on the release rod were selected from a standard spring assortment from Ace Hardware and cut down to fit the project

In order to cut the groove, I mounted the bar in the chuck of an electric drill that I clamped to my workbench and used a hacksaw blade to carefully groove the piece, guiding the hacksaw blade with my fingers. The set on the hacksaw blade is a little wide, but the depth of the groove, not the width, is the important part. That said, the groove doesn't have to be very deep. I recommend you cut the groove in small increments and keep trying the retaining clip until it fits snuggly.

Figure 6-19 also shows the configuration of the springs. The bottommost spring will rest on the phenolic block and keep the copper contact off the contact areas until the release rod is depressed. It should be small enough that it does not touch either side of the copper contacts. (Remember, the hole in the square contact is 1/2-inch wide, so the spring needs to be smaller than that.) When depressed, the top spring overrides the lower spring and keeps the upper part of the switch in firm contact with the lower sections. You can make fine adjustments of the spring length on a small grinder, or you can use a Dremel tool or a hand whetstone.

Now, measure 4 inches from the same end, and file the release groove in the lower half of the 1/4-inch release bar using a small needle file.

Figure 6-20 shows both the needle file used and the shape of the release groove. (You might want to use a hacksaw first to make the groove, and shape the upper side with a file.)

Figure 6-20: A close up of the release groove in the bottom of the release rod (top) and the small needle file used to make the groove (bottom)

The shape of the release groove is not overly critical, but make sure there is a slight bevel on the top side. The swirls you see on the shaft were made with 400-grit sandpaper I used to smooth the rod so it slides smoothly through the phenolic block and hole in the case and contact piece.

Making the Release Lever and Pylon

The release lever is made of a small piece of light-gauge steel sheet metal, approximately 0.060 inches thick. Use the pattern in Figure 6-21 to cut the lever. In this book's online resource files, you can find a PDF file of the lever template. Because the template is pretty small, you might try bonding the template to a blank piece of similarly sized stock, holding it in a pair of pliers (vise grips work well), and shaping the piece on a grindstone. Once you've bonded the template to the stock, you can also use a file to shape it.

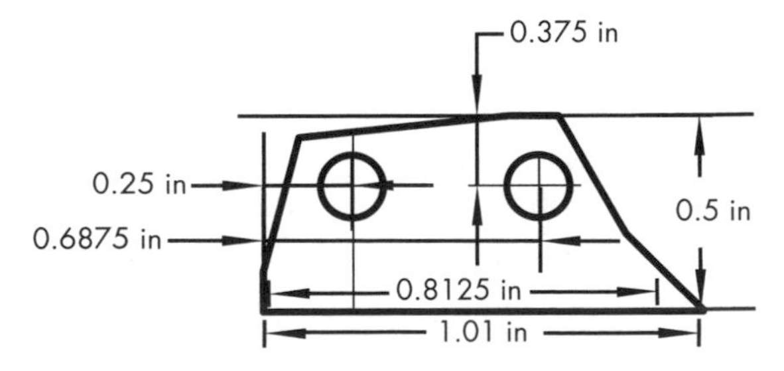

Figure 6-21: The release lever template

With a pair of tin snips, I cut a rectangle first and then cut the shape of the release lever. I firmly clamped the piece in place with a vise so I could finish shaping it with a file. Mild steel files easily with a satisfying feel. (You could also shape the lever with a small grinding wheel or a Dremel tool with an abrasive wheel.) Make the piece a little oversized at first, as indicated in the template, as it may require some adjustment in the final assembly. The lever should be under pressure from the tension spring, and should securely hold the release rod so that the square copper piece is firmly in contact with the both sides of the bottom copper contact assembly. Both pivot holes in the lever—one to attach to the solenoid and one to attach to the release pylon—are drilled with a #30 drill.

The pylon is for mounting the release lever to the enclosure, and it can be made of any scrap brass or aluminum you may have around. The pylon in Figure 6-22 was made from a 3/8-inch diameter brass rod. I reduced the top section's diameter because I didn't want the pylon to rub on the release rod, but that probably was not necessary.

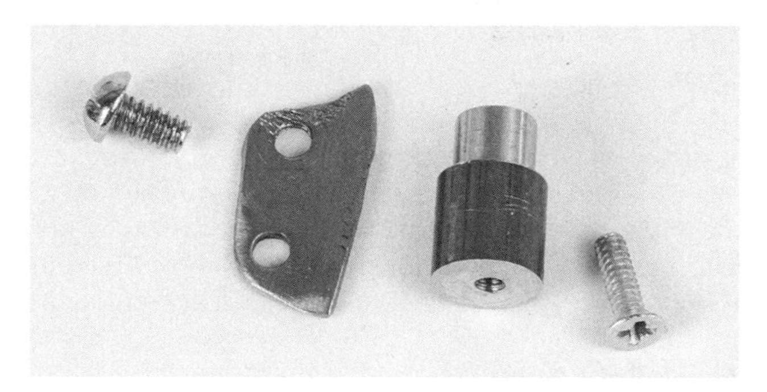

Figure 6-22: The release lever, the pylon, and two screws

Cut the pylon to about 0.61 inches long so that when mounted inside the enclosure, the end attached to the release lever hits the release rod just below the center of its diameter. This dimension is dependent on how precisely the release rod hole is drilled into the center of the top of the enclosure. Run the release rod through the hole, and measure how far from the back of the enclosure its center is. The pylon should be a little shorter than that.

You can make the top of the pylon smaller by chucking it into a drill like a bit and spinning it on a file. When you finally assemble the Battery Saver, you may also have to adjust the height of the pylon slightly to accommodate the thickness of the release lever and altitude of the release rod if the hole was not drilled perfectly straight.

Drill a #43-sized hole near the center of the pylon all the way through, and tap for a 4-40 screw from both sides. One side will mount to the back of the enclosure, and the other will hold the release lever. To make sure the screw doesn't go down too far and tighten against the release lever, I tapped down only about 3/16 inches so the screw bottomed out and jammed. Figure 6-23 illustrates what the tapped pylon should look like.

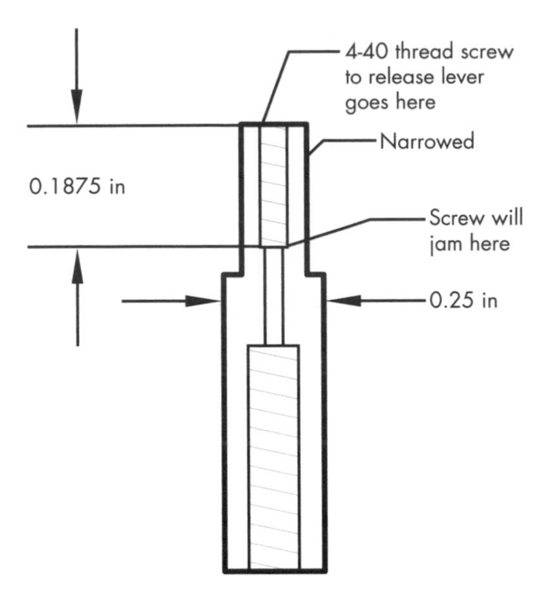

Figure 6-23: Inside the pylon

You can adjust either the length of the tapped hole or the length of the screw to make sure the release lever is free to move but secure, using the screw head as a bearing surface. For fastening both the release-lever bearing screw and the pylon mounting screw, I recommend using an anaerobic adhesive (such as Loctite Threadlocker) to secure the threads.

Assembling All the Parts

Now, we're ready to start putting all the parts from Figure 6-13 together and fastening them where they belong. The order of assembly is not overly critical, but more a matter of common sense. I'll go through it step-by-step:

1. Start by screwing the copper contact pieces into the phenolic support. Make sure the screw heads are below the surface of the copper. This is critical to assure good contact with the contact bar. Prior to assembly, as an added measure to improve the contact area, I sanded both the contact pieces and contact cap with a 400-grit sandpaper. To assure flatness, I put the sandpaper on a flat surface and rubbed the copper on it.
2. Slide the phenolic support into the enclosure, and fasten it with the four 4-40×1/2-inch screws.
3. Insert the release rod through the enclosure, and then thread it through the pressure spring, then through the copper contact, and finally through the release spring and down through the phenolic base. Insert the e-clip. This may be a little tricky. You'll have to hold the clip with a pair of needle-nose pliers.
4. Measure for the location of the pylon, and mount it to the enclosure with a 4-40×1/2-inch flathead screw. The pylon should be inside the enclosure, and the screw should be threaded in from the outside.

5. Screw the release lever to the pylon, and check by hand that it engages the release groove and secures the release rod. This is a little tricky, and you may have to adjust the release lever by grinding or filing it a little to fit in the release rod's groove. Once you have the lever in place, you can operate it with your fingers and assure that it locks tightly to the release-rod groove.
6. Install the spacer on the solenoid.
7. Temporarily mount the solenoid to the enclosure, and screw the release lever to the solenoid. To do this, I first unfastened the release lever from the pylon and fastened the other end to the solenoid plunger. I then applied double-sided tape to the bottom of the spacer, juggled everything in place, and loosely placed the screw through the release lever and into the pylon.
8. Exercise the release mechanism with the solenoid held in place. Once the release rod holds the top switch contact firmly down, mark the holes for the solenoid spacer.
9. Drill and tap holes in the enclosure to securely mount the solenoid and spacer to the enclosure. Alternatively, fasten the solenoid with double-sided adhesive, such as 3M outdoor double-sided adhesive.
10. Run the wires that will connect the solenoid to the Pro Mini circuit under the release rod, and connect them to the solenoid using the connectors.
11. Run the ground wire from the Pro Mini circuit under the release rod and out through the slotted opening for the copper bar. (It's the dark wire sticking out of the enclosure on the left in Figure 6-14.) You may want to either drill small holes in the phenolic for the wires or carve small grooves to run the wire through, using a Dremel tool and a small circular saw blade like the one shown in Figure 6-16. You can also make a groove with a file or hacksaw.
12. Attach the red positive voltage wire with the lug to the copper switch pole at the last hole you tapped in "Preparing the Copper Contact Assembly" on page 173. Use a 4-40×3/8-inch roundhead screw.
13. Mount the shield using small spacers. If you can't find a solder lug, you can solder-tin the wire, form it so it fits around the screw snugly, and then tighten the screw.

Compare your Battery Saver to the finished device in Figure 6-14 to make sure everything looks right. When the battery voltage reaches the trigger point, the Arduino triggers the solenoid, which releases the lever. Freed from the release lever, the rod pops up from the spring tension of the lower spring. The lower spring holds the contact piece above the copper mounted to the phenolic block, opening the circuit. To reset the Battery Saver, just push the rod back down again. Before doing a final test, check the latching of the release lever and rod several times.

Installing the Battery Saver into a Vehicle

Connecting your Battery Saver to your vehicle takes only minutes. First, disconnect both the positive and negative terminals from your battery. Then, connect a short battery cable from the positive side of the battery to the input side of the Battery Saver. Take the output side of the Battery Saver and connect it to the cable that originally connected to the battery. Connect the black wire to the negative terminal that will be reconnected to the battery.

Mounting the entire enclosure will depend on where and how your vehicle's battery and/or battery box are located. In many cases, the Battery Saver can simply hang from the battery cables. In other applications, I've used a heavy-duty cable tie to wrap around the entire battery and Battery Saver to hold it in place. In some cases, double-sided Velcro works well.

Operating the Battery Saver

In operation, once the Battery Saver is installed (see Figure 6-24), restore the ground connection to the battery and hook that ground connection to the Battery Saver. Then, set the Battery Saver by depressing the reset button—that is, the top of the release bar—until the unit is armed. You should hear or feel a click as you depress the reset button and the release lever engages.

Figure 6-24: The Battery Saver installed in my Boston Whaler runabout. The battery hooks up to the battery post and to the positive connection for the motor and accesssories. The ground lead of the battery was fed through the back of the battery box. The reset plunger is readily accessible.

Normal Operation

As long as the battery is fully charged and above the threshold voltage, the "on" indicator will blink at the rate established in the `timer()` function, which is approximately once every 2.2 seconds. When the battery voltage drops below the threshold level, which is set at approximately 11.9V via variable `B` in the sketch, the indicator LED begins flashing rapidly. When the LED sequence finishes, which is after about 3 minutes, the voltage will be checked once again. If the voltage is still below the threshold, the battery is disconnected; otherwise, it just returns to normal operation.

When you reset the Battery Saver after it has shut off, it will resume operation. After reset, if the battery voltage is above the threshold, the indicator will blink at the usual rate of once every 2.2 seconds. Often when the drain is removed from a battery discharged to some level, the battery recovers somewhat on its own after a relatively short time. If, however, the voltage is below the threshold level, the indicator will blink rapidly for the timeout period, as indicated earlier, to allow for the operator to start the boat, tractor, or other vehicle. You can set the timeout period by changing the value of `j` in the `timer2()` function.

The maximum value of the `j` variable is set at 1,800 in the initial sketch, and incrementing or decrementing `j` will add or subtract 100 milliseconds from the total timeout period. Thus, to set the timeout period to 5 minutes, you would set the maximum value of `j` to 3,000.

Setting the Threshold Voltage

The threshold voltage is established by resistors R1 and R2 with values of 10,000 and 5,600 ohms, respectively (see the schematic in Figure 6-6). These are set up as a voltage divider. According to the voltage divider calculation, whether you use a calculator or work it out with the formula, the voltage will be approximately 4.31V for a 12V input. Thus, you can calculate the exact threshold voltage at which you want the device to shut off the battery current by setting the threshold trigger point, which is `B` in the sketch, to whatever value you wish. While I calculated the theoretical value of the threshold, I experimented and found that a value of 387 establishes the shutoff threshold point at about 11.9V.

ON BATTERY TYPES

I experimented with several batteries and loads and found setting `B` to 387 consistently leaves at least half the energy in the battery after shutdown to restart the engine. On several different batteries, ranging from the small battery on a portable generator to large-capacity batteries for starting a truck, the same value seemed to work well.

That said, I have little experience with deep-cycle batteries and know that they have very different discharge parameters. If you want to attempt to use the Battery Saver for such batteries, take a look at their discharge rates and voltages, and set the threshold accordingly.

Protection from the Environment

Unfortunately, the Battery Saver is not weatherproof, and most applications call for it to be used in somewhat hostile environments. There are, however, a couple of possible solutions. On a variety of vehicles, including boats and tractors, I have wrapped the device in a plastic bag and tightly wrapped wire ties around the cables where they enter and exit the device.

But that's not terribly attractive, and I took some abuse from the distaff side of the family, so I applied Permatex silicon RTV sealant around all the openings where the copper bar comes through—but not where the reset (brass) rod comes through—and sealed all the screws.

For the reset button, I had difficulty finding a protective covering nipple (you would be surprised at what I found on the web), so I settled for affixing the top of an eye dropper to the enclosure with the silicone sealant. This also keeps the reset bar—which is hot to 12V when depressed—from shorting out.

Applying Cool Amp

While the copper-on-copper contacts work well, I burnished them with very fine sandpaper (400 grit) before assembly. Even though the untreated contacts have been used on a number of versions of the Battery Saver and have never failed, I decided to use Cool Amp, a simple-to-use silver-plating compound, for this version.

For very little cost, I was able to silver-plate the contact area of the contact bar and plate and thus reduce their resistance. Figures 6-25 and 6-26 show the difference between unplated and plated copper.

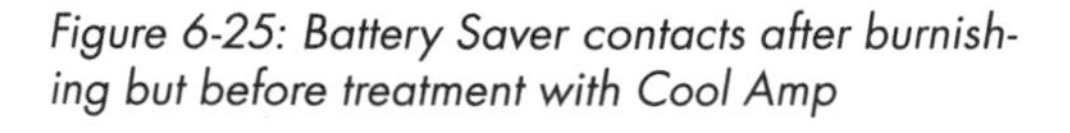

Figure 6-25: Battery Saver contacts after burnishing but before treatment with Cool Amp

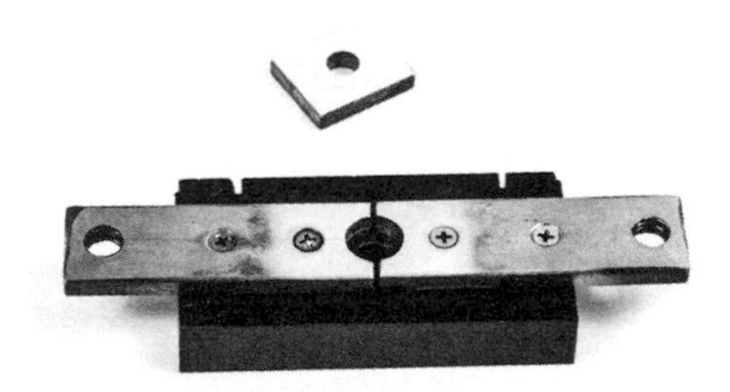

Figure 6-26: Battery Saver contacts after silver plating with Cool Amp

While this additional lowering of resistance is probably not necessary, all the current for the vehicle passes through this contact, so I figured it wouldn't hurt to be on the safe side. Further, I have used the Battery Saver in marine applications around saltwater, and the copper parts have acquired a green patina while the silver-plated areas have not.

I have used Cool Amp on a number of contacts, from motor-starting contacts to heavy-duty relay contacts, and it works well. You can learn more about Cool Amp at *http://www.coolamp.com/*.

NOTE *Copper does not oxidize too rapidly and the most common oxide of copper is highly conductive, which is one reason it is commonly used for current switching. Silver has the same properties—only better.*

MY ORIGINAL IDEA

I hold a US patent (4,149,093, now expired) on a device similar to the Battery Saver. The patent drawing is shown in Figure 6-27. Notice how similar it is to this project, despite being made decades earlier. It had essentially the same function, even though microcontrollers hadn't been invented yet!

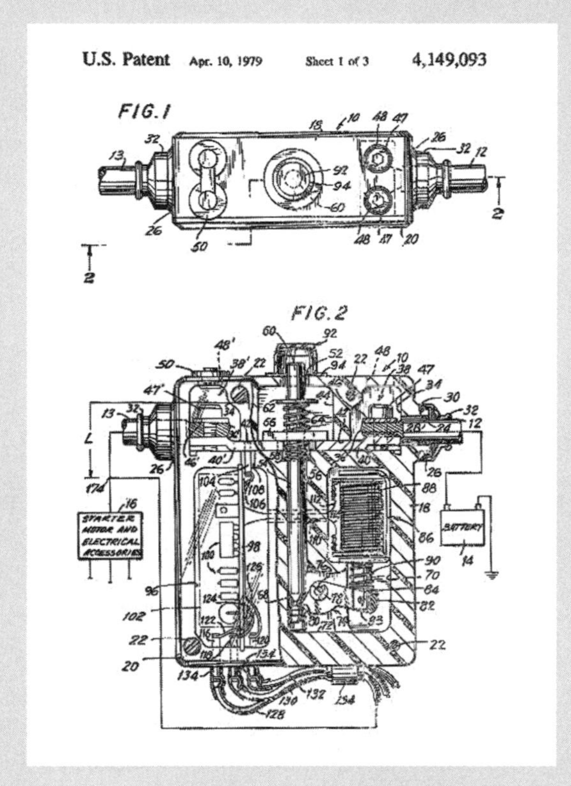

Figure 6-27: A patent drawing of the Battery Saver's predecessor

7

A CUSTOM PH METER

Microcontrollers are used in many, if not most, commercial and scientific instruments. They are accurate and versatile, which makes them a relatively low-cost solution to a variety of measurement requirements. This project combines an Arduino microcontroller with a commercial probe and some analog circuitry to construct an accurate meter that measures *pH*, the relative acidity or alkalinity of a solution.

There are three basic ways of measuring pH. This project involves using a pH meter and probe. The other approaches are litmus paper indicators, which you might remember from high school chemistry, and colorimeters, the traditional swimming pool maintenance kit. The latter is usually a kit of chemical reagents with a comparison color chart. Of the three, a pH meter is by far the most accurate.

But what does pH measure, exactly? The pH value describes the activity of hydrogen ions in aqueous solutions. The higher the activity of hydrogen ions, the more acidic the solution is and the lower the pH is. Less activity of hydrogen ions (and greater activity of hydroxide ions) results in a higher pH.

The pH scale is logarithmic. A difference of one pH measurement unit represents a tenfold increase or reduction of hydrogen-ion activity in the solution. This explains how a solution's aggressiveness rapidly increases with the distance from the neutral point on the pH scale.

Figure 7-1 shows the finished Custom pH Meter, and if you wonder why measuring pH is useful, "Why Measure pH?" on page 187 discusses several answers to that question.

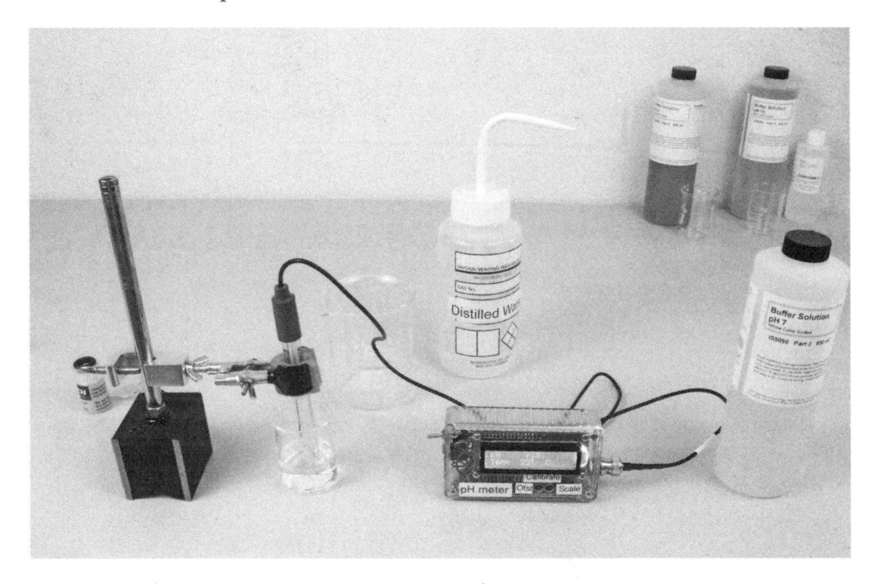

Figure 7-1: The Custom pH Meter in an actual measurement environment

Why Build Your Own pH Meter?

Commercial pH meters run the gamut of prices from low-cost portable units under $60 to full-fledged laboratory instruments costing several hundred or several thousand dollars. Relatively low-cost pH meters can do the job, but all have drawbacks, such as marginal accuracy, relatively short lifetime, calibration issues, and consistency.

A variety of pH meter kits are also available, including many stamps, some designed for the Arduino. (A *stamp* is a small circuit board with the critical circuitry to perform some function, minus the processor.) I have not had a chance to sample these kits, but they tend to be pricey, running close to or over $100. And they still require a power supply and packaging.

While this project doesn't propose to offer a full-scale laboratory instrument, it provides a good, workable pH meter and gives a lot of insight into what actually comprises a pH meter. I've made every attempt to tune the circuit for optimal performance, but you may find further adjustment helpful, so if you want to try something different after seeing the circuit, go for it.

WHY MEASURE pH?

In the past, pH was a relatively obscure measurement, confined to the laboratory bench and industrial environments (for quality control, process control, measuring and controlling waste effluents, and so on). However, more people have started using scientific measurement in areas that have traditionally relied on rote instructions or trial-and-error experimentation, like home winemaking and beer brewing, hydroponics, home agriculture, hydroculture, and baking. All of these applications can benefit from accurate pH measurement, and they don't even include the more mundane task of managing the chemistry of your swimming pool, koi pond, fountain, or aquarium.

For example, in baking, dough needs a low pH to rise. The pH of foods also impacts two of the four tastes: low-pH, or acidic, foods tend to taste sour, while higher-pH, or more alkaline, foods taste bitter. Lemon juice is an example of sour, and broccoli rabe or dark chocolate can be considered bitter. In home gardening, pH is an important soil characteristic for particular crops. Simple adjustments in pH can make aquarium water clear and reduce scum deposits on the glass sides, and a balanced pH in ponds keeps fish healthy and reduces algae. And the list goes on.

Required Tools

Soldering iron and solder

Drill and drill bits

Keyhole saw

Center punch

File

2-56 tap

Heat gun or hair dryer (for heat-shrink tubing)

Parts List

You'll need the following parts to build your Custom pH meter:

One Deek-Robot Pro Mini Arduino clone (Other Arduinos should work with the project in general, but not with the shield template provided for this book. I used an Arduino Nano clone for the breadboard because of the built-in USB interface. In the completed unit, I switched to the Pro Mini to conserve space.)

One LM35 (D) temperature sensor

One Texas Instruments TL072 dual op-amp (The pinout is shown in Figure 7-2.)

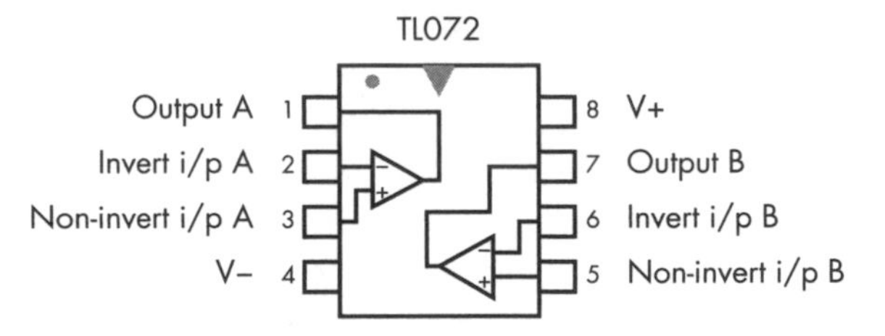

Figure 7-2: TL072 pinout

One 10-turn, 1-megaohm trimmer (R7)
One 10-turn, 10-kilohm trimmer (R4)
One BNC male connector
One LM7805 voltage regulator
One LMC7660 power inverter (The pinout is shown in Figure 7-3.)

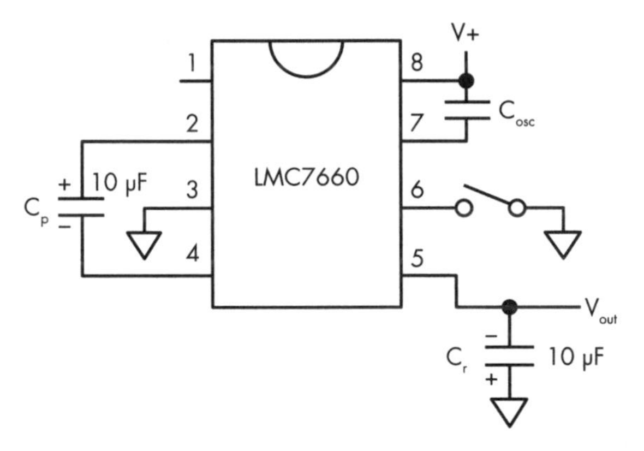

Figure 7-3: LMC7660 pinout

One 16×2 LCD
One I^2C adapter, if not included with the LCD
One 5.1V Zener diode
One 1 μF ceramic capacitor (C2)
Five 0.1 μF ceramic capacitors (C1, C6, C7, C8, C9)
One 0.01 μF ceramic capacitor (C5)
One 22 μF tantalum capacitor (C10)
Two 10 μF tantalum capacitors (C3, C4)
One 10-kilohm, 1/8 W resistor (R5)
Three 10-kilohm, 1/8 W resistors (R1, R2, R10)

Two 1-kilohm, 1/8 W resistors (R8, R9)
One pH probe
Four 4-40×1/2-inch screws
Eight 4-40 nuts and washers
Four 2-56×1/2-inch screws
28- or 30-gauge hookup wire
One Hammond 1591 BTCL plastic enclosure
Heat-shrink tubing

Downloads

Sketch *pHMeter.ino*
Cover template *pHCover.pdf*
Side template *pHBoxSide.pdf*
Shield *pHMeter.pcb*

About the pH Probe

At the heart of the Custom pH Meter is a pH probe. This measures the activity of hydrogen ions in a solution, which in turn determines the acidity or alkalinity of that solution. A basic pH probe, like the one in Figure 7-4, comprises two elements: a reference electrode and a measurement electrode. I won't go into the chemistry of the probe or the exact mechanism of a pH probe's operation, but I will describe its output and interface to the circuitry that provides the readout.

A pH probe produces a voltage proportional to the pH of the solution the probe is immersed in. The pH range starts at 0, which is the most acidic, and goes up to 14, which is the most alkaline. The probe delivers an output voltage from approximately –420mV to +420mV, representing an increment of roughly 60mV per unit of pH. A neutral pH of 7.0, at mid-scale from 0 to 14, is represented by 0.0mV.

The nature of the probe's output makes the Custom pH Meter's basic function in this project relatively straightforward: it needs to read and display a voltage. But there are a few other things the Custom pH Meter circuit has to account for. First, while commercial probes are built to the highest standards, they can be off by some nominal amount and require adjustment. Second, we're dealing with relatively small values, so to maintain accuracy, components and circuits have to be carefully selected. Further, as probes are used and age, they tend to change slightly, requiring recalibration.

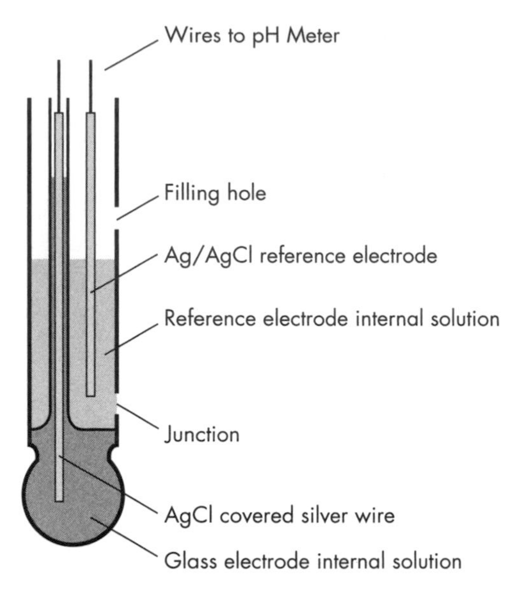

Figure 7-4: A simplified drawing of a pH probe

Finally, the pH probe has a very high electrical impedance—perhaps 10 to 100 megaohms or higher. Practically speaking, a high impedance means that despite the voltage level, there is very little energy available to change the state or condition of another device, so the circuit needs to amplify that signal. This requires a specialized input circuit involving an op-amp, which is designed to minimize noise while handling the high-impedance signal. Today's semiconductors are up to the task, and as I discuss in "Some Notes on IC Selection" on page 196, I checked out several op-amps to find one that seemed to offer the best combination of performance and price. Of course, while a good op-amp is important, the circuitry feeding input to the op-amp must also be as efficient as possible to achieve an accurate reading without introducing noise that could affect the sensitive output of the pH probe.

TAKING CARE OF YOUR pH PROBE

While your pH probe probably came with instructions, there are a couple of things you can do to increase its useful life. First, of course, follow the manufacturer's instructions. Second, unless the manufacturer specifies differently, when storing your probe, immerse the business end in a 3-molar solution of potassium chloride (KCl), as shown in Figure 7-5. You should be able to buy such a solution wherever you purchase your pH buffer solutions at a very modest cost. (A *buffer solution* is a mix of relatively weak acidic and alkaline chemicals that produces a specific pH.) You can also compound your own by dissolving about 22 g of KCl in 100 mL of distilled water.

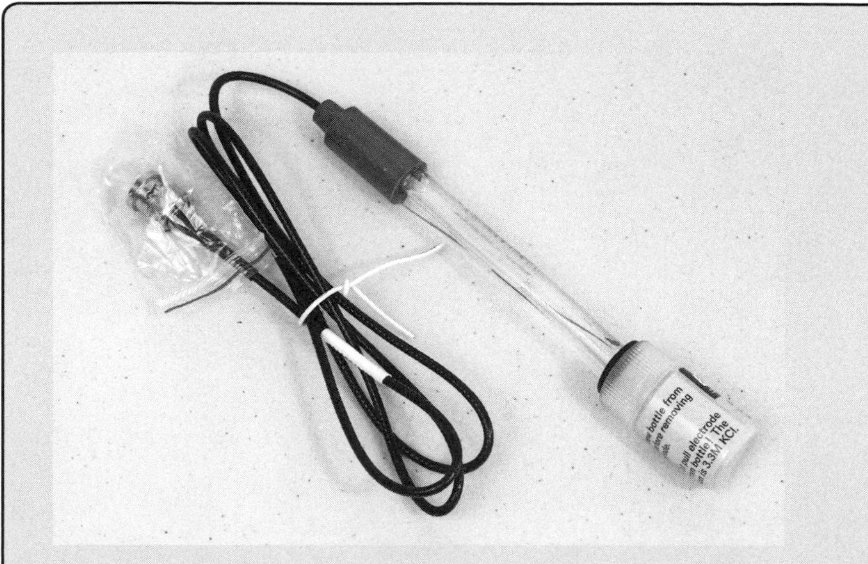

Figure 7-5: The pH probe I used for this project. The tip is protected in a small vial of KCl solution with a rubber seal.

Even for relatively short storage times, it is best to keep the probe in a pH 7 buffer solution rather than in air or water. Between samples or when moving the probe to a different buffer solution, make sure to rinse the probe carefully. Most manufacturers suggest rinsing with distilled water. You can gently blot the excess water off the probe, but most manufacturers caution against rubbing or wiping the electrode bulb for fear of creating an error due to polarization.

When calibrating your probe prior to a measurement, the manufacturers of even top-of-the-line pH meters suggest calibrating with a buffer closest to the expected pH of the sample. For example, if you suspect the pH value of the sample to be around 9, use a pH 10 buffer solution to calibrate the instrument.

The Schematic

The Custom pH Meter circuits shown in Figures 7-6 and 7-7 comprises a dual op-amp, a voltage inverter to supply ±5V to the op-amp, a voltage regulator, a temperature IC, an Arduino, and an I^2C LCD. You have the option to build your Custom pH Meter with an Arduino Nano or an Arduino Pro Mini. I had an easier time building the breadboard with a Nano, but my final product (and therefore, the shield PCB file provided for this chapter) uses the Pro Mini to conserve space.

The rest of this section describes the reasoning behind the design decisions made in creating this schematic.

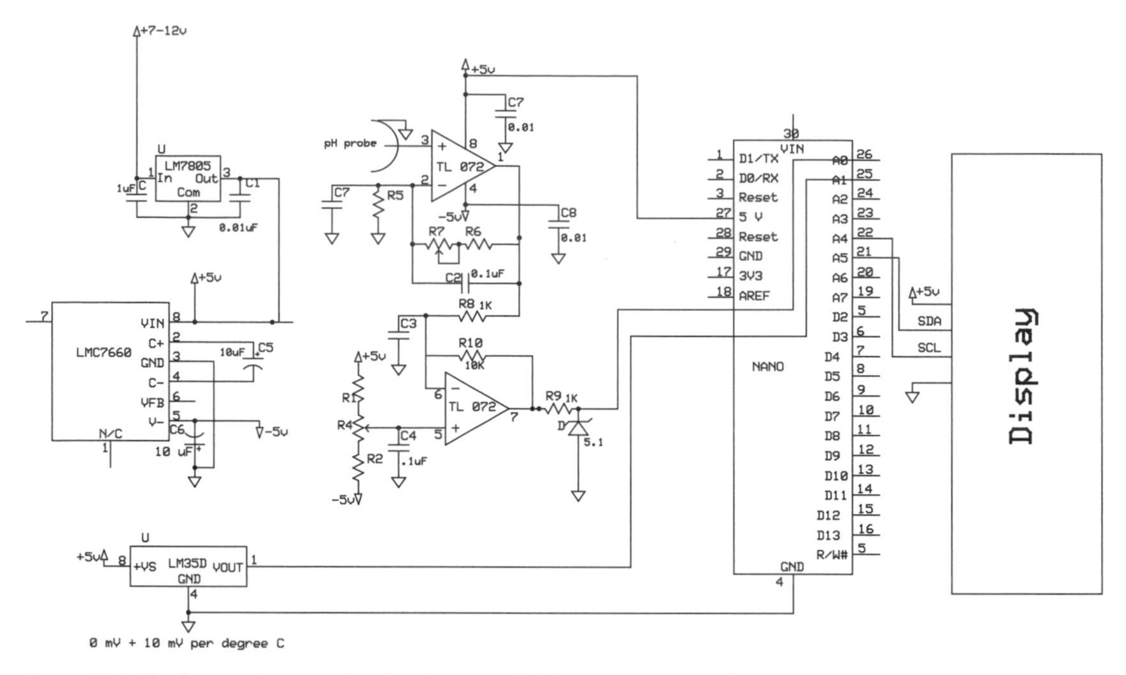

Figure 7-6: The basic schematic for the Custom pH Meter, using an Arduino Nano

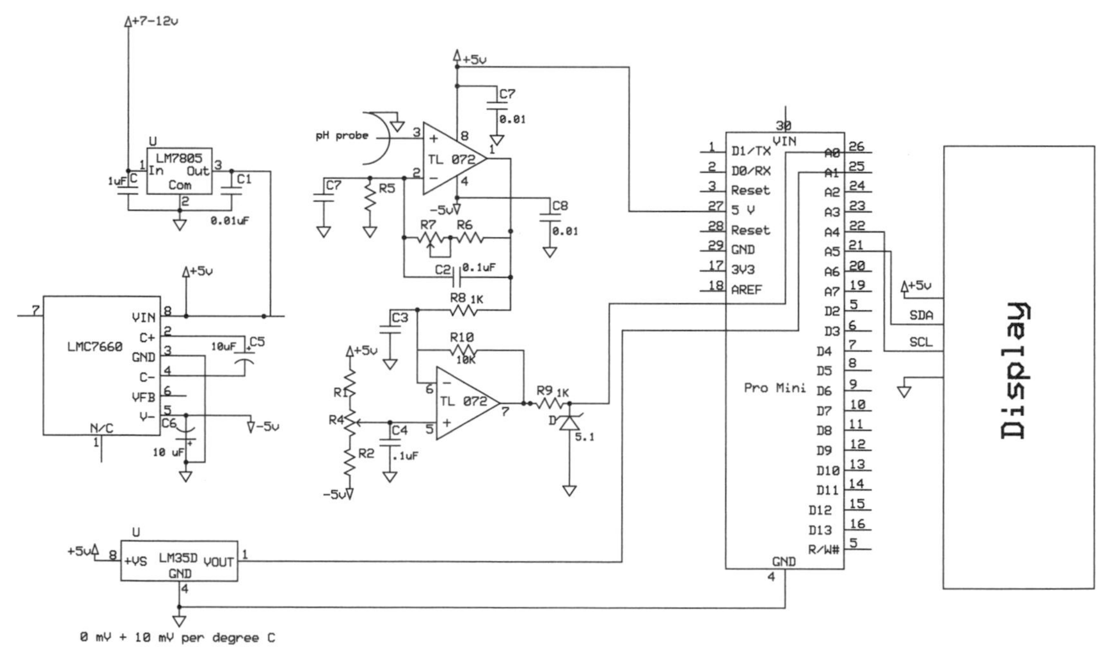

Figure 7-7: The basic schematic for the Custom pH Meter, using a Deek-Robot Pro Mini

Integrating the High-Impedance Probe

Recall that the pH probe delivers a DC output voltage that swings from –420mV to +420mV, giving approximately 60mV per unit of pH. This output is delivered at very high impedance, and the circuit must accept the probe's high-impedance input without adding spurious signals, reduce the impedance to a manageable level, and amplify the input so it can be read by the analog inputs of the Arduino. The circuit also has to provide a way to adjust the voltage that goes to the display in order to calibrate the probe in terms of both offset and gain (see "Offset and Gain" on page 194 for a crash course).

To handle the high-impedance probe output, the input of the op-amp must have a very high impedance, typically in the teraohm (1×10^{12} ohms) range, to read any voltage at all. The input of the op-amp must also have a low input current (the two go together); this is typically around 10 picoamps (1×10^{-12} amps), though some op-amps offer input current below 25 femtoamps (1×10^{-15} amps). It's also good if the op-amp has very low *drift* (that is, tendency to change output with no change in input).

General Design Notes

The Custom pH Meter is designed to work from a power supply of 9V, selectable between a battery and a plug-in module via a power switch (see the schematics in Figures 7-6 and 7-7). Because the input from the AC source could be suspect, the Custom pH Meter uses an external voltage regulator rather than the regulator built into the Pro Mini. An LM7805 with bypass capacitors at both the input and output worked well in previous projects, and this project uses the same regulator. This regulator supplies positive 5V to the inverter, the op-amp circuitry, and the Arduino. The power switch is a three-position switch, where the center is off, one position selects the battery, and the other selects AC power.

Because the pH probe provides an output of ±420mV, this circuit has to be able to handle a *bipolar* (above and below ground) voltage. The simplest way to achieve that is to use an op-amp with positive and negative supplies and a ground in the middle, which in turn requires a power supply that can provide those voltages. The LMC7660 voltage inverter is the solution: it converts the positive 5V from the voltage regulator to +5V and –5V, with ground in the middle. Thus, the op-amp can handle the input signal as long as it doesn't go above +5V or below –5V.

NOTE *Most voltage inverters are very similar to the LMC7660 and require minimal external components—in this case, only two capacitors. This circuit uses tantalum capacitors because of their compact size and reliability, but electrolytic capacitors could be used.*

OFFSET AND GAIN

To demonstrate how offset and gain work, Figure 7-8 illustrates the continuum of voltage used in this project, from –5V to +5V.

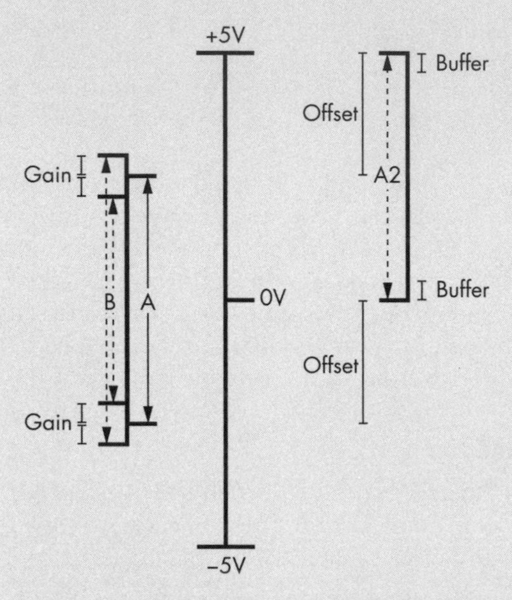

Figure 7-8: An illustration of what happens to the voltage when changing the gain and offset on the Custom pH Meter

Imagine that when the voltage from the pH probe is amplified, there is a voltage range A. In this illustration, voltage A represents half the supply voltage, ranging from –2.5V to +2.5V for a total of 5V. If the gain is adjusted, the voltage range will still center on 0V but will increase or decrease to some voltage range B. Adjusting the gain will always increase or decrease the lower and higher ends of the selected voltage sector equally, provided the original sector is within the total voltage range.

The entire voltage sector can also be shifted within the continuum by adjusting the offset, as illustrated in Figure 7-8. The range of voltage A2 still encompasses the same total voltage as range A, but its minimum and maximum voltages are different. In this instance, the center of the voltage range is shifted from 0V to 2.5V, such that voltage A2 swings from 0V to 5V.

This offset can be any voltage within the supply range, but in practice, it's best not to run the voltage to the voltage rails (the maximum and minimum of the supply voltage). Instead, leave some buffer between the rails and the voltage range a project needs.

"The Sketch" on page 205 shows that the Custom pH Meter maps the voltage within the limits of the supply to the Pro Mini (0V to +5V) to be an average of the gain selected. The final gain and offset adjustments for this project are made using the prepared buffer solutions and the potentiometers in the circuit.

The Custom pH Meter uses a Deek-Robot Pro Mini Arduino clone because this clone is small and inexpensive; however, a Nano could work if you make your own shield-printed circuit board. While the Pro Mini does not include a USB interface, there are a variety of ways to program it with little effort. See "Connecting and Programming an Arduino Pro Mini" on page 8 if you've never used this particular Arduino before.

The I^2C interface for the display comprises only two wires—clock and data—in addition to power and ground. The I^2C protocol can also be used with several I^2C devices at the same time, if required.

The Op-Amp Circuit in Detail

In the op-amp circuit, there are resistors and capacitors to minimize the effect of spurious signals and to couple the circuits. The amplifier circuit has two stages, which are both included in the single package of the op-amp: the first handles the high impedance from the probe and offers gain adjustment, while the second is a buffer that provides the offset, both for calibration and to accommodate the 0 to +5V analog input required by the Arduino. Each stage has a 10-turn trimmer potentiometer. The trimmer in the first stage is 1 megaohm, and it sets the gain; the trimmer in the buffer stage is 10 kilohms, and it is used for the offset.

The first stage provides most of the gain (the output is roughly six times the value of the input), which is adjustable via a negative feedback resistor (R2) and a potentiometer (R3). The adjustment range is a bit wider than required, but it works out quite well. Initially, I tightened up on the range but found that for some probes, a smaller range makes calibration difficult.

The second stage op-amp circuit uses a fixed feedback resistor of 10 kilohms (R5), while the noninverting input uses a combination of two resistors of 10 kilohms each (R1 and R2) and a 10-kilohm potentiometer (R4) to provide the offset adjustment. This stage provides a small amount of gain in addition to the offset adjustment to allow the pH probe voltage to center on 0mV (pH of 7) and swing between –420mV (pH of 0) and +420mV (pH of 14). In addition, this buffer stage changes the scale from a negative and positive voltage to a positive-only voltage for the Arduino.

Using the buffer stage to both provide the calibration offset and convert the plus-minus voltage swing works out conveniently with little, if any, discernable downside. Optionally, a separate reference voltage could be generated, but that would add additional components and offers little advantage over offsetting the voltage from the buffer stage and referencing the voltage to ground.

At the output end of the buffer, a Zener diode (D1) and resistor (R9) are added to protect the Pro Mini from any overvoltage condition. No protection was added to protect against a negative voltage to the analog input pin; however, during initial setup and experimentation, the analog pin accidentally received a negative voltage many times with no adverse effects.

Some Notes on IC Selection

Before settling on an IC for any project, it is a good idea to test multiple ICs to see which works best in your situation. When sampling chips, I suggest keeping notes on the pros and cons of each chip. For example, these are my notes on op-amp possibilities for the Custom pH Meter:

TL072 Worked well; a good all-around solution

TLC2262 A good all-around solution; a toss-up between this and the TL072

OPA129 Worked well, but not available in DIP

LMC6001 Worked well, but a little pricey for no advantage, at around $20

LMC6042 Probably would have worked, but was difficult to set up

LMP7702 Probably would have worked, but a little pricey and was difficult to set up

After trying several op-amp circuits in the public domain with mostly disappointing results, I used a generalized circuit to test each of these op-amps in turn, and for each test, the circuit required a certain amount of tuning to work. This tuning included changing the circuitry to stabilize gain and minimize spurious signals and stray voltages. The Texas Instruments TL072P op-amp proved the best option, and once I made that selection, I adjusted the circuit further to optimize it for the Custom pH Meter. The TLC2262 also would have worked well; I used it in some prototype samples.

The other op-amps I sampled might have worked as well, or almost as well, if optimized like the TL072; however, that would have been time consuming for a marginal or zero gain. The final Custom pH Meter circuit represents a best effort within self-imposed limitations, like budget. For example, a top-shelf op-amp, like the Texas Instruments OPA627/637, probably would have worked well, but the chip alone had a price tag between $25 and $50, depending on the version. That would have brought the total budget for the project to well over $100, a self-imposed limit. The decision to continue the project itself was already problematic because of the probe's cost ($36 at the time of writing); however, I believe the probe's capabilities warrant the expense.

Preparing the LCD

Before you build the circuit on a breadboard, make sure the LCD is prepared for prototyping. Though the LCD used in this project can be purchased with the I^2C adapter board, I have often had to buy the LCD and the adapter board separately, as was the case this time. When you buy them separately, the adapter board usually comes with header pins installed, and all you should have to do is insert them into the display and solder them. "Affixing the I^2C Board to the LCD" on page 3 describes this process.

> **ON LCD BACKLIGHTS**
>
> When I initially wired the I^2C board to the display board for the Custom pH Meter, I cut off the cathode (K) header terminal on the display board. The anode and cathode headers allow you to provide a voltage to power the display's backlight. The idea behind severing the connection was to include a separate switch to turn the backlight on and off to preserve battery current. It turned out the display was not readable without the backlight except in extremely bright light, so I abandoned that effort and manually rewired the backlight. You can experiment with other displays to try to find an ambient-light readable display.

The Breadboard

Like most of my Arduino projects, the Custom pH Meter began with a breadboard (see Figure 7-9). Despite the somewhat ragged appearance, the breadboard iteration worked well.

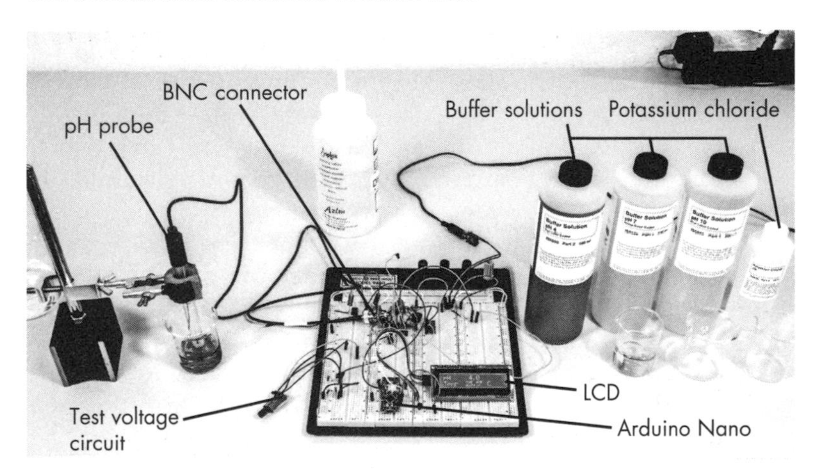

Figure 7-9: This circuit on a breadboard served as a proof of concept for the Custom pH Meter.

In addition to the basic Custom pH Meter circuit, I added a separate circuit (visible in the upper-left area of the breadboard in Figure 7-9) to supply a continuous, variable ±500mV test voltage so I could check the circuit and do some preliminary calibration prior to testing with the probe itself. This test circuit, shown in Figure 7-10, comprises a separate voltage inverter, a pair of voltage dividers, and a potentiometer to vary the voltage. You may want to set up this small circuit on a separate breadboard and use it to do the preliminary adjustment of the finished unit.

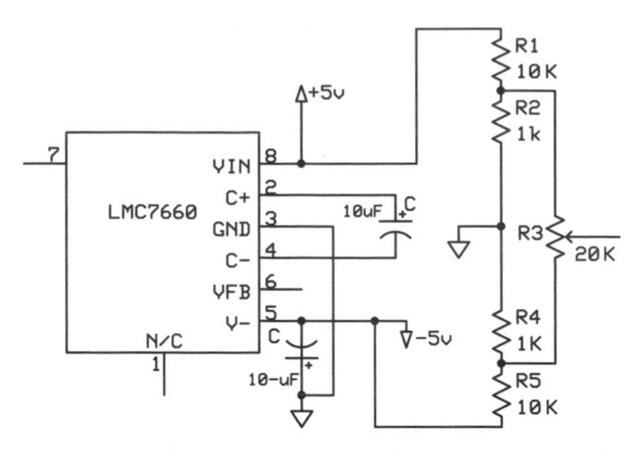

Figure 7-10: This circuit provided the test voltage for the Custom pH Meter.

In Figure 7-9, the pH probe is held by an inexpensive burette clamp attached to an old machinist's magnetic gauge holder. The probe was too thin to fit in the holder, so I wrapped some foam around the probe to clamp it.

Preparing this breadboard turned out to be a little messier and more convoluted than usual because there's a lot on the board. As shown in Figure 7-9, I used a large breadboard that included four vertical breadboards and a strip across the top for positive and negative rails. Initially, I used an Arduino Nano clone to build the Custom pH Meter on the breadboard. In the finished version, however, I suggest using a Deek-Robot Pro Mini board to reduce the size. Both Arduinos use the same 5V, 16 MHz Atmel 328 processor and other components. In keeping with the requirement to have the shortest possible connection to the input of the op-amp, the BNC connector is situated to provide a relatively direct connection to the noninverting input of the op-amp. The voltage regulator, an LM7805, is located in the upper right of the breadboard and is powered with either a 9V alkaline battery or a 7.5V to 12V wall adapter.

Here are the steps I took to construct the breadboard:

1. Connect all vertically oriented positive and negative rails (marked by red and blue stripes, respectively) to the horizontal positive and negative rails across the top of the breadboard.
2. Mount the power inverter chip (LMC7660) on the upper-left side.
3. Mount capacitor C3 (10 µF) between pins 2 and 4 of the LMC7660. (Make sure to observe polarity: connect the positive side to pin 2 and the negative side to pin 4.)
4. Mount capacitor C4 (10 µF) between pin 5 of the LMC7660 and the blue negative rail. (Make sure to observe polarity: the plus side of the cap goes to the blue negative rail.)
5. Connect pin 3 of the LMC7660 to the blue negative rail.
6. Connect pin 8 of the LMC7660 to the red positive rail.

7. Insert the voltage regulator (LM7805) into the right-hand section of the breadboard.
8. Connect pin 1 of the LM7805 to a blank row that will accept the incoming 9V or 7.5V voltage. (It can accept from +7V to +12V, as shown in the schematic.)
9. Connect pin 2 of the LM7805 to the blue negative rail.
10. Connect pin 3 of the LM7805 to the red positive rail.
11. Connect capacitor C2 (1 μF) from pin 1 of the LM7805 to the blue negative rail.
12. Connect capacitor C1 (0.1 μF) from pin 3 to the blue negative rail.
13. Insert the TL072 IC in the breadboard; I placed it in the second vertical section. Observe all antistatic precautions while handling the chip.
14. Make the connections to the TL072 as short as possible to eliminate possible spurious signals.
15. Connect capacitor C5 (0.01 μF) from pin 8 of the TL072 to the blue negative rail. Make the connection as close to the chip as possible to minimize effects of spurious signals.
16. Connect pin 8 of the TL072 to the red positive rail (once again, using as short a jumper as possible).
17. Connect resistor R5 (47 kilohms) from pin 2 of the TL072 to the blue negative rail.
18. Connect the center lead of the BNC input jack to pin 3 of the TL072 with as short a wire as possible.
19. Connect the ground of the BNC to the closest spot available on the blue negative rail. (I used a panel-mount BNC connector and screwed a piece of stiff wire to the flange so I could mount it very close.)
20. Connect pin 4 of the TL072 to the negative voltage of the LMC7660 (pin 5).
21. Insert one lead of capacitor C6 (0.1 μF) to pin 4 of the TL072, as close to the chip as possible.
22. Insert the other end of capacitor C6 to the closest spot available on the blue negative rail.
23. Insert the outside pins of potentiometer R7 (1 megaohm) between pins 1 and 2 on the TL072.
24. Insert a short jumper from pin 2 to pin 3 of potentiometer R7. The potentiometer will work with only the two pins (center and one end) used. For convention and stability, I usually connect the center to the pin not being varied.
25. Insert capacitor C7 (0.1 μF) from pin 1 to pin 2 of the TL072.
26. Insert resistor R8 (1 kilohm) from pin 1 to pin 6 of the TL072.
27. Connect one lead of capacitor C8 (0.1 μF) to pin 6 of the TL072 and the other lead of capacitor C8 to ground.
28. Connect resistor R10 (10 kilohms) between pins 6 and 7 of the TL072.

29. In an open area of the breadboard (as close to the TL072 as possible), insert potentiometer R4 (10 kilohms).
30. Connect one side of resistor R2 (10 kilohms) to pin 1 of potentiometer R4.
31. Connect the other side of resistor R2 to negative 5V (pin 5 of the LMC7660).
32. Connect one side of resistor R1 (10 kilohms) to pin 3 of potentiometer R4.
33. Connect the other side of resistor R1 to the red positive rail.
34. Connect pin 2 (the center pin or slider) of potentiometer R4 to pin 5 of the TL072.
35. Connect one lead of capacitor C9 (0.1 μF) to pin 5 of the TL072 (as close to the pin as possible) and the other lead to the blue negative rail.
36. Insert the Arduino Nano into the breadboard. I placed it in the second row from the left toward the bottom of the board so the USB connection would be easily available.
37. Connect 5V from the Nano (pin 27) to the red positive rail.
38. Connect the ground of the Nano (pin 29) to the blue negative rail.
39. Connect one side of resistor R9 (1 kilohm) to pin 7 of the TL072.
40. Connect the other side of resistor R9 to A0 (pin 26) of the Nano. You may have to identify a blank space on the breadboard and then use a jumper wire.
41. Connect the anode of the Zener diode (D1) to pin A0 on the Nano.
42. Connect the cathode of the Zener diode to the blue negative rail.
43. Connect the positive lead of capacitor C10 (22 μF) to A0 of the Nano.
44. Connect the other lead of capacitor C10 to the blue negative rail.
45. Make a four-wire cable for the LCD with wires for plus, minus, SDA, and SCL. (See "Connectors Used in This Book" on page 18 if you've never made a cable.)
46. Connect the positive and negative wires of the LCD harness to the red positive and blue negative rails, respectively.
47. Connect the SDA pin from the LCD to A4 (pin 22) of the Nano.
48. Connect the SCL pin from the LCD to A5 (pin 21) of the Nano.

In Figure 7-9, at the upper left of the board, you will see four resistors and a potentiometer on three wires. That's the probe voltage simulator circuit in Figure 7-10. To wire up the simulator circuit, make the following connections:

1. Connect the open end of resistor R5 (10 kilohms) to negative 5V (pin 5 of the LMC7660).
2. Connect the open end of resistor R1 (10 kilohms) to the red positive rail.

3. At the juncture of resistor R1 (10 kilohms) and resistor R2 (1 kilohm), connect pin 1 of potentiometer R3 (20 kilohms).
4. At the juncture of trimmer R4 (1 kilohm) and resistor R5 (10 kilohms), connect pin 3 of the potentiometer R3 (20 kilohms).
5. Connect the center lead (pin 2) of potentiometer R3 (20 kilohms) to pin 3 of the TL072.

The temperature sensor is not included in the breadboard.

Finally, calibrate the Custom pH Meter. I suggest calibrating first with the simulator circuit and then with the actual probe, as described in the next section.

Calibrating the Custom pH Meter

Calibrating the Custom pH Meter for the first time may be a little trying, but it shouldn't take long to get the hang of it, and it should work well afterward. First, set both the scale and offset potentiometers as close to the middle of their ranges as possible. Because pH 7 is neutral, start there, put the probe in the pH 7 solution, and adjust the offset until the display reads 7.00.

Next, clean the probe, place it in the pH 4 solution, and adjust the scale trimmer until the display reads 4.00. After that, clean the probe again, and try the pH 7 solution again; the reading should remain close to center. If it is off center, then adjust the offset to exactly 7.0 again, and repeat the process. This time, it should require only a small adjustment to set the scale to pH 4.0. Now, check the display with the probe in the pH 10 buffer solution, see how far off the reading is, and adjust the scale trimmer accordingly. Repeat this process until the readings match all three buffer solutions. (After about three or four tries, adjusting both scale and offset, I got it to line up perfectly.)

When you know the meter works, I suggest resetting it and calibrating it again for practice. I was able to do it quite a bit faster the second time, with only two repetitions. I rechecked my pH meter several times over a period of about three weeks, and it seemed to stay in calibration; you should have similar results.

NOTE *As a preliminary test, I also used the test voltage circuit in Figure 7-10 to perform initial calibration. You will also need a digital voltmeter. I started with 0V and adjusted the offset potentiometer until the LCD showed a pH of 7. Next, I adjusted the test voltage circuit to output 180mV on my digital voltmeter and turned the scale potentiometer until the LCD showed a pH of 10. I then adjusted the test voltage to –180mV and adjusted the scale until the LCD showed a pH of 4. After only a couple of tries, I had good results, so I disconnected the test supply and replaced it with the probe. This time, I was able to calibrate the meter in only a single try.*

When your meter is built and calibrated, try testing it on some common household products, like these:

- Coca-Cola Classic: pH 2.5
- Orange juice: pH 2.8
- Coffee: pH 5.0
- 5 percent ammonia solution: pH 11
- Clorox bleach: pH 11.9

IN AN ANALOG FRAME OF MIND

Some of us dinosaurs still like analog readouts, and for those holdouts mired in the 20th century, I have included provisions in the schematics in Figures 7-6 and 7-7, the sketch, and the final shield PCB file for using an analog readout. This was a bit of an afterthought and it's optional, so the series resistor for the meter is not included in the shield; however, the Pro Mini pin connections to the readout are. The single required resistor can be mounted on the rear of the meter movement. In Figure 7-11, I simply connected the meter to the breadboard circuit.

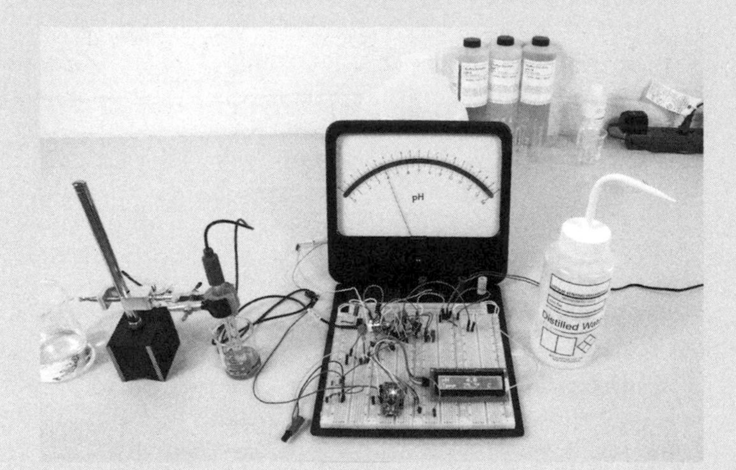

Figure 7-11: The Custom pH Meter breadboard circuit, with both the digital display and a 20 mA meter movement

I've collected analog meters and movements over the years; the meter pictured is a Simpson 20 mA meter movement. To drive the meter, I simply used the PWM (pulse-width modulation) output from pins 5 and 6 on the Pro Mini, hooked directly to the meter with a resistor in series. In the sketch, I centered the meter on 0V so it uses a positive and negative voltage.

I ended up using a variable resistor to set the analog meter's minimum and maximum to –420mV and +420mV, respectively. This eliminated the problem of attempting to set the gain and offset to match the digital readout. However, that in no way affects the accuracy of either readout, and the digital and analog readouts match. They also track identically through the entire pH range.

For many meters, the case can be removed to easily place a different scale, as has been done in Figure 7-11. The face on this project's meter was prepared with a laser printer to make the scale reflect pH value and show an mV scale for reference. A nice scale can be made using a drawing program such as Corel Draw or Adobe Illustrator, and you can purchase full sheets of adhesive-backed label stock to adhere it to the original meter plate. Just be careful not to damage the needle or movement in the process.

About the Effects of Temperature

Thus far, I have not addressed the issue of temperature, but a solution's pH value is temperature-dependent, as illustrated in Figure 7-12.

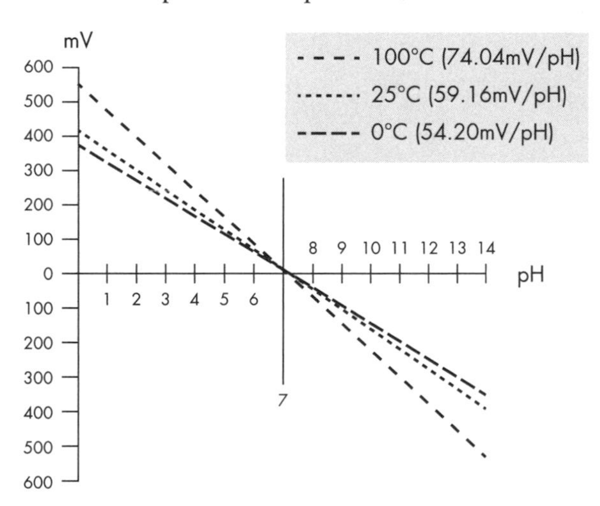

Figure 7-12: How pH varies with temperature

The effect of temperature on pH at or around room temperature (25°C) is nominal. In fact, according to the chart, the difference in pH from 25°C to 0°C would measure as little as half a pH unit at both pH extremes. At 100°C, however, the difference is more pronounced and could be as much as 1.5 pH units.

Given that information, the question becomes exactly how to handle the effect of temperature on pH.

Adding a Temperature Sensor

The first step to address the effects of temperature is to include a temperature sensor in the circuit so the Custom pH Meter knows what the temperature is. One of the easiest and most widely used temperature sensors in Arduino land is the LM35, which according to the data sheet outputs a linear 10mV/°C at 25°C, with half-degree accuracy. You can see this sensor in the schematics in Figures 7-6 and 7-7.

I hooked up the chip and included it in the sketch, but I was somewhat concerned about the accuracy. Unfortunately, I didn't have a National Institute of Standards and Technology (NIST) temperature standard to go by, so I compared the Arduino sensor readings to a glass scientific thermometer, a bimetal dial thermometer, and a Radio Shack digital thermometer. None of these agreed with each other or the Arduino.

Checking Accuracy

The obvious way to calibrate the temperature sensor was to check it against two temperature values I really knew: the freezing and boiling point (at sea level) of distilled water. Because I live about 12 feet above sea level, the altitude would not be a problem. The result of my ice water and boiling tests indicated that the sensor was indeed off by about 2 percent. This was probably due to something in the external circuitry, such as the reference voltage, and is compensated for in the sketch. The temperature IC and connections were protected in a heat-shrink tube to eliminate any moisture getting to the connections, as described in "Construction" on page 211.

According to the manufacturer for the glass electrode I used in this project, the error caused by temperature can be calculated as follows:

$$\text{Error in pH} = 0.003 \times (\text{Calibrated temperature} - \text{Current temperature}) \times (\text{Neutral pH} - \text{Actual pH})$$

For example, if the electrode is calibrated at room temperature (25°C) and is measuring a sample around pH 4 at around 5°C, you would calculate the error as follows:

1. Calculate the temperature difference: 25°C – 5°C = 20°C
2. Calculate how far away from neutral the pH is: 7 pH – 4 pH = 3 pH
3. The total error is: $0.003 \times 20 \times 3 = 0.18$ pH.

As of this writing, I have not attempted to integrate the temperature reading into the sketch to adjust the pH automatically. The Custom pH Meter does, however, display the temperature on the LCD, so you can decide whether or not it's worth adjusting. As you can see in Figure 7-12, the effect is minimal and can easily be approximated from the chart or calculated with tests similar to mine.

Chances are that most of your measurements will be at or near room temperature. Temperature compensation is generally required only in severe environmental and industrial environments. If you are taking pH readings at extreme temperatures, you may want to include the formula in your sketch.

If you're curious, you can read more about how to compensate for pH probe reading errors due to temperature at http://www.qclscientific.com/electrochem/phtemp%20comp.html.

The Sketch

The Custom pH Meter sketch, like many others in this book, comprises parts of other sketches and examples. I've included comments throughout that describe how the most significant pieces work. This unit was tested on and used with the Arduino IDE version 1.0.5-r2.

```
//Custom pH Meter Sketch
//Smoothes both temperature and pH

#include <Wire.h>
#include <LiquidCrystal_I2C.h>
/* Visit http://playground.arduino.cc/Main/I2cScanner for code
   you can run to figure out your LCD's I2C address if 0x27 doesn't work. */
LiquidCrystal_I2C lcd(0x27, 16, 2); //16x2 display
//There are a couple of libraries out there. The one I used was
//simply Liquid Crystal_I2C for a generic type display.
const int numReadings = 10;
const int numReadings2 = 20;
const int meterOut1 = 5;
const int meterOut2 = 6;

float readings[numReadings];      //The readings from the analog input
float readings2[numReadings2];

int index = 0;                    //The index of the current reading
int index2 = 0;
float total = 0;                  //The running total
float total2 = 0;

float average = 0;                //The average
float average2 = 0;

int pHpin = A0;
int tempPin = A1;
int meterdrive1;
int meterdrive2;

int pHvalue = 0;
float val;
float val2;
float tempC;
float temp2;
```

```
void setup() {
  lcd.init(); //You may have to use a different command, depending on the
              //library you use
  lcd.backlight();
  pinMode(meterOut1, OUTPUT);
  pinMode(meterOut2, OUTPUT);

  //Initialize serial communication with computer:
  Serial.begin(9600);

  //Initialize all the pH and temperature readings to 0:
  for(int thisReading = 0; thisReading < numReadings; thisReading++) {
    readings[thisReading] = 0;
  }
  for(int thisReading2 = 0; thisReading2 < numReadings2; thisReading2++) {
    readings2[thisReading2] = 0;
  }

  //Configure the reference voltage used for analog input to 1.1V
  analogReference(INTERNAL);
}

void loop() {

  tempC = analogRead(tempPin);
  temp2 = tempC/9.31; //My calibration factor was 9.31, as determined
                      //by the boiling water and ice tests

  pHvalue = analogRead(pHpin);

  val = map(pHvalue, 0, 1023, 0, 1400);
  val = constrain(val, 0, 1400);

  val2 = val/100;

  meterdrive1 = map(average, 0, 14, 0, 255);
  meterdrive2 = map(average, 14, 0, 0, 255);

  analogWrite(meterOut1, meterdrive1);
  analogWrite(meterOut2, meterdrive2);

  //Subtract the last reading:
  total = total - readings[index];

  //Read from the sensor:
  readings [index] = val2;

  //Add the new reading to the total:
  total = total + readings[index];

  //Advance to the next position in the array:
  index = index + 1;

  //If we're at the end of the array...
  if(index >= numReadings)
```

```
    //...wrap around to the beginning:
    index = 0;

  //Calculate the average:
  average = total / numReadings;

  //Subtract the last reading:
  total2 = total2 - readings2[index2];

  //Get readings from the temperature sensor:
  readings2 [index2] = temp2;

  //Add the temperature reading to the total:
  total2 = total2 + readings2[index2];

  //Advance to the next position in the array:
  index2 = index2 + 1;

  //If we're at the end of the array...
  if(index2 >= numReadings2)
    //...wrap around to the beginning:
    index2 = 0;

  //Calculate the average:

  average2 = total2 / numReadings2;

  delay(1);                //Delay between reads for stability

  lcd.setCursor(0,0);
  lcd.print("pH");
  lcd.setCursor(4,0);
  lcd.print("          ");
  lcd.setCursor(7,0);
  lcd.print(average,2);  //Truncate to two decimal places
  lcd.setCursor(0,1);

  lcd.print("Temp  ");
  lcd.print(average2*.98,1); //Error calculated from empirical measurement
  lcd.print((char)223);      //Print the degree symbol
/* This may vary depending on display. One display used ((char)178) for the
degree symbol.*/
  lcd.print("  C");

  delay(600);

}
```

The basic pH measurement functionality is straightforward: it reads an analog value from the analog output of the op-amp circuitry and feeds that value to an analog input pin of the Pro Mini. The pH and temperature are read every time the main loop runs and then stored in `pHvalue` and `tempC`, respectively. My first version of the sketch printed these directly to the LCD.

But when I laid out the circuit on the breadboard and adjusted the components, I noticed that the output was a little jumpy. The pH value jumped around by two or three tenths of a pH unit, plus or minus some core value. For example, the reading might fluctuate from a pH of 4 to 4.1, then to 3.9 and back to 4.

Smoothing the pH and Temperature Output

I went back to the drawing board. I played with the circuit, trying to find where the jumpiness was coming from, and failed to nail it down. Then, because the pH was unlikely to change quickly, I decided to average a few readings. While that stabilized the reading, the drawback was that the more samples I took, the slower the reading.

However, I didn't think that was a problem, as some expensive commercial pH meters I've used took some time to stabilize, very likely for the same reason. But there was still room for improvement.

Fortunately, there is a useful sketch on the Arduino website written by David Mellis and subsequently modified by Tom Igoe that uses an array to smooth a signal. (You can see the original sketch in full at *https://www.arduino.cc/en/Tutorial/Smoothing/*.) I used this example as a model to smooth out the pH voltage in this project's sketch. I experimented with several different values and found that somewhere between 5 and 10 samples worked well. I set `numReadings` equal to 10, and that resulted in a minimal drag on stabilization period, smoothing things out fairly well. The sketch shown in this book stores the result after smoothing in the `average` variable, which is printed to the LCD at the end of the main loop. In addition, I continued to fine-tune the circuit, so the sketch required less and less averaging.

Notice that the same smoothing technique has been employed in the part of the sketch that handles input from the temperature-sensing circuit. (The `average2` variable contains the smoothed temperature reading result.) This was necessary for the same reason smoothing was needed for the pH voltage: even the temperature sensor output was a bit jittery. My first suspicion was that perhaps the Arduino Pro Mini and its voltage reference was causing hiccups in both the pH and temperature voltages. However, I hooked the temperature sensor directly up to my multimeter with a well-filtered power supply and experienced the same disruptions. In the end, the smoothing approach solved the problem.

NOTE *While the smoothing approach used in this sketch worked well, that isn't the only approach you could take. For example, a moving average approach could also work well.*

Centering an Analog Meter

If you choose to use an analog meter, the Arduino will need to drive the positive and negative sides of the meter. The sketch maps the meter drive to reverse the PWM values on two pins as follows.

```
meterdrive1 = map(average, 0, 14, 0, 255);
meterdrive2 = map(average, 14, 0, 0, 255);
```

To obtain `meterdrive1`, the average value (the average of pH values measured) is mapped from 0 to 14, while `meterdrive2` is the same average value mapped from 14 to 0. Both mappings use the `map()` function from Arduino's preloaded libraries.

The `map()` function is a useful tool that lets you map a number from one range to another. The syntax is as follows:

```
map(value, fromLow, fromHigh, toLow, toHigh)
```

The `map()` function can be used to shift a set of values or, as in this case, to reverse the values going from 0, 14 and from 14, 0. If you want to use an analog meter and don't want to reset the indicator to the center, you can simply use either output pin 5 or pin 6 (leave the one you don't use open) and change the value of the resistor to result in a correct reading.

A NOTE ON SIGNIFICANT FIGURES

The following line of code prints the pH to the LCD, showing two decimal places:

```
lcd.print(average,2);  //Truncate to two decimal places
```

An earlier draft of the sketch called `lcd.print(average,1)` instead, showing only one decimal place, but when I was trying to minimize the jitter on the Custom pH Meter, I changed the display code to include two decimal places for finer granulation. For the most part, the pH reading remained extremely stable even to the second decimal place.

In the final sketch, I kept two decimal places, but to be honest, I'm not sure how meaningful or accurate the second decimal place is. It has a slight tendency to drift as the probe sits in the solution, which I believe is normal. I dutifully researched and learned a lot more about *significant figures* (the digits in a measurement that actually have meaning) than I ever wanted to know, but I was still left without a definitive answer.

Here's the bottom line: all of the literature I perused regarding pH discussed pH in terms of integers—or at best, to the tenths position. Only in some references to scientific and industrial applications was the hundredths position even used at all.

For most practical applications, you can change the sketch to use a single significant digit if you prefer. I do also strongly recommend that you use only a single significant digit during preliminary calibration, as the additional digit could be confusing. If you must (as much for ego as anything), you can put the second significant digit back, but know that its accuracy is suspect and, in my experience, it doesn't really buy you anything.

The Shield

The Custom pH Meter shield, shown in Figure 7-13, is designed to minimize noise from the pH probe to the input. The Pro Mini and the LCD in this project can generate a little electrical noise; thus, all of the active analog input components are at one end of the PCB, while the Pro Mini and interface to the display are at the opposite end of the board. The inverter IC and associated components are located under the Pro Mini to conserve space.

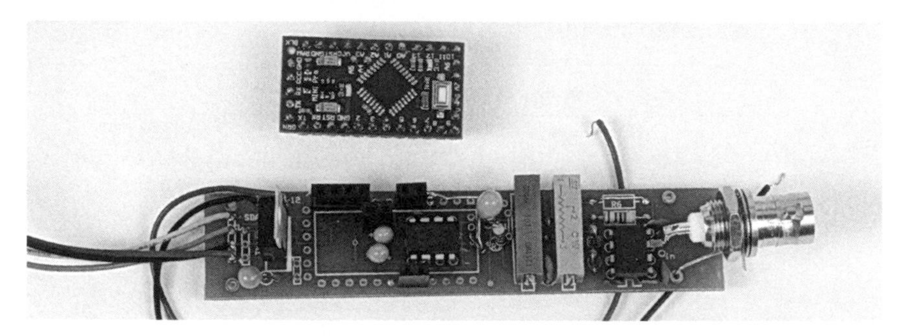

Figure 7-13: The shield PCB has headers soldered in place only for the pins this project uses on the Pro Mini. The voltage inverter and capacitors are located under where the Pro Mini will plug in.

For this project, I decided to use a double-sided circuit board. This made the PCB layout a lot simpler than trying to squeeze everything on one side, and it allowed the amplifier, buffer stage IC, and associated components to be arranged in close proximity. You can see the layout file for this shield, which you can download with the rest of this book's resource files, in Figure 7-14.

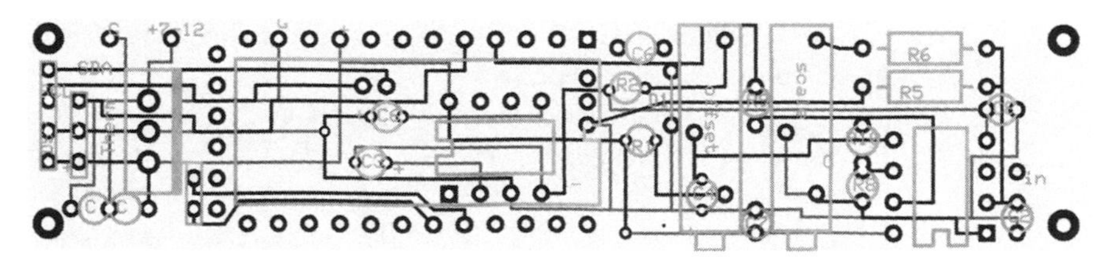

Figure 7-14: The top traces in the shield layout are the darkest, the bottom traces are second darkest, and the silkscreen layer is the lightest. Notice that the silkscreen layer shows boxes around various components.

Figure 7-15 shows the completed printed circuit board before and after population. When populating this board, make sure to take precautions to prevent static electricity damage to the TL072. Because of this chip's very high input impedance, it is particularly sensitive to static discharge from handling. I used a socket to hold the op-amp so in case it got damaged, replacing it would not be a major job.

Figure 7-15: The shield PCB before and after population. The voltage inverter and associated components are under the Pro Mini. The populated version is shown with the display connected.

Also note that many resistors on this PCB are mounted vertically to save space and reduce lead length and circuit-board-trace lengths. I used 0.100-inch female headers to mount the Pro Mini, which leaves plenty of room for the components underneath. It is not necessary to fully populate the board with headers to fit all the Pro Mini's pins; you just need enough to mechanically support the Pro Mini and provide the necessary electrical connections. I found it helpful to place headers at the very end of at least one side to align the pins and simplify my aim when plugging in the Pro Mini board.

Construction

Be sure the sketch is loaded onto the Arduino, and solder all components to your PCB now, including wires for power and ground and for the jack for the (optional) temperature sensor. Place the op-amp into its socket now as well, but bend pin 3 of the op-amp so that it sticks out. You will need to be able to access pin 3 in a later stage of the construction process.

When the Custom pH Meter circuit is soldered and the sketch is loaded onto the Arduino, only one step remains in terms of actual construction: putting everything inside a protective box. This section describes some suggestions for an enclosure and for how to mount the circuit board inside. Figure 7-16 shows the finished enclosure.

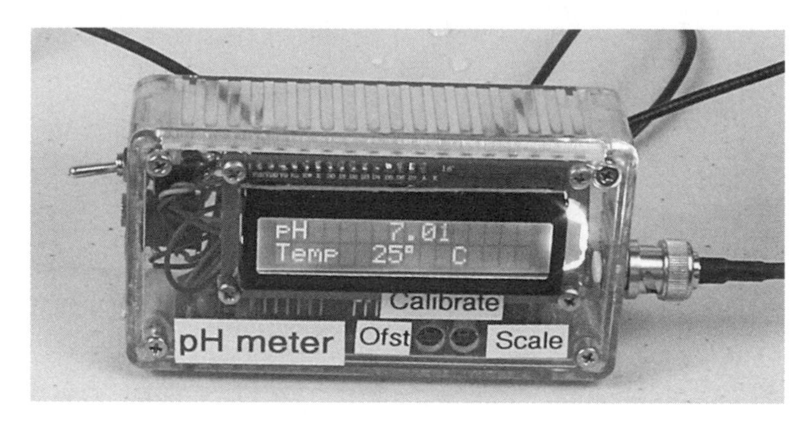

Figure 7-16: The finished Custom pH Meter in the enclosure. This close-up illustrates the positioning of the holes for the offset (labeled Ofst*) and gain (labeled* Scale*) calibration adjustments.*

The Custom pH Meter Enclosure

Your choice of enclosure will depend on how you want to use the Custom pH Meter, whether or not you elect to include an analog meter, and the level of portability required. I selected a standard ABS clear plastic box with outside measurements of approximately 1.3×2.45×4.4 inches. The space was somewhat tighter than I planned, but I was able to squeeze in the printed circuit board, display, switch, connectors, and battery.

Making Room for the Display

This project's enclosure is a Hammond Manufacturing case, model 1591 BTCL. Figure 7-17 shows a drawing of the top of the plastic box, marked with lines for cutting the display hole and drilling the mounting holes.

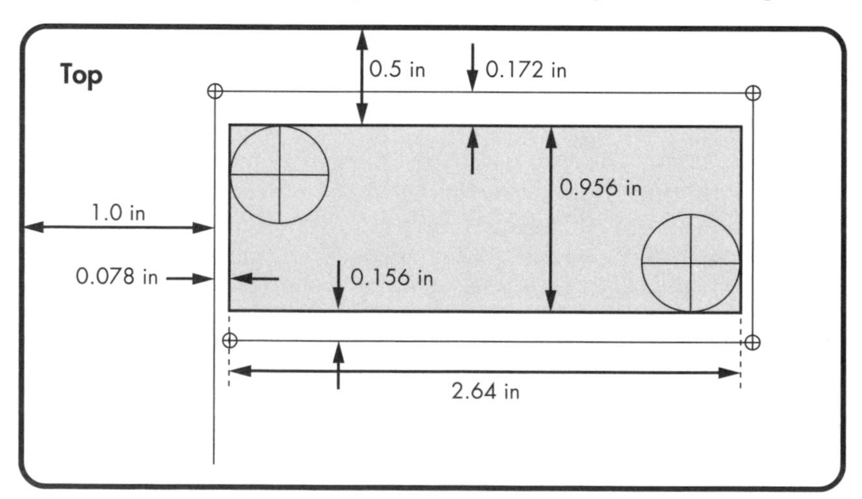

Figure 7-17: Template for the cover of the enclosure, showing an opening for the 16×2 display (shaded area) and where to drill mounting holes

You can download this drawing with this book's resource files (see *https://www.nostarch.com/arduinoplayground/*) and use it as a template for center punching the holes.

I needed as much vertical room as possible inside the enclosure to accommodate the battery, so cutting out the top of the enclosure was necessary. Be careful, though: some displays have slightly different footprints, so measure yours and check it against the drawing first. If there's a discrepancy, adjust the measurements.

The ABS plastic the enclosure is made of cuts easily, so cutting out the display hole shouldn't pose any major problems. Before you get started, clamp the enclosure securely to a piece of scrap wood attached to a workbench or table.

WARNING *When you use a relatively large bit to drill into a thin layer of ABS plastic, the bit will tend to grab the plastic. Do not hold the enclosure by hand.*

To follow this template, first drill the two big holes in the opposing corners of the display area, using the punch marks as centers. I found 1/2 inches to be a good size for these holes, but just make them big enough to accommodate the saw blade you're going to use to make the cutout. You may want to drill the centers of the 1/2-inch holes with a smaller drill first to make sure they are on center. Then, use a keyhole saw to cut out the smaller rectangle.

To make sure there is enough room for the battery, I suggest mounting the display off center, as shown in Figure 7-16, with the measurements indicated in Figure 7-17. When your display hole is cut, drill the smaller holes for mounting the LCD with a #30 or 1/8-inch drill to accommodate the 4-40 screws.

Drilling Holes for Other Hardware

Once the top is prepared, drill holes for the BNC connector (A), on/off switch (B), power input switch (C), and optional temperature jack (D) in the two smaller sides of the main body of the case, according to the template in Figure 7-18. The hole for the BNC connector is 3/8 inches, while the holes for the switches and the temperature jack are all 1/4 inches.

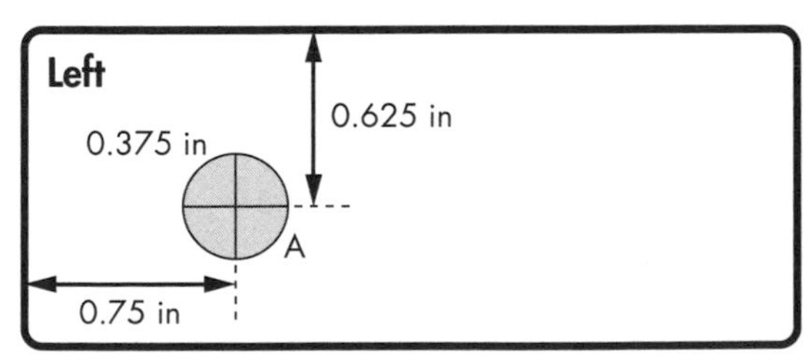

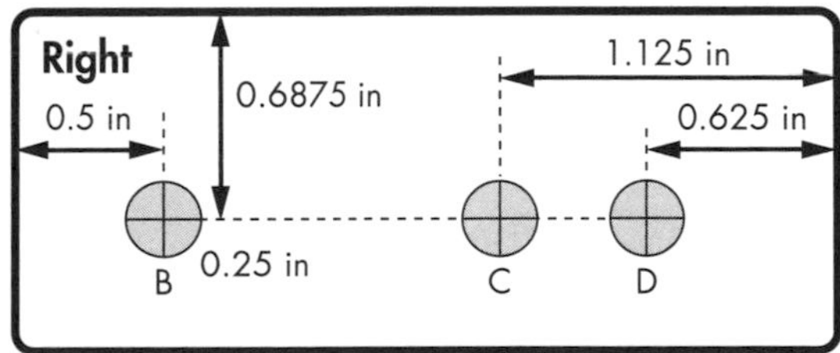

Figure 7-18: Approximate layout for the holes for the BNC connector (A) on the left side of the enclosure, and the on/off switch (B), battery/AC switch (C), and optional 3.5 mm temperature sensor input jack (D) on the right side.

Mounting the Circuit Board

Finally, choose where to mount the printed circuit board to the inside of the case. For both pH meters that I made, I held a nonpopulated shield PCB inside the case and marked the locations of the four mounting holes on the plastic; you can use a printout of the board layout in this book's resource files as a pattern if you prefer. With your four holes marked, gently center punch and drill them. Be careful not to crack the plastic enclosure, and make sure they line up with the mounting holes indicated on the PCB. I suggest using a #41 drill bit (about 3/32 inches) to drill these holes.

If you use the shield design included with this book, you may also need to drill the circuit board's mounting holes to the correct size. Conventional manufacturing practice would advise using a spacer, screw, and nut to hold the board inside the enclosure, but instead, I drilled and tapped the circuit board itself. I used a #50 bit to drill the circuit board in all four corners and tapped it with a 2-56 tap. I then used 2-56 screws to fasten the board in place. While this may not be a recommended practice for all applications, the board was light enough that it worked out well.

Furthermore, I would not expect a finished pH meter to be subjected to much physical punishment. With that in mind, I just put a strip of foam on the noncomponent side of the board and screwed the board to the side of the case—but not too hard. If you expect the unit to experience repeated and extreme vibration, you can put a dab of Thread-Lok or another anaerobic adhesive on the screw threads before assembly.

Once the board is mounted, mark the access holes for the gain and offset trimmers. Place the cover on the enclosure, look straight down through the cover, and use a Sharpie marker to mark the locations of the two trimmer access holes. Remove the top from the enclosure, and drill 1/4-inch holes corresponding to the screw heads on the trimmers (see Figure 7-16).

Installing the Other Hardware

Now you can install the BNC connector, on/off switch, power switch, and (optional) temperature jack. I found it easiest to mount the switches and jack inside the enclosure first and solder the wires to them later. The same technique works well with the BNC connector, though you may wish to solder the ground wire beforehand. I took a piece of 22-gauge solid wire, wrapped it three-quarters of a turn around the base of the connector, tightened it with the retaining nut, and soldered it to a ground tap on the PCB (see Figure 7-13).

NOTE *While I did not bring out the signals for the analog meter for my personal build, if you build the shield, you'll see the two connections marked with a small box around them.*

Connecting the Probe to the Op-Amp

"Integrating the High-Impedance Probe" on page 193 describes special considerations required for using the pH probe as an input to the Custom pH Meter. It turns out that even high-quality FR-4 circuit board material can cause some current leakage, from dirt or moisture on the surface or other contamination. To minimize leakage, connect the input from the IC directly to the probe. In order to do this, place the IC in close physical proximity to the BNC connector inside the case. Then, instead of soldering pin 3 of the op-amp to the board or plugging that pin into a socket, bend it out and wire it directly to the BNC connector (see Figure 7-19).

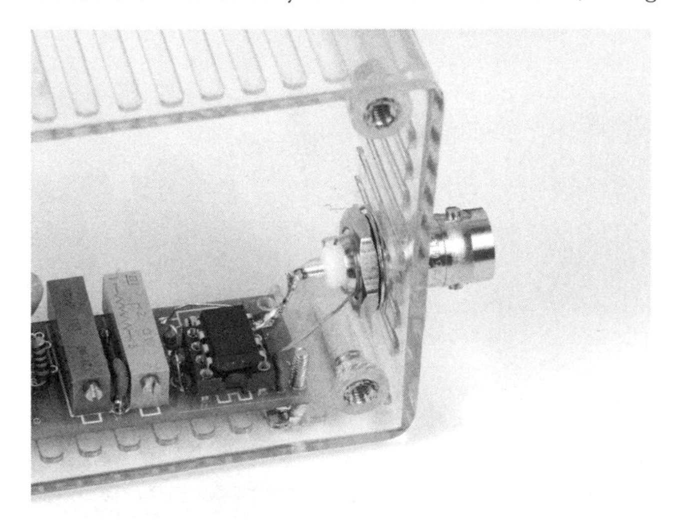

Figure 7-19: To minimize spurious signals, the connection from the BNC connector to the IC pin is soldered directly.

Because this is the only high-impedance part of the circuit, some other precautions mentioned in the op-amp's data sheet (such as ground-isolation rings around the other inputs) are not required. However, this project's shield PCB does keep critical traces as short as possible and places components in close proximity.

Connecting the Temperature Sensor

For the optional temperature sensor, you will need to mount a 3.5 mm three-conductor jack inside the enclosure; the drilling hole for this is marked D in Figure 7-18. Wire this jack to the temperature IC connections on the PCB, which are the three holes between the display connections and regulator IC in Figure 7-20. The sleeve of the connector is ground, the ring is positive, and the tip is the output of the temperature-sensing IC.

Figure 7-20: The three connections for the temperature sensing IC are just above the word Therm.

You can protect the IC itself from liquid immersion by encapsulating it in a short length of heat-shrink tubing with sealant, as shown in Figure 7-21. This heat-shrink tubing is readily available online. The sealant on the inside of the tubing should make a completely waterproof seal for the IC. To seal the IC, just insert it into the tubing and heat with a heat gun or a hair dryer on high heat until it is completely sealed.

Figure 7-21: This LM35 temperature IC is completely encapsulated in a short length of heat-shrink tubing with a sealant thermal gel. The thermal gel can barely be seen where the wires enter the tubing.

Once all hardware is mounted inside the enclosure and all components are soldered to the PCB, screw the LCD to the cover and close it all up to finish. You might also want to label your potentiometers and switches, as shown in Figure 7-16, for ease of operation later.

8

TWO BALLISTIC CHRONOGRAPHS

This project is a device for measuring the velocity of a projectile. It originally measured the velocity of pellets from airsoft and paintball guns, and it evolved to be capable of measuring projectile velocities from BB and pellet guns before finally measuring velocities of over 3,000 feet per second (fps) from higher-powered weapons. The main intention of this project is not to measure the velocities of traditional firearms, but this project does have that capability, and the end of this chapter describes how to use it to measure the velocity of a 9 mm bullet.

The Ballistic Chronograph was meant to be simple, but it turned out a little more complex than originally planned. The result is two projects: the Full Ballistic Chronograph and a more diminutive and simpler device I call the Chronograph Lite (see Figure 8-1).

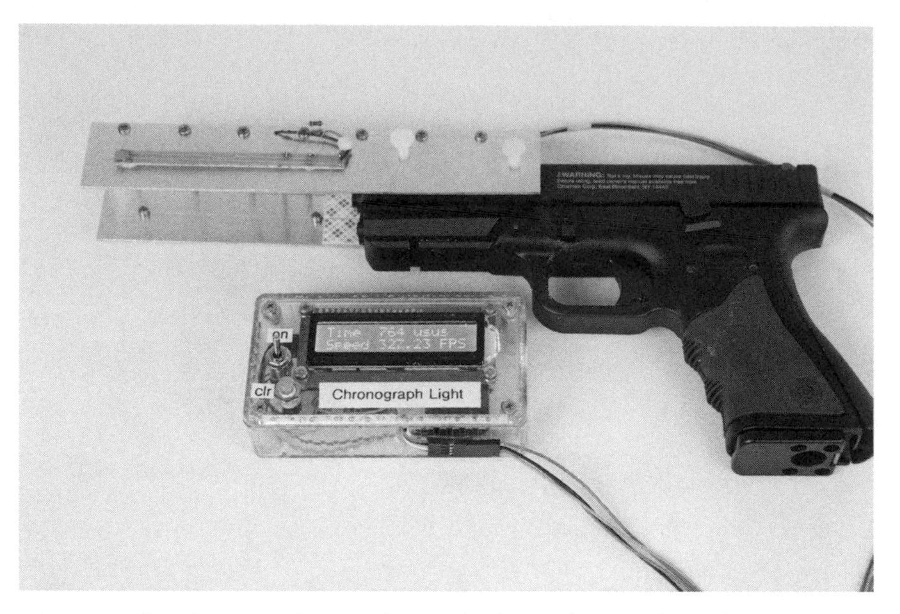

Figure 8-1: The Chronograph Lite with a projectile-acceleration channel attached to a 0.177 caliber pellet gun

I've attempted to make each Ballistic Chronograph system both flexible and accurate. The flexibility comes from separating the sensor elements from the readout and permitting different types of sensors to measure different devices—and producing different readouts with the same acceleration channel.

In this project, you will use some components not too frequently encountered in Arduino projects, such as a crystal oscillator to provide precise timing (outside of the crystal oscillator used in the Nano), infrared LEDs, phototransistors, logic gates, a 12-stage digital counter, and a digital-to-analog converter (DAC) to help perform the counting function.

A WORD OF WARNING

With deference to Jean Shepherd's *A Christmas Story* (in which everyone warns Ralphie, "You're going to shoot your eye out!"), remember that any firearm is inherently dangerous, and many air-powered weapons can fire at lethal force. Whether you test an air-powered device or a weapon using high-powered bullets, use extreme caution. The Full Ballistic Chronograph and the Lite version were developed, tested, and made primarily for lower-powered weapons using CO_2 and air power to accelerate projectiles. Though the device is capable of measuring bullets from traditional firearms, such as the 9 mm pistol mentioned earlier, it was not developed or tested for that application. I strongly recommend that you not attempt to use the device you build in this chapter in such applications.

What Is a Ballistic Chronograph?

A device for measuring the velocity of a high-speed projectile exiting a firearm is generally known as a *ballistic chronograph*. The term *chronograph* was co-opted from the horological community and is now widely used to describe instruments for measuring the speed of bullets, arrows, darts, and so on.

This chapter proposes two versions of the Ballistic Chronograph: one offering the ability to accurately measure very high-speed projectiles and a Lite version offering a little less precision but a far simpler implementation. Though I refer to the simpler build as the "Lite" version, it is by no means unsophisticated.

Commercial Chronographs

There are several commercially available chronographs, most of which are intended for high-powered pistols and rifles. Chronographs are usually placed on the ground or a table in front of the shooter. Most of the popular commercial devices depend on ambient sunlight for operation and, therefore, don't work indoors or on overcast days. And while they are modestly priced, they are not really cheap.

Chronographs vary from simple two-wire devices (still in use and believed by some to be the most accurate) to relatively elaborate units with digital memory, average velocity calculations, and other features. The two-wire approach simply uses two thin strands of wire (36- or 40-gauge wire will do) stretched between two pairs of contacts accurately spaced apart. The projectile is shot and breaks the first wire to start a timer, and then, if you have good aim, breaks the second wire to stop the timer. The time between breakages is calculated to provide a speed value in feet or meters per second. The very early chronographs were built with a clock, which had readouts of ones and zeros displayed in a bank of LEDs. The binary number had to be translated to a decimal number and then calculated with the distance between the wires to get the velocity.

Measuring Muzzle Velocity

Now that you know what types of prebuilt chronographs are out there, let's take a look at the physics of the device. A projectile leaving the muzzle of a weapon has a velocity imparted to it by some propellant, such as air, CO_2, or the gas created by the rapid oxidation of the fuel in gunpowder. The projectile travels down the barrel and exits the muzzle. The speed of the projectile as it exits is called *muzzle velocity*.

The muzzle velocity of air-powered guns tends to vary depending on several factors, including the charge of the propellant, cleanliness of the barrel, and projectile-to-barrel matchup. Some air rifles can be pumped to almost 3,000 psi (pounds per square inch) to fire larger projectiles at relatively significant velocities. These larger air guns have relatively low muzzle velocities in the sub-1,000 fps range, but they pack a real punch. Compared to conventional air rifles, which shoot 0.177-inch pellets that pack between 15 and 25 ft-lbs (foot-pounds) of power, these larger-bore rifles offer between 500 and 700 ft-lbs of power.

Ideally, you would want to measure the velocity as close to the end of the barrel as possible. However, this can be difficult, and some chronograph makers claim that the velocity is not attenuated much in the first several feet (or even yards) of travel. On the other hand, there is little doubt that air resistance is a significant factor, and the projectile will slow at least somewhat in the first few feet—especially in the case of larger projectiles.

This Project's Approach

As in the two-wire ballistic chronograph systems, we're trying to measure the time it takes for a projectile to travel a fixed distance. But instead of breaking thin wires, this project takes advantage of an infrared light source and light-sensitive receiver, as illustrated in Figure 8-2.

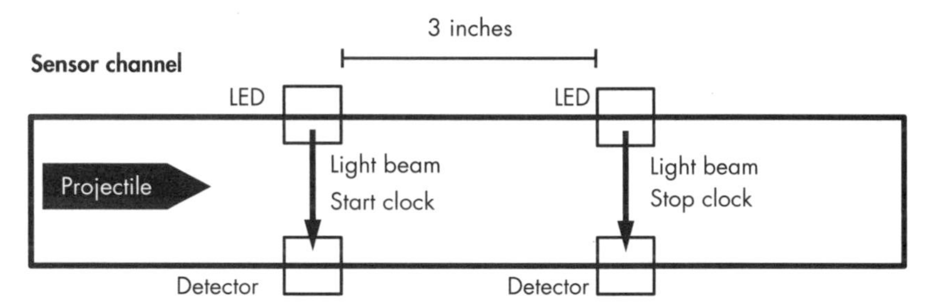

Figure 8-2: The basic principle in measuring the speed of the projectile is to have it break a beam of light to start a clock and then break another beam of light to stop the clock.

Two pairs of LEDs and IR sensors are arranged so that the IR sensor normally detects the light source. But when the projectile breaks the light beam of the first pair, the sensor goes dark and changes its electrical state. The processor senses this change and starts a timer. When the projectile interrupts a second source/receiver pair, the timer stops. The two sets of light sources and receivers are set an accurate distance apart so the time of travel can be relatively easily calculated into projectile speed.

For a simple example, say the beams of light are set a foot apart. A projectile interrupts a beam of light and starts the clock; when the projectile interrupts the second beam of light, the clock stops. If the microcontroller's timer measured 1 second, the velocity would be 1 foot in 1 second, or 1 fps.

This system can be used with a variety of projectiles and provide a digital readout on an LCD. Unlike other approaches, this device separates out the sensor bank from the electronics such that, if desired, different sensors can be swapped in and out for different firearms or even different applications. For instance, you could set up a sensor channel and do some basic physics experiments by dropping small objects through it and recording their velocity.

The Chronograph Lite

First, we'll take a look at the Chronograph Lite, which is simple to construct and has only a handful of parts.

Required Tools

Soldering iron and solder

Drill and drill bits (1/2, 1/4, and 1/8 inches)

Philips head and slotted screwdrivers

Saw (keyhole or saber saw)

Parts List

One Arduino Pro Mini or clone

Two IR LEDs, about 650–850 nm

Two IR photosensors (I used the Honeywell Optoschmitt SA5600.)

NOTE *Some users have had trouble matching the IR LEDs with the Optoschmitt photosensors. If you run into this problem, try the Honeywell SE3450/5450 or equivalent. Another option is to use two Adafruit IR Break Beams (part #2167) instead of the separate LEDs and sensors. The IR Break Beams will work for the Chronograph Lite, but the output must be inverted for the full version.)*

One 270-ohm, 1/8 W resistor

(Optional) Two 10-kilohm, 1/8 W resistors (if using phototransistors rather than Optoschmitt photosensors)

One channel holding two LED/sensor pairs

One 16×2 LCD

One I^{2}C adapter, if not included with the LCD

One x4 adapter housing (see "Connectors Used in This Book" on page 18)

Four female pins for housing (see "Connectors Used in This Book" on page 18)

One SPST switch

One momentary NO switch

One 9V battery connector

One 9V battery

One Hammond 1591 BTCL enclosure or equivalent

Two 7 1/2 × 1 1/2 × 0.06–inch aluminum pieces

One 1 3/8 × 7 1/2–inch piece of 1 3/8-inch acrylic sheet

Two #10-24×1-inch nylon screws

Two #10×24 nylon nuts

Four 4-40×1/2-inch screws

Eight 4-40 nuts

Four 4-40 washers

Assorted 28-gauge hookup wire

Downloads

Sketch *ChronographLite.ino*

Templates *PanelCutoutLite.pdf, PanelCutout.pdf, AccelerationChannel.pdf*

The Schematic

Outside the Arduino board, the circuitry for this project is not very complex. The schematic in Figure 8-3 uses the I^2C bus to power the LCD, two connections for the photosensors, and two connections for the clear switch.

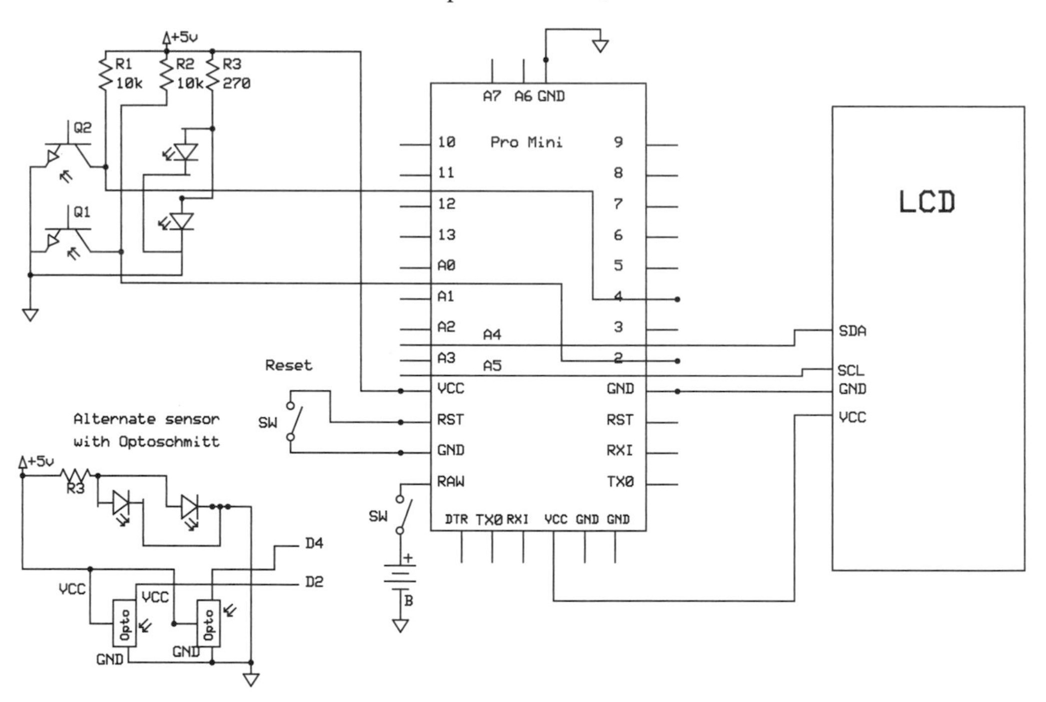

Figure 8-3: The schematic of the Chronograph Lite. The primary schematic shows the phototransistors and the alternate section shows the Honeywell Optoschmitt sensors (bottom left).

Building a Test Bed

Figure 8-4 shows the test bed that was used to prove the concept and develop the sketch. I suggest you build your own and install your LEDs and photosensors into it before building the breadboard.

For this test bed, I cut two pieces of cardboard approximately 2×6 inches and punched them to fit two pairs of IR LEDs and phototransistors that were spaced 3 inches apart. I then screwed the cardboard to a 1-inch-thick piece of wood, though you could glue or staple it if you prefer. When you build yours, be sure that each phototransistor is directly opposite an IR LED in the channel.

After installing the IR LEDs and phototransistors into the test bed, I recommend preparing them for the breadboard as follows:

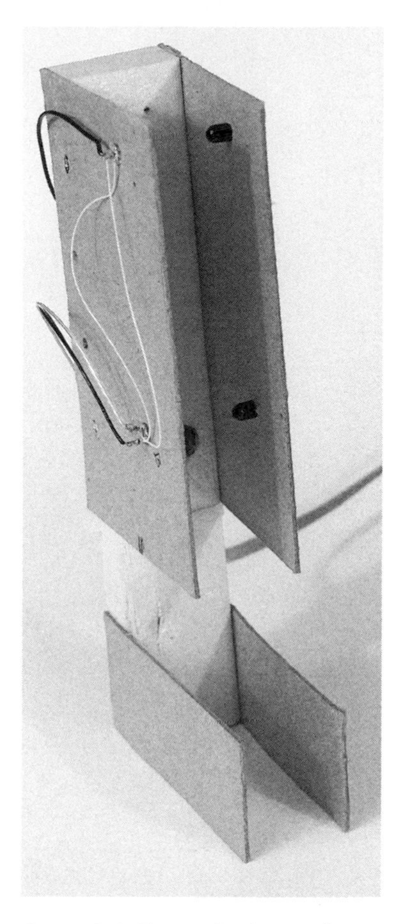

Figure 8-4: The test bed I initially used to check out the chronograph concept

1. Connect the two IR LED anodes with a piece of wire by soldering or wire-wrapping.
2. Solder one 24-inch length of wire to the combined LED anodes. If you're using solid-core wire that fits in a breadboard, you can just strip the other end of the 24-inch wire. If you're using stranded-core wire, attach a male crimp pin to the end of the wire.
3. Connect the two IR LED cathodes with a piece of wire by soldering or wire-wrapping.
4. Connect the two phototransistor emitters with a wire by soldering or wire-wrapping.
5. Connect the combined IR LED cathodes to the combined phototransistor emitters; I suggest soldering a long wire.
6. Solder a 24-inch length of wire to the combined LED cathodes and phototransistor emitters, and finish the other end of the wire with a male crimp pin, as you did in step 2.
7. Solder a 24-inch length of wire to each phototransistor's collector, and finish the other end of the wire with a male crimp pin, as you did in step 2.

In Figure 8-4, the LEDs and light sensors (these will be phototransistors or Honeywell Optoschmitt sensors depending on your choice) are placed in holes punched in the cardboard of the acceleration channel. I used a relatively small hole punch so that friction would hold them in place. You could instead glue them with hot glue or contact cement. I used a 24-inch length of four-conductor telephone wire to connect the channel to the breadboard, but any wire will do. In the completed version and in other prototypes, I just used four lengths of 30-gauge twisted wire because it was more flexible. Alternatively, you could build the final sensor channel now.

The LEDs and phototransistors will need to be wired to the Arduino, as indicated in the schematic diagram in Figure 8-3. I wired the LED anodes to the power supply through a 270-ohm resistor (R3). In the case of the phototransistors, I set them up so that the emitters were grounded and each collector went through a 10-kilohm resistor (R1 and R2) to the positive of the power supply for an open-collector configuration. Thus, if the beam of light were interrupted, the phototransistor would conduct, and the voltage at the collector would drop.

NOTE *If you use the Honeywell Optoschmitt SA5600/5610, the 10-kilohm resistors (R1 and R2) are not required, as they are included in the SA5600/5610 chip. The wiring of the Optoschmitt sensors is shown in the lower left of the schematic in Figure 8-3.*

Rather than shooting up the office with live paintballs, BBs, or pellets, I set up the gig vertically so a projectile could be dropped through the light beams to test the system. This method meant that the velocities measured didn't approach those of a projectile leaving a weapon's barrel, but it was good enough for an initial proof-of-concept experiment. The higher the target was dropped from, the higher the recorded velocity—that is, if your aim is good. (Remember $s = (1/2)at^2$, where s is displacement or distance, a is acceleration due to gravity, t is time, and initial velocity is zero.)

If ambient light causes problems in testing, an additional piece of cardboard can be taped to the top (side) of the two pieces of cardboard to shade the sensor, though I didn't find this was a problem in any of the experiments I conducted.

The Breadboard

The next step is to build a breadboard, as shown in Figure 8-5. For this, we'll use the Chronograph Lite schematic in Figure 8-3. The most complicated part of the circuit is wiring up the photosensors and LEDs, which are not plugged directly into the breadboard but rather need to be installed in the sensor channel, as described in the previous section.

Whether you have already made the finished channel or are using the cardboard prototype, you will need to connect the sensors and LEDs in the channel to the breadboard—or, for that matter, to the completed unit—with four wires: positive, ground, first sensor, and second sensor.

Figure 8-5: Photosensors and LEDs in the early prototype attached to the breadboard via discrete wires. To test the unit, a coin was dropped through the channel.

Here's how to wire the breadboard:

1. Connect the red positive rails together and the blue negative rails together. Do not connect the red positive rail and blue negative rail to each other under any circumstances—it will result in a short circuit and damage to components.
2. Insert the Arduino Pro Mini or clone in the breadboard, leaving a fair amount of room—about four or five rows—at one end.

3. Connect the VCC pin on the Mini to the red positive rail.
4. Connect the GND pin on the Mini to the blue negative rail.
5. Take two 10-kilohm resistors (R1 and R2) and connect one end of each to the red positive rail. (Note that these are not required if you are using the Optoschmitt SA5600 photosensor.)
6. Connect the other end of resistor R1 to pin D4 on the Mini via a jumper wire.
7. Connect the other end of resistor R2 to pin D2 on the Mini via a jumper wire. (If you use the Optoschmitt photosensor, you can connect pins D2 and D4 directly to the output pins on the photosensors, as shown in the bottom left of Figure 8-3.)
8. Connect the collector pin of phototransistor Q1 to pin D2 on the Nano using the attached 24-inch wire.
9. Connect the collector pin of phototransistor Q2 to pin D4 on the Nano using the attached 24-inch wire.
10. Connect one end of the 270-ohm resistor R3 to the red positive rail.
11. Connect the other end of resistor R3 through one of the 24-inch lengths of wire to an empty row on the breadboard.
12. Connect the combined anodes of LED 1 and LED 2 to the row where you connected resistor R3 in step 9 via the attached 24-inch wire. Refer to Figure 8-3 to see how the LEDs are wired together and to the breadboard.
13. Connect the 5V pin and the GND pin on the Mini to VCC and GND on the LCD, respectively.
14. Connect pin A4 on the Mini to the SDA connection on the LCD.
15. Connect pin A5 on the Mini to the SCL connection on the LCD.

    Now you're ready to enter the sketch.

The Sketch

Now to write the sketch to make things work. Here is the sketch for the Chronograph Lite:

```
//Lite version of chronograph, using Optoschmitt sensors

#include <Wire.h>
#include <LiquidCrystal_I2C.h>

unsigned long start_time = 0;
unsigned long stop_time = 0;
unsigned long time_of_flight = 0;
float velocity = 0;
```

```
LiquidCrystal_I2C lcd(0x3F, 20, 4);

void setup() {
  Serial.begin(9600);
  pinMode(2, INPUT);
  pinMode(4, INPUT);
  lcd.init();
  lcd.backlight();
}

void loop() {

❶ while(digitalRead(2) == 1) {
    //Waiting for first sensor to trip
  }
❷ start_time = micros();

❸ while(digitalRead(4) == 1) {
  }
❹ stop_time = micros();

❺ time_of_flight = stop_time - start_time;

  Serial.print("  time of flight         ");
  Serial.println(time_of_flight);

  velocity = 1000000*.25/(time_of_flight);


  lcd.clear();
  lcd.print("tm of flt ");
  lcd.print(time_of_flight);
  lcd.print(" us");
  lcd.setCursor(0, 2);
  lcd.print("Speed FPS    ");
  lcd.print(velocity);
}
```

The sketch is pretty straightforward. After setting up the variables and inputs, a `while` loop waits for the first sensor to be interrupted by checking the condition `digitalRead(2) == 1` ❶. When tripped, the clock is started with `start_time = micros();` ❷, and another `while` loop counts until the second sensor is activated (because the second sensor is plugged into pin D4, this `while` loop checks whether `digitalRead(4) == 1` ❸). When the second sensor is activated, the clock is stopped with `stop_time = micros()` ❹.

The sketch then calculates the time that lapsed between the first sensor and the second with `time_of_flight = stop_time - start_time` at ❺. Once the sketch makes its calculations, it provides instructions to display the results on the LCD screen.

All you need to do now is load the sketch, stand the test channel up as in Figure 8-4, and drop a projectile like a marble through the cardboard channel. To begin, power the Mini with the programmer. Alternatively, you can use a separate regulated 5V power supply connected to the 5V terminal on the Mini, or you can connect a battery to the VIN port of the Mini and use the Mini's on-board regulator. Do *not* connect a 9V battery to the 5V supply rails—it could burn everything out.

When you're satisfied that the sensors are working, you can attach your temporary sensor channel to a real air pistol if you'd like to test the circuit further (see Figure 8-6).

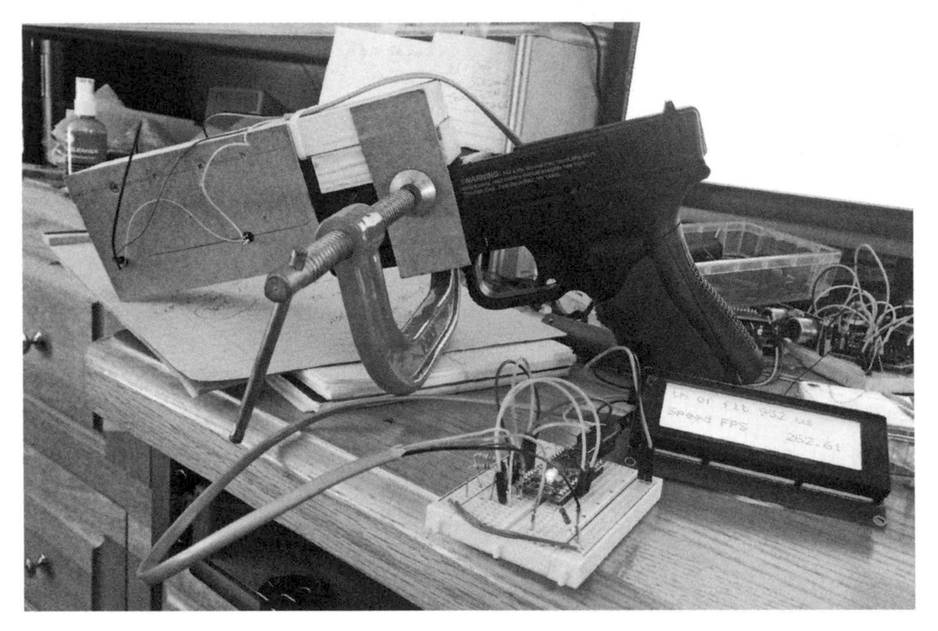

Figure 8-6: Photoelectric sensors in an early prototype, attached to a Crossman 0.177 caliber air pistol. The setup was obviously primitive, using a C-clamp to hold the channel to the weapon. Because the test channel was so simple, I set it up on my desk.

If this is enough for your needs, you can package this up and skip everything in "The Full Ballistic Chronograph" on page 233. The Chronograph Lite should work well for games and low- and medium-velocity weapons (less than 600 fps), such as pellet guns, BB guns, airsoft weapons, and so on.

Despite satisfactory performance, however, I was bothered by the fact that, while the Arduino can count microseconds, it can deliver results only as multiples of four, so in reality the resolution is only 4 microseconds. That is why I developed the Full Ballistic Chronograph project. If this bothers you, too, and you don't plan to package up the Chronograph Lite, you can skip to "The Full Ballistic Chronograph" on page 233 now.

TESTING THE CHRONOGRAPH LITE WITH A PROJECTILE SIMULATOR

To test the Chronograph Lite or Full Ballistic Chronograph for errors, I made a simulator to simulate the effect of a projectile traveling through the two sensors rather than shooting up my work area with pellets or paintballs. I primarily used the simulator in development—it is not necessary for the completion or use of either chronograph in this chapter—but it provides some insight into how to turn on and off relatively high-speed signals.

I used a square-wave generator (see Chapter 9 if you want to create your own) and made a very simple breadboard simulator. The schematic of the simulator, shown in Figure 8-7, includes only the turn-on and turn-off functions and is driven by the square-wave generator.

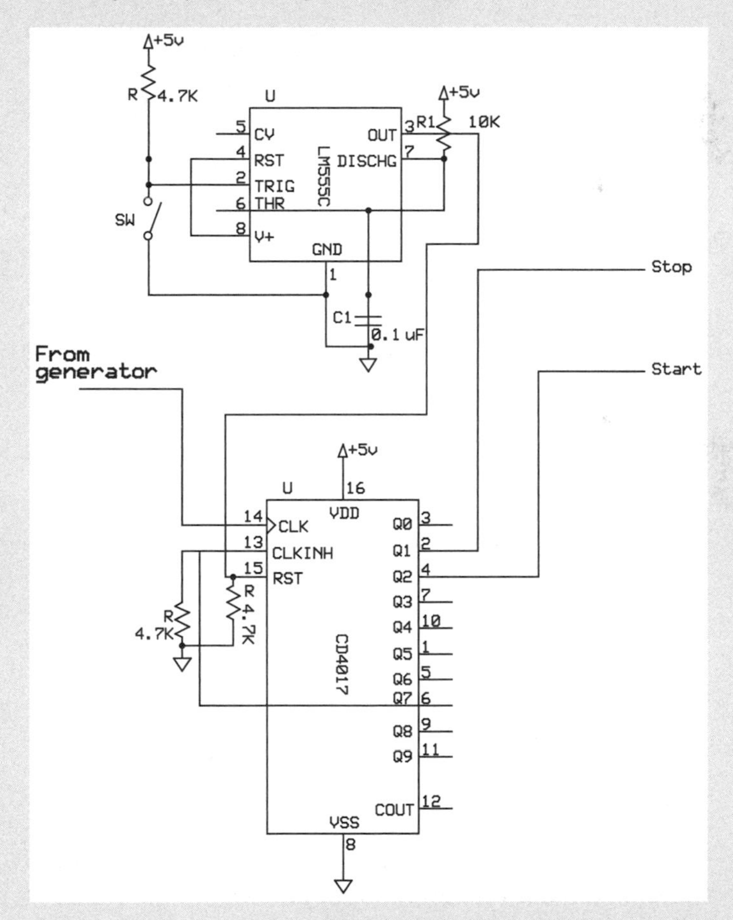

Figure 8-7: Schematic of the simulator used to simulate the sequential firing of the Optoschmitt sensors. It is used in conjunction with a square-wave generator. Resistor R1 and capacitor C1 can be adjusted for satisfactory debounce, but the values shown worked well.

(continued)

The simulator receives a clock signal from the square-wave generator. On initiation, depressing the switch (labeled SW in Figure 8-7) on the simulator begins the sequence of start and stop signals from the CD4017 decade counter. A manual switch (SW) fires the simulator after a debounce from a NE 555 timer. The function of the simulator is to turn the connections to the photosensors on and off just as if a projectile were traveling through the start and stop LED/photosensor pair. Figure 8-8 shows a breadboard for the simulator with the finished board and square-wave generator.

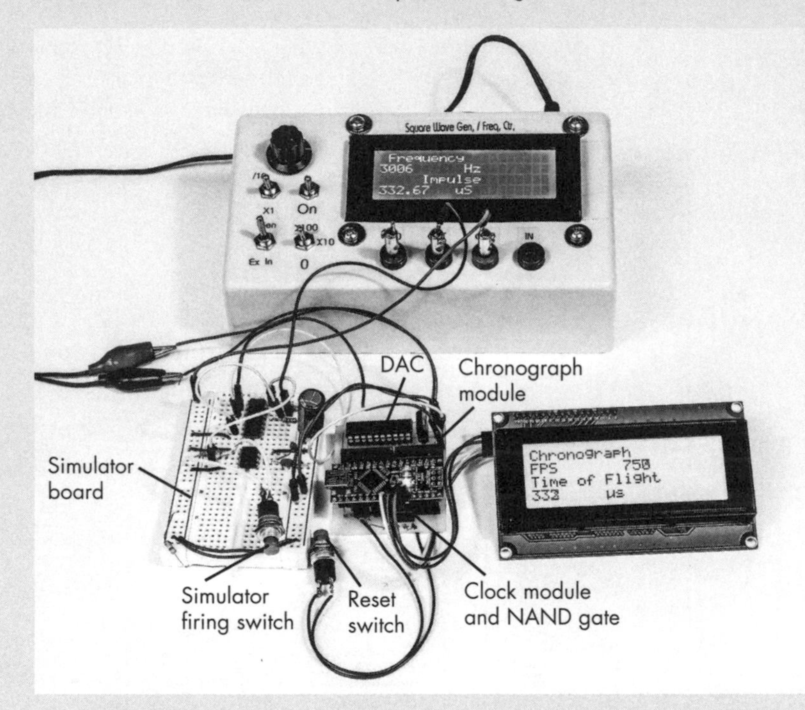

Figure 8-8: The simulator breadboard and the square-wave generator hooked up to the finished prototype board for the Full Ballistic Chronograph. The simulator works equally well with either the Chronograph Lite or the Full Ballistic Chronograph.

Construction

To complete the Chronograph Lite project, all you have left to do is to package the Mini, display, battery, and appropriate switches in an enclosure, leaving a connector exposed to connect the unit to the sensor channel. Unlike most of the other projects in this book, I did not use a shield for the Chronograph Lite, because the wiring to the Mini was sufficiently straightforward that it did not require one. I used a Hammond ABS plastic enclosure 1591 BTCL, as indicated in the parts list. See Figure 8-9 for the completed Chronograph Lite.

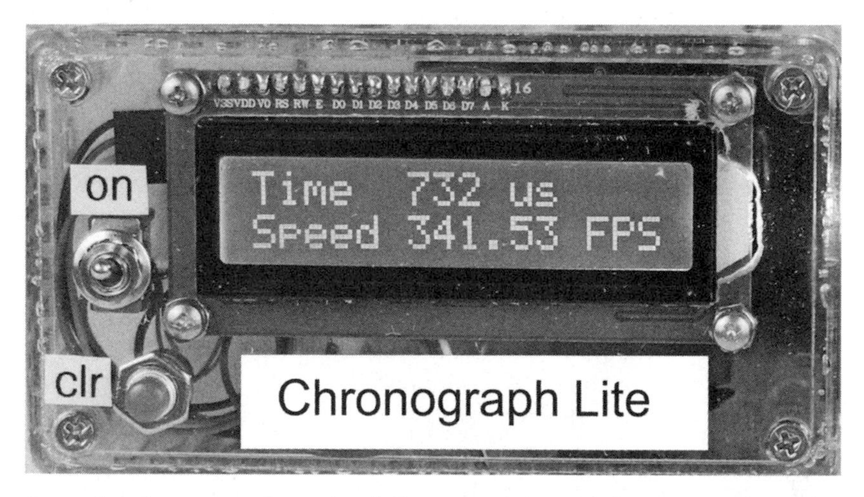

Figure 8-9: Front view of completed Chronograph Lite. A hole is cut into the enclosure to the right of the screen to allow space for the backlight protrusion.

Figure 8-10 shows the template for the enclosure. You can download a PDF of this drawing from *https://www.nostarch.com/arduinoplayground/* and use it to mark and center punch the enclosure for the holes.

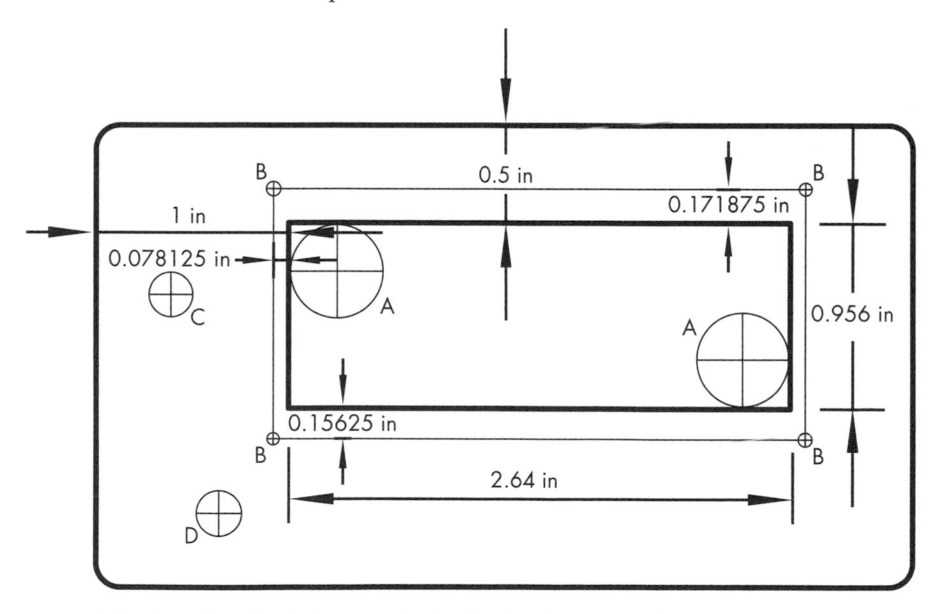

Figure 8-10: Template for holes and display for the Chronograph Lite

I prepared the enclosure for the Chronograph Lite as follows:

1. Carefully mark, center punch, and drill 1/2-inch holes for the corners of the display (A), 1/8-inch holes for the mounting holes for the display (B), 1/4-inch holes for the on/off switch (C), and 1/4-inch holes for the clear switch (D).
2. For the LCD screen, mark the edges of the 1/2-inch holes (A). Draw lines connecting the edges so you have a rectangle to cut out. (You can use a Sharpie marker and clean excess markings later with alcohol.) Drill the holes and cut the opening along those lines using a keyhole or saber saw.
3. There's a slight protrusion in the middle of the LCD on the right-hand side (facing up); this is part of the backlight assembly. You can cut a hole to accommodate for this, as I did in Figure 8-10, or you can leave the edge straight and use spacers to keep the protrusion from hitting the enclosure.
4. Mount the display and fasten it in with 1/2-inch-long 4-40 mounting screws and nuts. If you made a cut out for the LCD backlight protrusion, you can mount the display directly. If you did not, use extra 4-40 nuts to space the display back from the face of the enclosure. If needed, add additional washers; 4-40 nuts can vary in thickness.
5. Mount the *on* and *clr* switches as indicated in Figure 8-9.

Now to wire up the Pro Mini. There is no shield, so we will solder directly to the Pro Mini board as follows:

1. Solder the wires for the I^2C connection. To make your life easier, use colored wire and create a code for yourself. Solder connections to the 5V (some clone boards may say VCC) and GND pins on the Nano. Then, solder 3-inch wires to pins A4 and A5 on the Nano. Connect the other end of these wires to a four-pin female connector. (See "Connectors Used in This Book" on page 18 for details on making Pololu connectors.) Connect the 5V and GND pins on the Nano to 5V and GND on the LCD. Connect pin A4 on the Nano to SDA on the I^2C board, and A5 on the Nano to SDL on the I^2C.
2. Connect the positive (red) wire of the battery connector to one side of the SPST switch. Connect the other side of the switch to the VIN terminal on the Nano (some clone boards may say RAW).
3. Solder the black (negative) wire from the battery connector to the GND pin on the Nano.

Finally, connect the Nano to the sensor channel as follows:

1. Prepare a four-conductor female Pololu connector with four color-coded wires approximately 3.5 inches long. Attach two wires (I suggest red and black) from this connector to the VCC and GND pins on the Nano.

2. Connect the remaining two connectors to pins 2 (D2) and 4 (D4) on the Nano.
3. Make a slot or hole in the side of the enclosure, and run the Pololu connector with the sensor channel connections through it (see Figure 8-11).

Figure 8-11: A slot in the enclosure for threading the four-pin sensor channel connector through. The connector is mounted with double-sided adhesive

4. Connect one side of the *clr* pushbutton to GND and the other side to the RST (reset) pin on the Nano.
5. Finally, connect the battery, screw on the top of the enclosure, plug in the sensor channel, and flip the switch to turn on the device.

You should be all set to use your Chronograph Lite. Go to "Final Setup and Operation" on page 252 for instructions on using the Chronograph Lite.

The Full Ballistic Chronograph

While the Chronograph Lite worked well and I used it to successfully measure projectile speeds, I had a nagging feeling that it could be better. If you're using the device for slow-speed projectiles—that is, 600 fps or less—the accuracy of the Chronograph Lite is more than enough. But the restriction to 4 microseconds of resolution resulted in what I perceived to be a fair amount of error in feet-per-second (fps) at higher speeds, so I decided to construct the Full Ballistic Chronograph.

Required Tools

Soldering iron and solder

Drill and drill bits (1/2, 1/4, and 1/8 inches)

Philips head and slotted screwdrivers

Saw (keyhole or saber saw)

Parts List

Assembling the Full Ballistic Chronograph is relatively simple. Here's what you'll need:

One Arduino Nano or clone

One 16×4 LCD

One I^2C adapter, if not included with the LCD

One PCB shield

One enclosure (Hammond 1591 BTCL)

Four 1/2-inch×4-40 screws

Four 4-40 nuts

One 3PDT toggle switch

Two momentary pushbutton switches

Four 0.100×4 female headers

Four female X4 shells

Sixteen (eight male, eight female) adapter pins

One 4 MHz crystal

One TI SN 74LVC1GX04 crystal-oscillator driver

One SOT23 adapter board

One HCT 4011 4-input NAND gate

One CD4013 dual D flip-flop

One CD4040 12-stage binary counter

One ADC DAC8562 digital-to-analog converter

One LM7805 voltage regulator

One NPN transistor 2N5172 (or equivalent)

Four 5-kilohm, 1/8 W resistors

One 1-megaohm, 1/8 W resistor

One 1-kilohm, 1/8 W resistor

One 1.5-kilohm, 1/8 W resistor

One 270-kilohm, 1/8 W resistor

One 4.7 MFD tantalum capacitor

Two 33 pF capacitors

One 0.01 µF capacitor

One 5 mm LED

Two IR detectors (I used the Honeywell Optoschmitt SD5610.)

Two IR LEDS, about 850–950 nm

28- or 30-gauge hookup wire

NOTE *For the Full Ballistic Chronograph, you will have to use the inverted version of the chip, the SA5610, or externally invert the signals. See the note at the bottom of Figure 8-3.*

Downloads

Sketch *FullBallisticChronograph.ino*

Templates *ChronoCover.pdf, AccelerationChannel.pdf*

PCBs *ChronoPCB.pcb, LEDHolder.pcb, SensorHolder.pcb*

Improving the Accuracy

There are several possible solutions for improving the accuracy of the chronograph. The Arduino Nano uses a 16 MHz clock, yet when configured using the Arduino Nano platform and IDE, it results in a 1 microsecond resolution (±2 microseconds), even though the period—the time between cycles—of a 16 MHz clock is 1/16,000,000 of a second, or 0.063 microseconds. While a processor could never resolve down to its own clock speed, it's probably capable of much better than 1 microsecond. Clearly there is some overhead in the current project—perhaps part hardware (the components in the Arduino board) and part software (the compiler and firmware part of the IDE)—that limits performance. Here are some ideas I had to improve accuracy, starting with one that didn't make it into the final project but that I think is educational.

Digging into Machine Code

One possible solution is to dig into the basic Atmel machine and AVR code. Without going into excruciating detail, AVR assembly is the functional language of the Atmel chip. The Arduino community has surrounded that with special code that lets the AVR run in the Arduino environment.

According to the ATmega328 data sheets, it's possible to directly address the individual timers on the ATmega328 and get the resolution required. However, looking into it, I saw that this method could prove overly complex and figured there had to be another way.

Creating a High-Speed Window

The time of flight of a projectile that we want to look at covers a range of roughly 90 microseconds (about 3,000 fps in a 3-inch distance) to

950 microseconds (about 260 fps in the same 3-inch distance) from fastest to slowest. Relative to the higher frequencies of some clocks, such as the 16 MHz clock of the processor, 90 microseconds is a fair amount of time.

One method for measuring the velocity is to open a timing window when the first beam of light is interrupted that lets a stream of high-speed signal through until the second beam is interrupted. While the window is open, the pulses in that signal will be counted; when the window is closed, the count will represent the time the window was open.

As an example, say the window opens and a signal of 10 cycles per second (cps) passes through until the window closes; 100 cycles are counted. For this illustration, the Arduino's clock is the high-speed signal. When you know the distance the projectile traveled, you can use some simple arithmetic to determine the time of travel and the speed: 100 cycles at 10 cps gives us 10 seconds. If the distance were 1 meter and 100 cycles were counted while the window was open, the speed would be 1 meter per 10 seconds or 0.1 m/s.

A single NAND logic gate can be used to make a window that can be opened and closed. A *logic gate* is simply an electronically controlled switch that outputs a voltage only under certain conditions, corresponding to a Boolean logic equation. *NAND* is the Boolean expression for "not AND," and a *NAND gate* outputs a voltage when its two inputs are not the same.

I sampled both a 74HC00 high-speed NAND gate and a standard CD4011BC gate, and the standard part works fine. There are several other parts that will work, too—what you're looking for is a part with a propagation delay (T_{PD}) under 100 nanoseconds.

Selecting a Counter

After deciding to take the window approach, the next thing to consider is how high you need to count. If you were to count in integers from 1 to 100, for example, you would need a counter that could count to 100, which would provide a resolution of 100. If you scaled that up, the counter could provide a range from 10 to 1,000 or from 100 to 10,000. If that range were the result of the calculation for fps, you would then have a resolution of only 1,000 fps (each increment would equal 100 fps) plus any included error, which we'll go into later.

So where should you go from here? To the parts bin, of course, to see what counters are available to count the signal passing through that window. When selecting a counter, you need to consider how fast it needs to be and how many pulses you want it to count. The tried-and-true CD4040, 12-bit, serial-in, parallel-out, digital counter seemed capable of doing the job. (The CD4040 worked at the 4 MHz frequency, but you could always use a faster one, like the 74HC4040 or 74HCT4040.) The CD4040 will provide a digital count from 0 to 4095, or 2^{12}.

Selecting a Clock Speed

Next, consider what signal frequency is needed in order to suit the range of projectile speeds. I started with the assumption that I wanted to achieve a range of roughly 300 fps to 2,500 fps with as much latitude on both ends as possible.

Further, while the counter will ideally count from zero to the maximum 4,095 counts, there is the possibility of some error. So rather arbitrarily, I chose to look at the total digital count between 400 and 4,000 to account for the possibility of error.

Given the number of cycles counted, the signal frequency, and the distance traveled, the velocity of a projectile can be found with the following calculations:

$$\frac{1}{\text{Frequency}} = \text{Time per Cycle}$$

$$\text{Time per Cycle} \times \text{Total Cycles} = \text{Time of Flight}$$

$$\frac{\text{Distance Traveled}}{\text{Time of Flight}} = \text{Velocity}$$

Let's go through the arithmetic for a projectile that travels 0.25 feet (3 inches) within 4,000 cycles of a 2 MHz signal:

$$\frac{1}{2\text{ MHz}} = 0.0000005\text{ s per cycle}$$

$$0.0000005\text{ s per cycle} \times 4{,}000\text{ cycles} = .002\text{ s in flight}$$

$$\frac{.25\text{ ft}}{0.002\text{ s}} = 125\text{ s in flight}$$

For a 4 MHz clock, a full 4,000 cycle count will amount to about 281 fps for the low end of the speed range. At the high end, given 400 cycles counted and a 2 MHz clock signal, you will measure 1,250 fps, and at 4 MHz, you can measure up to 2,500 fps.

You can be creative with your frequency. If you elect to use a 2 MHz clock, it will provide maximum resolution in the very low-speed range. If, on the other hand, you select a 4 MHz clock, you will be in the middle of the resolution range. An 8 MHz clock will provide a very good resolution in the fast range (faster than any conventional weapon) but will curtail performance at the lower-speed range.

Because I anticipated that the bulk of speeds I needed to measure would fall in the middle of the counting range, a clock around 4 MHz sounded good. I was not anticipating many occasions when velocities would be in the sub-300 fps range, and at the high end, it looked like accuracy could be maintained to well over 5,000 fps (a digital count of just under 200, which might be stretching it a little but seemed to work well in simulations).

If your projectiles remain in the sub-300 fps range, I suggest revisiting the Chronograph Lite. If, for some reason, you want to stay in the lower fps range but require maximum accuracy with perhaps multiple digits, build the Full Ballistic Chronograph with the slower clock rate. You can simply swap out the 4 MHz crystal for a 2 MHz crystal and adjust the sketch to slide the range down to the lower area.

Adjusting the Clock Speed

To address the speed of the clock (the signal that is gated to the counter), the most accurate method by far is to use a crystal-controlled oscillator, which generally has errors only in the sub-50 parts per million range. I configured a 4 MHz crystal with the TI SN 74LVC1GX04 crystal-oscillator driver experimentally and it worked well, so I used one in the final project.

While I did look at, review, and test single-chip oscillators, such as the Maximum stand-alone oscillator (7375), it was not quite as stable as the crystal-controlled version.

Designing the Full Ballistic Chronograph

Now, we have the means to clock the signal into the 4040 counter, but we need to figure out how to display the velocity on the LCD. One method would be to use a different counter with a serial output that would be clocked directly into the Nano. Another possibility would be to take the parallel data from the CD4040, serialize it with a shift register, and feed the result to the Nano.

However, I took a different direction, as illustrated in the block diagram in Figure 8-12. I decided to use a 12-bit digital-to-analog converter (DAC) to accept the parallel digital signals and convert them to a single analog value. DACs and their counterpart, analog-to-digital converters (ADCs), are used in digital music, TVs, and a host of other areas where an analog input needs to be digitized, manipulated, transferred, stored, and eventually output to return an analog signal. I thought this would be a good opportunity to introduce the capabilities of digital-to-analog converters.

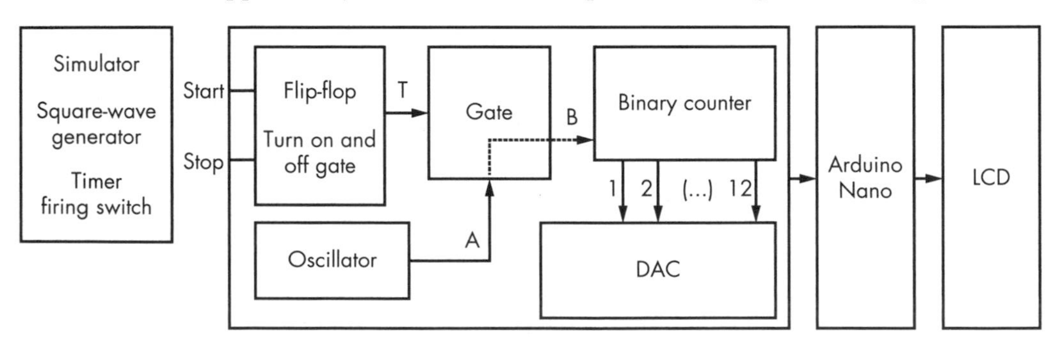

Figure 8-12: Block diagram of the Full Ballistic Chronograph

The process in Figure 8-12 depicts the operation of the chronograph using a simulator. In actual operation, the simulator would be replaced with the two LED-sensor pairs. The simulator, under control of the firing switch, initiates a start signal that remains active until the second stop switch is activated after a period determined by the square-wave generator. This essentially simulates the projectile passing through the first and then the second pair of sensors.

When the start switch is activated initially, it turns on the *flip-flop*—a bi-state device that turns on with the activation of the start switch and remains on until the stop switch is activated. The flip-flop feeds the trigger of the gate. When the trigger (T) is inactive—that is, when it's set to a logical 0—the signal from the oscillator at input (A) cannot go through the gate to output (B). When the trigger is activated (set to a logical 1), the gate allows the signal from the oscillator (A) to travel through the gate to output (B) and eventually to the input of the binary counter. The binary counter counts the number of pulses that pass from the oscillator through the gate and stops counting when the gate closes.

The outputs of the binary counter are fed to the DAC. They represent binary numbers from 0 through 4,095—that is, 0 through $2^{12} - 1$. The DAC converts these digital values to a single analog value. The technique of this conversion depends on the type of DAC used; for example, in the DAC8562 used here, an R-2R resistor ladder is switched, and a transistor is used to yield the output. (For complete information, look up the data sheet from Analog Devices on the DAC8562.)

The output of the DAC has a scale of 0V to 4.095V corresponding to the digital inputs. This output is then directed to one of the analog inputs on the Arduino Nano, which provides the inverse function of the DAC and converts the analog signal back into a digital format that the Nano can handle. The Nano takes that signal and, following instruction from the sketch, adjusts the value to represent the velocity in fps for the time it takes the projectile to travel the 3 inches. The Nano finally sends that data to the LCD, which displays the velocity of the projectile and travel time.

The Schematic

Figures 8-13 and 8-14 show the schematic diagrams for the completed Full Ballistic Chronograph. Note the extra gates at the bottom. I included these in the schematic because they are available to you in the NAND gate and flip-flop IC packages suggested for this project, but my design does not use them. If you want to add functionality, they are available.

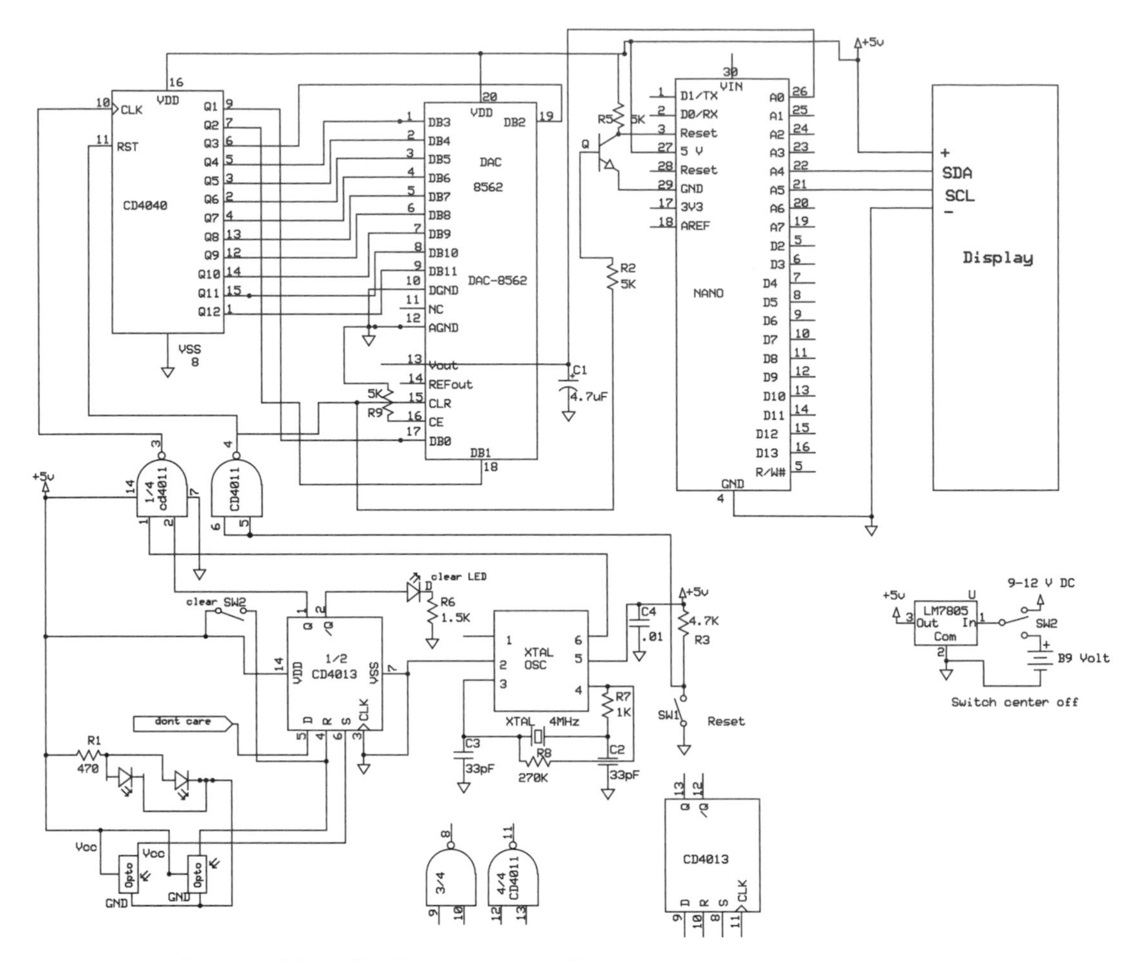

Figure 8-13: Schematic of the Full Ballistic Chronograph

Another thing included in the schematic that we haven't covered is the reset button. In the Full Ballistic Chronograph, I included a button to trigger the reset rather than having it reset automatically. I could have set it so that the result was displayed on the LCD for a fixed period of time before the system reset, but it might have turned out that the number was erased before users had time to record it, or users may have found themselves sitting idle while it timed out. I decided a reset button would be more convenient.

Because resetting the microcontroller wasn't going to upset the sequence of things, I chose to use a hard reset on the controller through transistor Q1. To reset the CD4040 and the DAC, I used the reset signal and then inverted it using one of the CD4011's four NAND gates with the two inputs tied together. SW2 manually closes the second set of sensors in case the first pair of sensors fires and not the second, and SW3 is the power and battery switch.

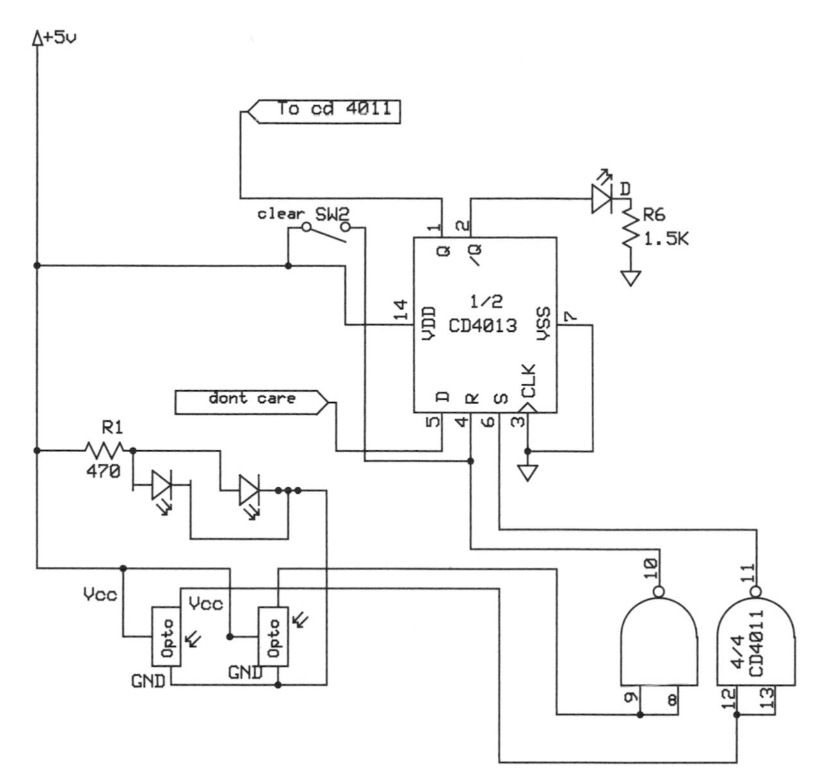

Figure 8-14: Inverters for Optoschmitt or Adafruit sensors if you use the SA5600 instead of the SA5610. This circuit uses the previously unused gates of the CD4011 NAND gate as logic inverters. They are not accommodated for in the PCB, so you will have to wire them by hand.

The Sketch

The sketch for the Full Ballistic Chronograph is relatively straightforward:

```
//Full Ballistic Chronograph

#include <Wire.h>
#include <LiquidCrystal_I2C.h>
LiquidCrystal_I2C lcd(0x27, 16, 2);
int DACpin = A0;
float DACvalue = 0;
float FPS;
float Time;

void setup() {
  lcd.init();
  lcd.backlight();
}
```

```
void loop() {
❶ DACvalue = analogRead(DACpin);
  Time = DACvalue*5/1023/4*1000;
  FPS = .25/Time*1000000;
  lcd.setCursor(0,0);
  lcd.print("Speed  ");
  lcd.print(FPS,0);
  lcd.setCursor(11,0);
  lcd.setCursor(0,1);
  lcd.print("Time   ");
  lcd.print(Time);

  lcd.setCursor(11,1);
  lcd.print(char(0XE4)); //To display the mu symbol, use 228 or 0XE4
  lcd.print("s");
}
```

In this sketch, the software receives an analog signal from the DAC at ❶ and converts it to a digital value. It then goes through a couple of quick mathematical operations to come up with the time of flight (`Time`), calculates the speed in feet per second (`FPS`), and finally exports those values to the LCD.

In designing electronic circuits, there are always tradeoffs between hardware and software. Many of these have to do with timing issues and built-in latencies in software-based approaches. In this instance, the tradeoff is the need for greater accuracy not available with the straight Arduino IDE approach without dropping to some level of native code. To avoid native code, the Full Ballistic Chronograph has more complex hardware than the Chronograph Lite.

The Shield

Unlike the Chronograph Lite, the Full Ballistic Chronograph is best built on a shield. The shield is a little more involved than some of the others in this book, but don't be intimidated. Figure 8-15 shows the actual traces of the shield, while Figure 8-16 shows the silkscreen image with the part placements and hole configuration. For this project, I opted for a double-sided board because the circuit was a little more complex than some of the others and because it allowed me to minimize the space required. The complete PCB file is available for download at *https://www.nostarch.com/arduinoplayground/*.

I attempted to keep the shield footprint to a minimum to make it possible for the user to squeeze the system into a small portable enclosure. As is, the finished Full Ballistic Chronograph fits easily in a 11×8×4 cm box.

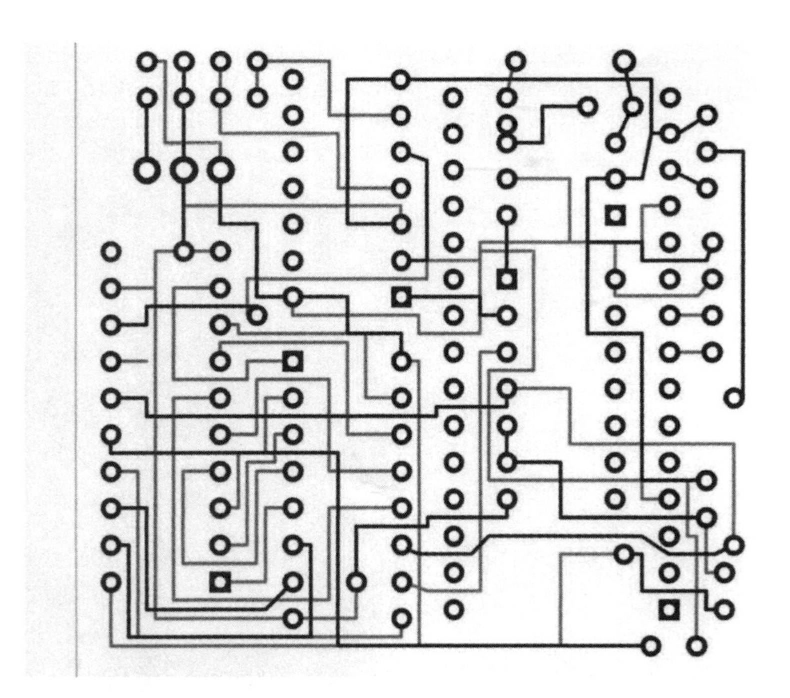

Figure 8-15: Trace patterns for the shield. The darker gray is the upper copper layer, and lighter gray is the lower layer.

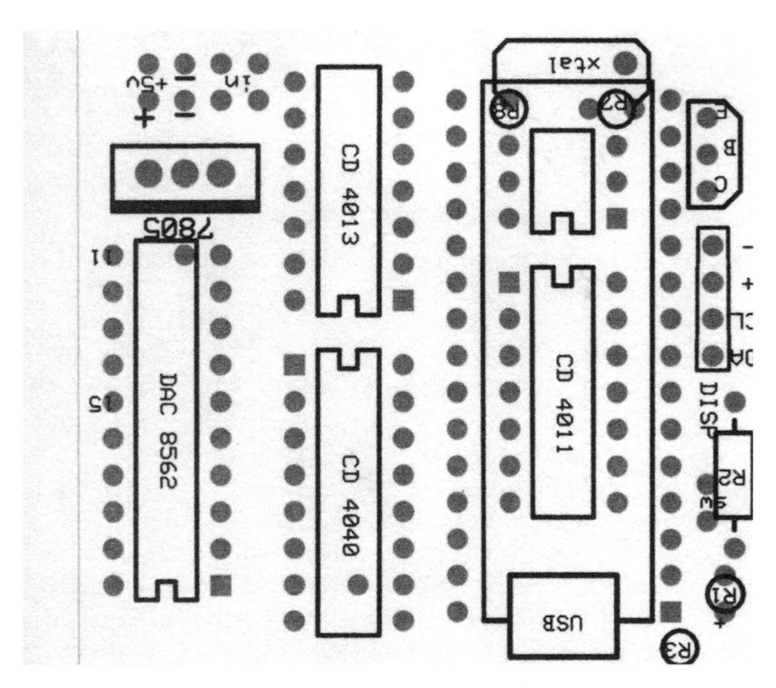

Figure 8-16: The component placement on the shield

This is another case where I opted to outsource the PCB construction after I had made and refined the first sample myself and made sure all critical connections could be soldered on both sides of the board. Figure 8-17 shows the raw board as it was received from the service bureau.

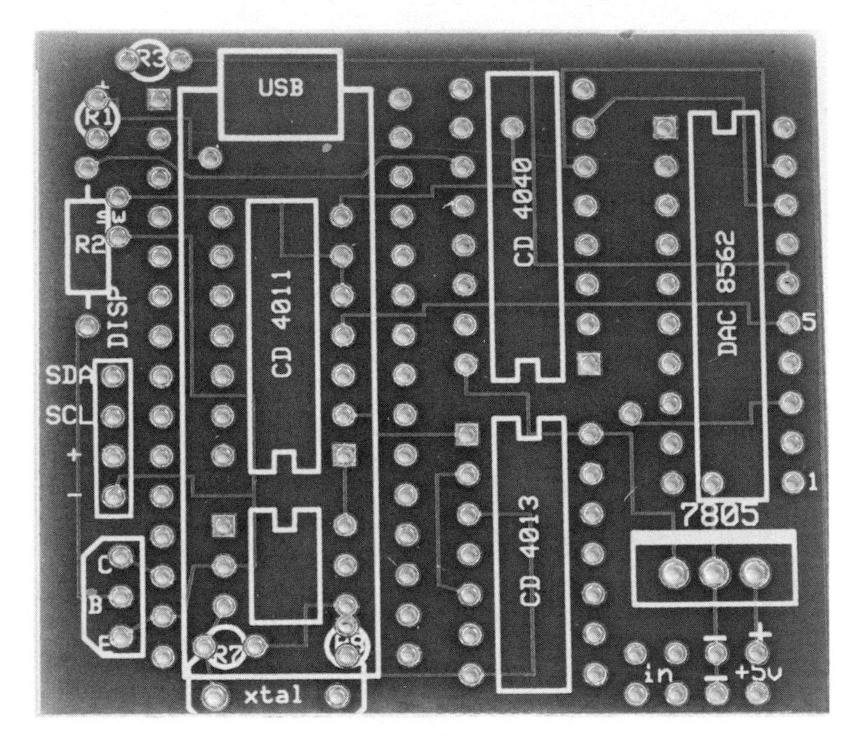

Figure 8-17: The Full Ballistic Chronograph circuit board before population

Soldering the Full Ballistic Chronograph

Once you have all your parts, follow this guide to build the Full Ballistic Chronograph:

1. Prepare the oscillator adapter board by soldering the headers in place. Solder the chip to the adapter using one of the approaches suggested in "Using SOICs" on page 20.
2. Begin populating the PCB. I usually like to start with the components that go under the Nano—in this case, the oscillator adapter board, resistors, crystal, and CD4011. Place them in the PCB, as indicated in Figure 8-15. Next, I like to include the headers that the Nano plugs into. Once again, it's not necessary to fully populate all the headers for the Nano. While on occasion I do use a full complement of headers, I tend to populate only those with connections, as well as a pair at the very top, in order to simplify alignment when plugging in the Nano. Additionally, there should be enough to mechanically support the Nano. Solder these headers in place now.

3. Populate the balance of the board, including the headers for the I^2C display and the sensor channel. Solder wire pigtails to the connections on the board for the reset and clear switches, LED, and positive and negative power supply. The lead sensor—the one that is interrupted by the projectile first—should be the one wired to pin 6 of the 4013, with or without the inverter circuit; the other sensor should be connected to pin 4.

Construction

Figure 8-18 shows the positions for the holes and cutout in the enclosure. You can download a copy of Figure 8-18 from *https://www.nostarch.com/arduinoplayground/* and use it as a template.

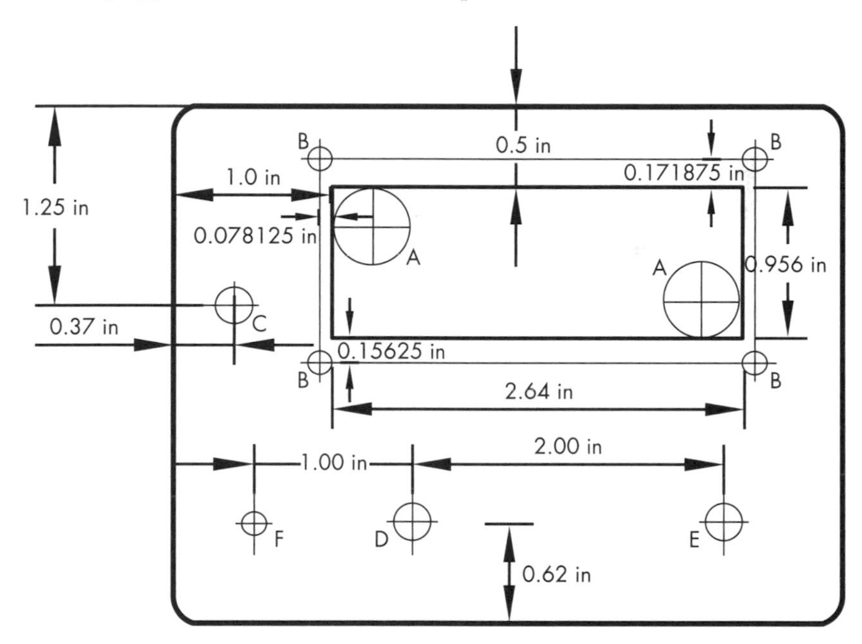

Figure 8-18: The holes and LCD cutout on the cover of the enclosure

Prepare the enclosure for the Full Ballistic Chronograph as follows:

1. Prepare the cover of the enclosure, as shown in Figure 8-18, by drilling 1/2-inch holes for cutting out the LCD (A); 1/8-inch holes for mounting the LCD (B); 1/8-inch holes enlarged with a reamer for a tight fit for the 5 mm LED (F); and 1/4-inch holes for the momentary clear switch (D), momentary reset switch (E), and on/off switch (C).
2. Mark the edges of the 1/2-inch holes (A) and connect lines tangent to the holes—you can use a Sharpie marker and clean excess marks later with alcohol.
3. Cut the opening for the display using a keyhole or saber saw.

4. There's a slight protrusion in the middle of the LCD on the right-hand side (facing up). This is part of the backlight assembly. You can cut a hole to accommodate this, as I did on both the Full Ballistic Chronograph and Chronograph Lite, or you can leave the edge straight and use spacers to keep the protrusion from hitting the enclosure. Even though I cut a space for the protrusion, I used a one-nut spacer anyway to space the connections on the top of the screen away from the front of the enclosure (see Figure 8-19).

Figure 8-19: The four-conductor female connector is mounted to the side of the enclosure using double-sided adhesive.

5. Mount the LCD's I^2C assembly to the front of the enclosure.
6. Mount the switches to the enclosure. Connect the switches and LED as shown in the schematic in Figure 8-13. Use pigtailed wires as indicated in step 3 of "Soldering the Full Ballistic Chronograph" on page 244.
7. Prepare cable assemblies to connect the shield to the I^2C adapter and the 4-pin female connector that connects the Full Ballistic Chronograph to the sensor channel. (See "Connectors Used in This Book" on page 18 if you've never built a connector yourself.)
8. Stick the shield to the bottom of the enclosure using double-sided adhesive.
9. Mount the battery holder using a 4-40 flathead screw or double-sided adhesive.

10. Make a cut in the side of the enclosure to feed the wires through to the connector for the sensor channel connection. The width of a single hacksaw blade is sufficient for 28-gauge wire.
11. Mount the connector for the sensor channel to the enclosure with double-sided adhesive, as shown in Figure 8-19.

The Sensor Channel

We built a sensor channel test bed earlier in the chapter, but now we'll build a more permanent sensor channel and look at the sensor and LED pair we'll use inside.

Building the Sensor Channel

The sensor channel is a U-shaped tunnel that fastens to the muzzle of a weapon and holds the photo detector/LED pairs that handle the switching. This channel can be constructed out of a variety of materials. I used a 3/8-inch section of acrylic and two pieces of 0.060-inch thick aluminum (see Figure 8-20).

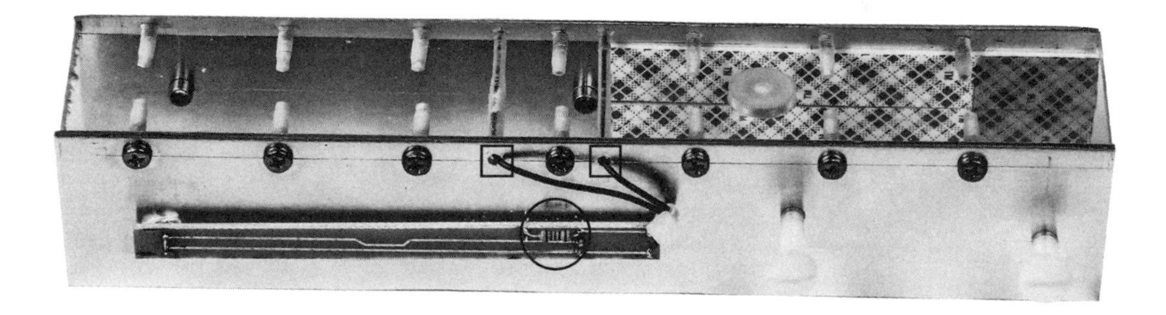

Figure 8-20: The completed sensor channel shown from the top (right side up). Note the feed-through holes in the acrylic for the positive and negative power supply connections to the LEDs (boxed). Also note the current-limiting resistor on the PCB holding the LEDs (circled). The cross-hatch area is foam taped to protect the weapon's slide from being scratched.

You could just as easily use mild sheet steel for the side pieces. The top piece, shown in Figure 8-20, could be any lightweight material, such as phenolic, Lexan, or another plastic to support the side pieces. I chose clear acrylic because it allowed me to see the gun without having to look down the barrel. You can get a wider look at the whole channel, including the sensor cable, in Figure 8-21.

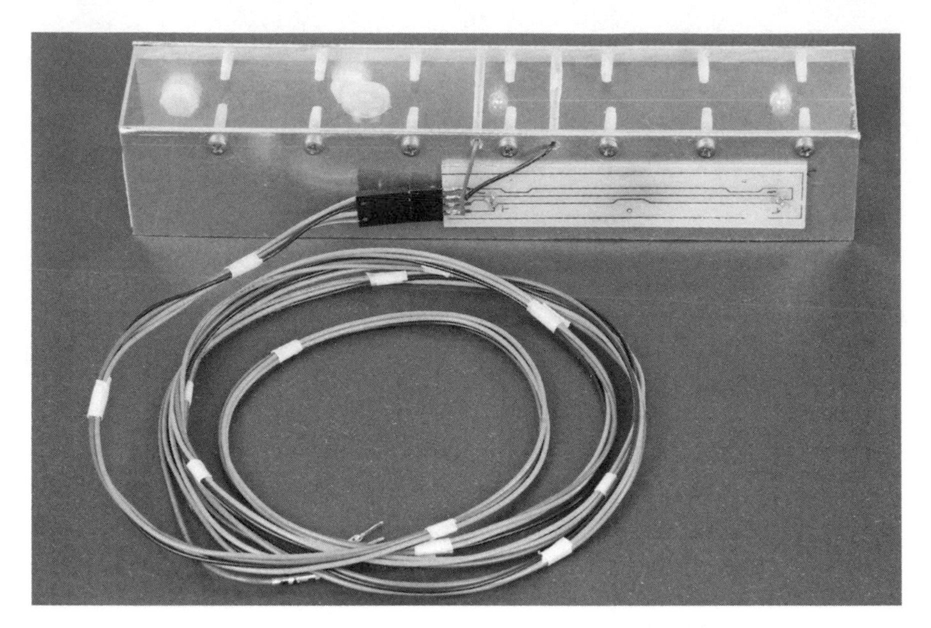

Figure 8-21: The channel with the cable attached and the PCB mounted so that the connector faces toward the back (where the weapon attaches)

There are two PCBs, attached to either side of the channel with double-sided tape, to hold the LEDs and photosensors. These PCBs are slightly different from each other, as shown in Figures 8-22 and 8-23. PCB files for these boards are available to download at *https://www.nostarch.com/arduinoplayground/*.

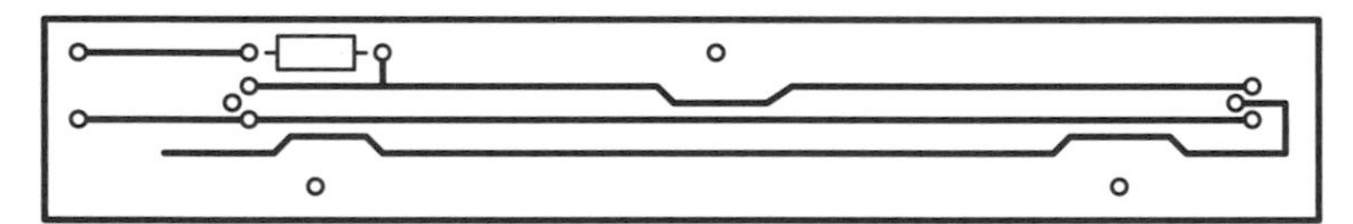

Figure 8-22: The pattern for the PCB that holds the LEDs and mounts to the sensor channel. Note the current-limiting resistor.

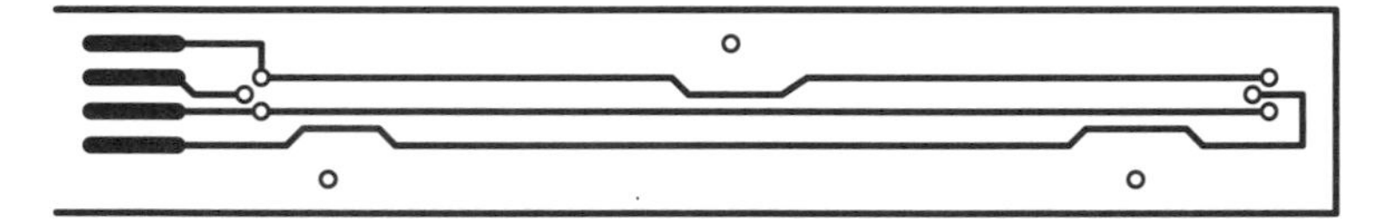

Figure 8-23: The PCB pattern for the phototransistor side of the sensor channel. Note the three pins for each phototransistor. The edge fingers are for soldering to headers that connect to the main processing and display board via an umbilical cable.

The acrylic top of the sensor channel measured 1 3/8 × 7 5/8 inches. I used a straight wooden dowel to line the barrel up with the sensors/LEDs. Note that there is a slight indentation, made with a 1/2-inch drill bit, in the top of the acrylic to allow for the optical sight of the Crossman T4 air pistol.

The aluminum sheets used for the sides measured 1 3/8 × 7 5/8 inches. I drilled the holes to fasten the aluminum sides to the acrylic top with a #30 drill bit and spaced the holes 1 inch apart. The acrylic was drilled with a #43 drill bit and tapped for 4-40 screws. See Figure 8-24 for drilling specifications for both the acrylic and aluminum pieces. The holes for the acrylic are drilled through the width.

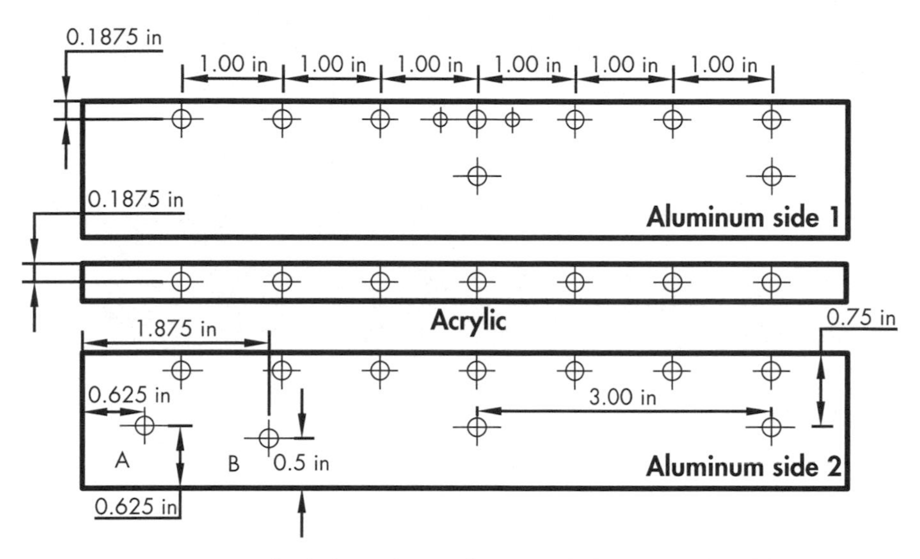

Figure 8-24: Dimensions for holes in the acrylic top and aluminum sides of the sensor channel. This template can be downloaded and used as a stencil for marking and center-punching holes.

In addition, as a feed-through for the wires from the LED side to the photo detector side, I drilled two #43 holes on either side of the fourth mounting holes on both the acrylic and aluminum pieces. The exact location of these holes is not critical.

In addition to the holes for fastening the acrylic, the aluminum required two holes on each side for the LEDs and photo detector pairs, and another two holes on one side for mounting to the barrel (slide) of the gun (A and B in Figure 8-24), drilled with a #25 drill and tapped for a 10-24 screw. Check out how big the IR LEDs are. Most are 5 mm, and a 3/16-inch hole is generally a close fit. The Optoschmitt sensor also fits snugly in a 3/16-inch hole. The holes for the LEDs and photosensors can be measured exactly 3 inches apart, or you can measure them yourself to match with the PCBs mounted on the side.

Depending on the weapon(s) you intend to use, you may want to adjust the positioning of holes A and B in Figure 8-24. The sensor channel can also accommodate more tapped holes for multiple weapons. To mount the sensor channel to the top of the gun, I used nylon screws with locking nuts. The nylon screws were able to tighten against the blued-steel finish of the pistol without marring it.

The screws for mounting the gun worked well with the Crossman T4 as well as on an older Crossman pellet gun (see Figure 8-25).

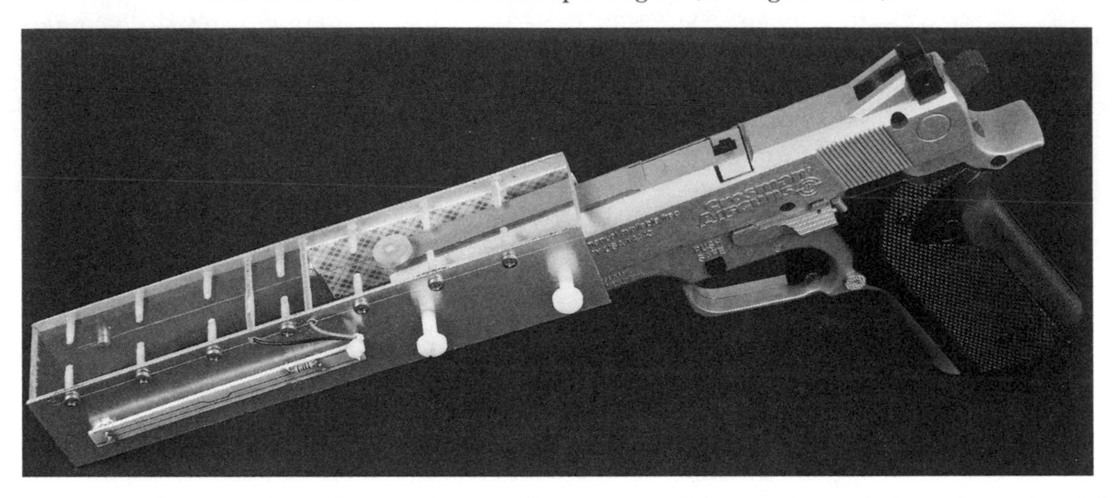

Figure 8-25: The sensor channel mounted on an older Crossman pellet gun. This angle shows the top (acrylic) side of the channel.

On the inside of the channel, I placed some double-sided adhesive foam tape (if you can find a single-sided adhesive foam tape, all the better) to give it a more snug fit and protect the weapon from damage. I left the protective covering on the other side of the foam so it would not adhere to the weapon or mar the finish.

Depending on the weapon you are using, you might want to add an extra layer of foam to pad the channel so the center of the barrel is closer to the center of the channel. But as long as the barrel is not so far on either side that the projectile could strike either the LED or photo detector, centering the barrel perfectly is not critical.

It is critical, however, to center the vertical adjustment so the LED/detector pairs line up with the trajectory of the projectile. To set this alignment, I used a straight wooden dowel of the same diameter as the bore of the barrel, inserted it partially into the barrel of the weapon, and then adjusted the position so it lined up with the LED/detector pairs. Once it's aligned, tighten the nylon screws to secure the channel to the weapon.

Optoschmitt Light Sensors and UV LEDs

In preparing the sensor channel, I sampled several different types of LEDs and detectors to see which offered the best price and performance. Units purchased on eBay (UV LED and phototransistor pairs) worked well, and I used them in early prototype versions. However, I continued to search for a sensor that I was sure would be fast enough and provide good sensitivity in a narrow field, which helps to exclude ambient light. After reviewing several samples, I chose the Optoschmitt SD5610 detector from Honeywell—so named, I guess, because it includes Schmitt-trigger circuitry (see Figure 8-26).

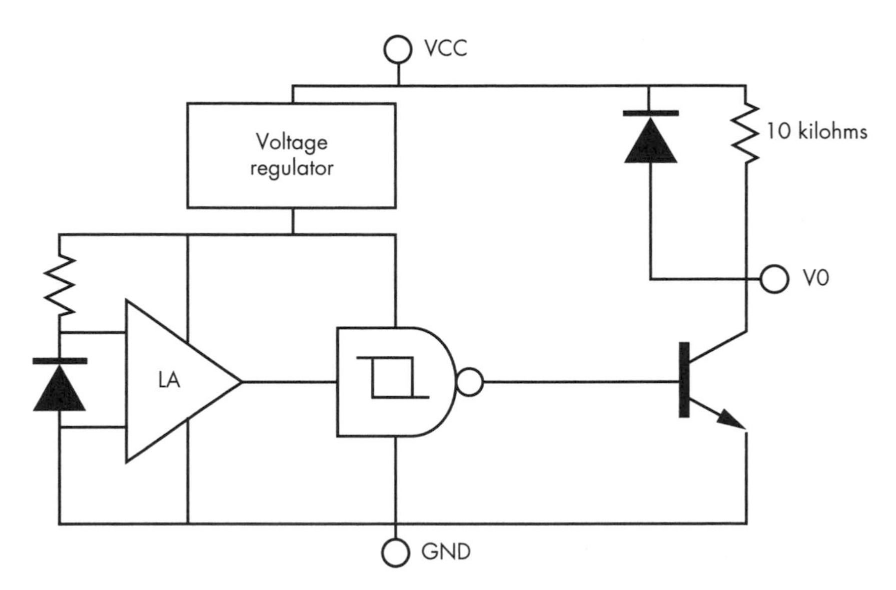

Figure 8-26: Schematic for the Optoschmitt SD5610 detector I used in the chronograph. Note that the 10-kilohm pull-up resistor is included, but the inverter function is not included in the schematic.

The Optoschmitt SD5610 sensor is a little pricey, but it features a 6-degree acceptance angle, which worked well for projectiles of all sizes, including very small and large ones. It also reduced the effects of ambient light.

According to the manufacturer, the photodetector consists of a photodiode, amplifier, voltage regulator, Schmitt trigger, and an NPN output transistor with a 10-kilohm (nominal) pull-up resistor (see Figure 8-26). The internal pull-up resistor eliminates the need for an external resistor in the circuit. Note that there are two versions of this device: the SD5600 and the SD5610. The SD5610 includes an inverter so that the output is low when the ambient light is above the turn-on threshold. Because I required the inverted output, I used the SD5610. The spectral sensitivity is greatest in the 800 to 850 nm wavelength—the area of most common IR LEDs. Additional information on the SD5600 series can be obtained at Honeywell's website.

For the LEDs, I just used regular IR LEDs that claimed output in the 850–950 nm range. I simply bought a bag of 50 units on eBay, and they work fine. Alternatively, SparkFun offers single units very cheaply.

The LEDs and photosensors should be soldered on the PCBs made for them and should fit snuggly into the holes. I fastened the PCB to the sides of the acceleration channel using some standard 3M double-sided adhesive tape.

Sensor Umbilical Cable

The cable I used to connect the sensor channel to the Full Ballistic Chronograph's PCB is made from four lengths of 30-gauge wire twisted together, fastened with masking tape, and connected to a female four-conductor Pololu connector at each end. These connectors are not polarized and do not have a detent, so they can be relatively easily unplugged or plugged in the other way. Before you plug them in, make sure to line up the color-coded wire.

NOTE *I initially attempted to use a length of four-conductor telephone cable, but it was too stiff and caused problems.*

The channel and completed chronograph can be connected easily. If you used colored wire, you'll know that the plug is connected correctly because it is not polarized.

Final Setup and Operation

Once you've finished assembling the unit and sensor channel, it's time to take it out on the range and give it a try. Both the Chronograph Lite and the Full Ballistic Chronograph have been designed to operate from battery power, so you don't need to plug them in. Set up the umbilical cable to connect the sensor channel to the chronograph unit, and then mount the channel to the weapon securely (see Figure 8-27).

Figure 8-27: The completed Full Ballistic Chronograph with the acceleration channel mounted to a Crossman 0.177 caliber pellet gun

Once the channel is attached to the weapon, carefully align the barrel of the weapon with the LED/detectors in the channel by using a straight dowel the same diameter as the bore of the barrel or by making a simple adapter that uses a straight length of tubing (see Figure 8-28).

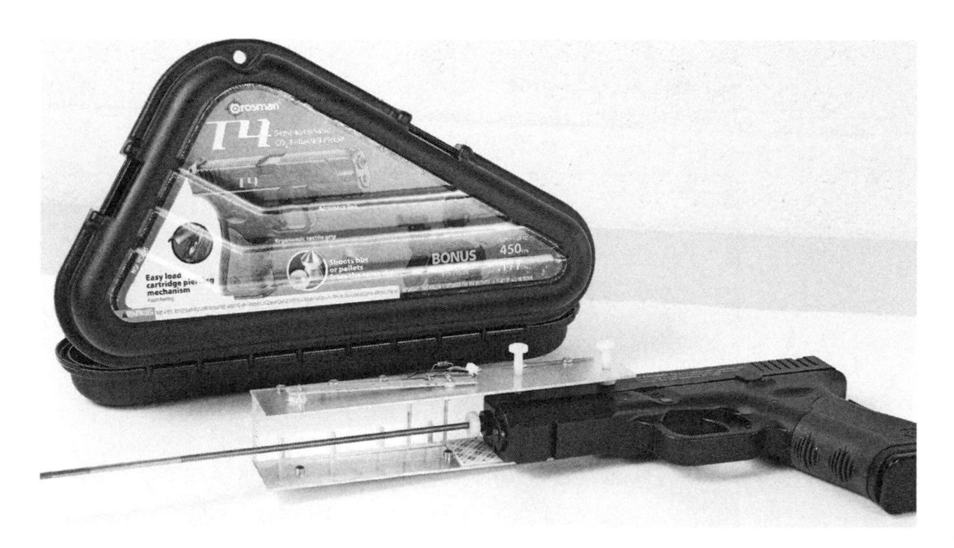

Figure 8-28: Using a small diameter brass rod with a Teflon end to set up proper alignment between the barrel of a Crossman T-4 CO_2 pistol and the sensor/detectors in the sensor channel

WARNING Always use caution when handling a weapon. *Do not look down the length of the barrel to align it. Look through the clear acrylic top of the sensor channel.*

Using the Full Ballistic Chronograph

After aligning the barrel as best as possible, turn on the Full Ballistic Chronograph and press the reset button. Aim carefully at your target and fire the weapon. The velocity of the projectile in feet per second and the time it took to travel the 3 inches between sensors should appear on the LCD screen. To make another measurement, simply press the reset switch and fire again.

If by any chance you fail to align the weapon in the channel correctly, there is a possibility that the projectile will interrupt the first set of photosensors and not the second. In this case, you can press the clear switch and then the reset switch to try again. The clear switch simply closes the connection for the second photosensor set.

The Full Ballistic Chronograph should give accurate readings from about 300 fps to well over 2,000 fps.

Using the Chronograph Lite

The Chronograph Lite operates in much the same way, only it automatically resets so no reset button is required. However, should the projectile fail to interrupt the second set of sensors/detectors, it will be necessary to clear the display by pressing the clear button. Essentially, this does the same thing as interrupting the second sensor/LED pair, but you should always use the clear button and never attempt to interrupt the second sensor/detector pair with an external object—especially your finger. Should the weapon accidentally fire, you could sustain a severe injury. The Chronograph Lite will provide accurate measurements from about 200 fps to well over 1,000 fps, but its accuracy tends to roll off as it approaches 700 fps.

HIGH-POWERED WEAPON TESTING

Unless you are experienced with firearms, I strongly recommend against using the Chronograph to measure high-powered weapons. That said, I did test the Full Ballistic Chronograph on a few of them.

Figure 8-29 shows the sensor channel mounted to a Smithfield XP/M 9 mm semiautomatic pistol. I tested the Chronograph with a number of cartridges, and the measurements came within a couple fps of the manufacturer's specification of the bullet. For example, the Remington JHP claims the bullet travels at 1,155 fps, and I measured about 1,152 fps. Other weapons also measured close to the published velocities.

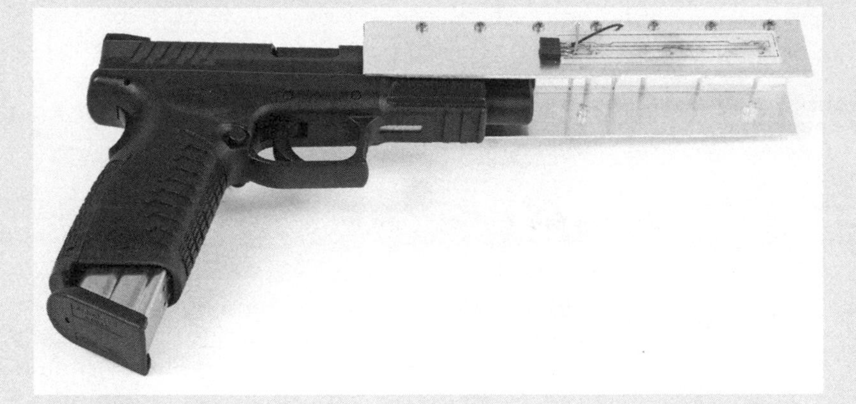

Figure 8-29: A Smithfield XP/M 9 mm pistol set up with the sensor channel for the Full Ballistic Chronograph. The magazine is intentionally inserted backward for safety reasons.

9

THE SQUARE-WAVE GENERATOR

Signal generators, also called waveform generators or function generators, create an alternating current (AC) voltage that can be used in a variety of electronic tests and diagnostic procedures. A *square-wave generator* like the one you will build in this chapter (shown in Figure 9-1) is an electronic lab instrument that creates a continuous sequence of equally spaced pulses of electricity that are on for a certain amount of time, switch off for an equal duration, and switch back on again, repeatedly.

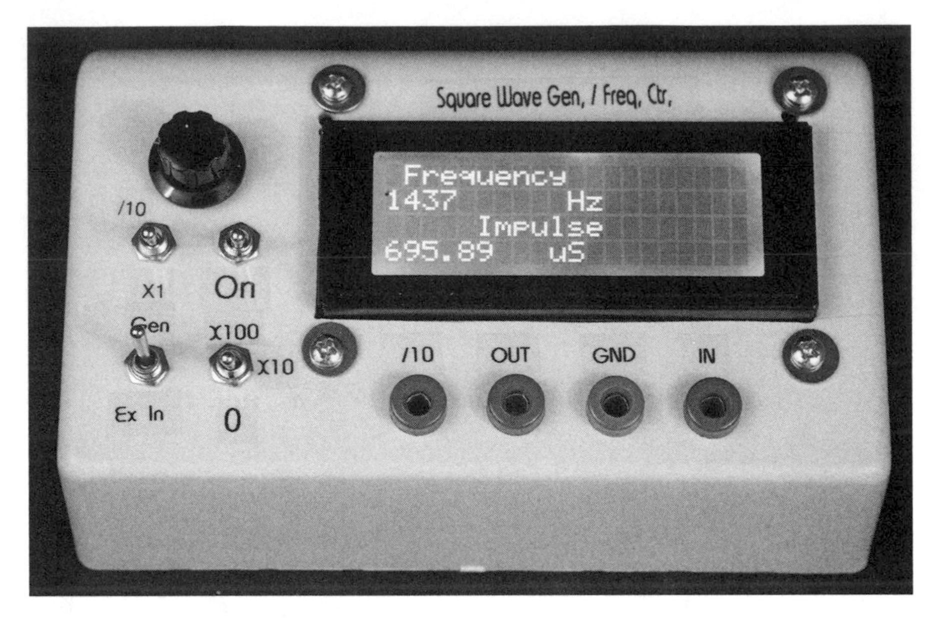

Figure 9-1: The completed Square-Wave Generator

Why Build a Square-Wave Generator?

Signal generators like this one are frequently used to perform diagnostic jobs, from evaluating the frequency response of components and subsystems to providing stimulus to systems under development. Some specific uses for signal generators include:

- Observing the integrity of an amplifier, attenuator, or other device
- Measuring timing characteristics of a circuit
- Simulating real-world on/off events

The signal generator part of this project is primarily a square-wave generator.

What Is a Square Wave?

What's a square wave and what's it good for, you ask? A *square wave* is an electrical signal that starts at zero voltage, rises to some level (its *amplitude*), stays at that level for some duration, returns to zero, and then repeats the process in a symmetric pattern.

The square wave is one of the fundamental wave forms in electronics, and it is in many respects the most useful, in part because it has both a DC component and an AC component. The DC component is the fact that it stays at a certain voltage for a period of time and then almost instantaneously transitions to a different level. The AC component is that it repeats this transition at a regular period. Figure 9-2 shows a square wave.

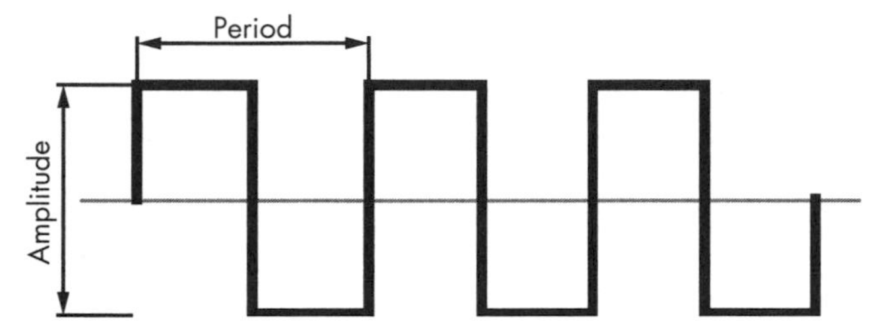

Figure 9-2: A typical square wave

The square wave has an amplitude and a period. The *period* of the wave is the duration of a complete cycle, and it could be in seconds, minutes, milliseconds, microseconds, and so on. The *frequency* is how many cycles occur in a certain period of time (one second is the accepted standard) and is therefore the reciprocal of the period. For a period of T, the formula to determine the frequency is $f = 1/T$.

Why Square Waves Are Useful

Square waves are particularly useful when developing and testing electronic products. For example, the clock in a microcontroller system is essentially a square wave. In some diagnostic and test procedures, the microcontroller's internal clock can be disconnected and replaced by an external signal generated by a signal generator—in this case, a square wave or sine wave work equally well. You can then test the microcontroller at different frequencies. For certain processors, it's often valuable to slow a processor clock during testing to see exactly where software glitches occur.

Other uses include sending a signal into a device being tested in order to tune the circuit to the proper value or checking a device's frequency response or integrity. A square-wave generator can also act as a pulse generator to test a variety of digital circuits. You will find this application useful in the Ballistic Chronograph in Chapter 8.

Because a square wave's voltage starts at zero and rises almost instantly, it can also be used as a switching voltage to turn circuits off and on at the frequency of the square wave. The frequency of the Square-Wave Generator in this chapter ranges from 1 kHz to around 30 MHz (with the divider switch included, it can go down to 100 Hz). This frequency range can be varied via a potentiometer, so you can turn things on and off at different rates. This allows the Square-Wave Generator to simulate almost any repetitive switching action, which is useful for cycling things on and off for life-test applications, too.

OTHER USEFUL WAVEFORMS

A square wave is only one common type of waveform, though. Probably the most common is the *sine wave*, or sinusoidal wave, in which the wave is a continuous curve and one cycle represents 360 degrees. Yet another wave type frequently encountered in electronics is the triangle wave. Both are depicted in Figure 9-3. The amplitude and period of a sine wave and a triangle wave are measured the same way as for a square wave.

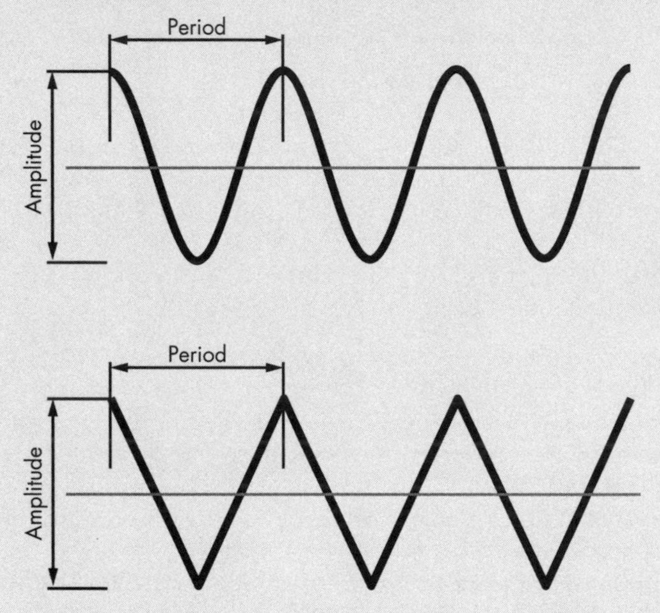

Figure 9-3: A sine wave (top) and a triangle wave (bottom)

Each of these waveforms has specific characteristics that make them useful in different applications. Sine waves and triangle waves are both important in electronic music projects, for example. This project, however, focuses on the square wave, and as you progress through this chapter, you will see it can be applied in a variety of ways.

A Frequency Counter

In addition to providing a signal generator, this project includes a *frequency counter*, which reads wave frequencies, so you can display the generator's output frequency on a digital readout. You can also use the frequency counter as a separate instrument on its own to measure frequency from an outside source and display it. This project also displays the period, or impulse time, of the wave.

In operation, the frequency counter receives an AC input signal and counts each pulse. After a certain number of pulses have been counted, the counter compares this against a clock signal, sometimes referred to as a *time base*, and displays the number of pulses per unit time—for example, pulses or cycles per second.

To assure accuracy, the counter compares the pulses against a clock, usually one that is crystal-based. For instruments that require the utmost accuracy, the crystal-clock assembly is a precision subsystem often placed in a temperature-controlled environment. The frequency counter in this project uses the 16 MHz crystal on the Arduino Pro Mini. In fact, the Pro Mini has most of the circuitry to implement a complete frequency counter with few additional components. It includes the clock registers for counting and just about everything else a frequency counter needs with no external components.

NOTE *Professional laboratory and desktop instruments selling for hundreds (or thousands) of dollars often provide a very wide frequency range, from under 1 Hz to several GHz, and offer anywhere from 6- to 10-digit precision. Their displays can be switched to read frequency or time (the time between pulses), too.*

To do all of this, the Square-Wave Generator takes advantage of the Arduino AT328 16 MHz, 5V Pro Mini, which is a smaller, lower-priced version of the Arduino Nano. The generator also includes special circuitry to divide a signal's frequency, allowing you to provide an output of very low frequencies at one end and allowing the frequency counter to read very high frequencies—above what the Pro Mini can normally handle—at the other.

Shortcomings of the Square-Wave Generator

While this project produces a square-wave generator and a frequency counter that perform well in many applications, it has shortcomings compared with professional laboratory and bench instruments. The generator has a less than 1 percent frequency error, which is good for hobby projects but doesn't match the tolerances of laboratory-grade and direct-digital-synthesizer (DDS) generator units. Those units often have errors measured in the part-per million (PPM) range. And the frequency counter uses the time base of the Arduino, which, while accurate, doesn't match higher-priced units with crystal ovens and other special circuitry.

This generator also doesn't offer the resolution of multiple digits. Many lab and bench instruments have resolutions that go into as many as 10 digits. That said, the instrument has worked well for me in a broad variety of Arduino and other projects, where a greater resolution wasn't required.

Required Tools

Drill and drill bits

Keyhole saw

Soldering iron and solder

File

Parts List

In addition to the Arduino Pro Mini, you'll need a Linear Technology oscillator chip and a small handful of other components. Here's the complete parts list:

One Arduino Pro Mini or clone (There are several available, and some have different pinouts—particularly for pins A4 and A5. Figure 9-4 shows the pinout for the particular clone that I used. Other units with different pinouts will work, but the connections on the shield may have to be changed.)

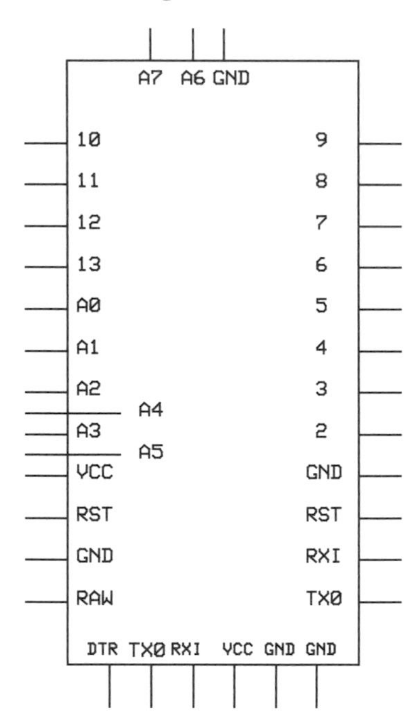

Figure 9-4: Pinout of the Deek-Robot Pro Mini Arduino clone

One LTC1799 oscillator chip and breadboard-compatible adapter board, like the 5-SOT-23 adapter board shown in Figure 9-5 (See "Using SOICs" on page 20 for tips on how to use a surface mount chip like this.)

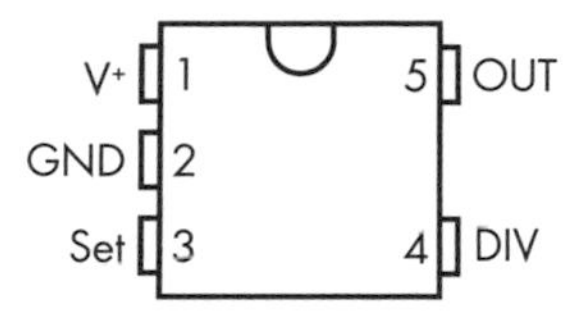

Figure 9-5: This adapter board has a complete variable frequency oscillator (1 KHz to 30 KHz) and CMOS buffer circuit.

One 250-kilohm carbon potentiometer

Two 0.1 μF ceramic capacitors

One LM7805 voltage regulator

Two SPDT center-off toggle switches

Two SPDT toggle switches

One HCT4017 decade counter IC, like the CD4017 B shown in Figure 9-6

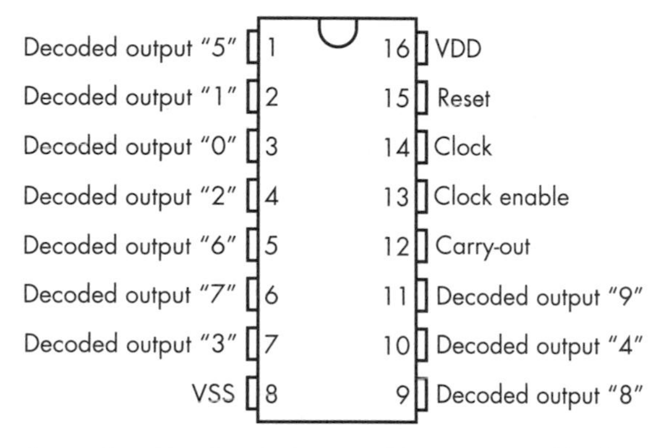

Figure 9-6: The CD4017 B is a CMOS counter/divider comprising a 5-stage Johnson counter with 10 decoded outputs. It is used here as a divide-by-10 counter.

One 1 μF electrolytic capacitor

One 10 μF electrolytic capacitor

One 20×4 LCD

One I^2C adapter, if not included with the LCD

One PCB shield (See "Downloads" on page 262 if you don't want to design your own.)

One Hammond 1595C sloped front enclosure (or equivalent)

One battery holder

One knob

Four 4-40×1/2-inch screws and washers

Eight 4-40 nuts

One piece of double-sided foam tape

One 9V battery

Assorted 28- or 30-gauge hook-up wires

(Optional) Four banana plug jacks *or* three BNC connectors

(Optional) One 3.5 mm jack

(Optional) One 9V, 100 mA, 110V wall power supply (for more information see "Battery Power" on page 278)

Note that I elected to use banana plug jacks for the I/O on the front panel, though this is a bit old fashioned and probably not the best practice. You could replace the two output connectors with a BNC connector as is used in the pH meter project, which is a little more pricey and eliminates the need for a ground, as the BNC connector has a center conductor and a shielded ground surrounding it. The banana jacks have only one conductor each.

Downloads

Sketch *SquareWave.ino*

Front panel template *SquareWaveEnclosure.pdf*

PCB foil pattern *Generator.pcb*

The Schematic

The Square-Wave Generator circuit in Figure 9-7 doesn't call for a lot of different components. However, before you start building, note that the Arduino Pro Mini has a very different pin configuration than the Nano. It is also worth noting that there are many versions of the Pro Mini available, so check the pinout of the version you buy. The particular Arduino I suggest in the parts list has the pinout detailed in Figure 9-4. Read "Important Notes on the Pro Mini" on page 263 for a description of key differences.

In the schematic, notice the switches: SW1, SW2, SW3, and SW4. SW2 provides the divide-by-10 display. SW3 allows you to use the frequency counter with an external source instead of the signal generator. SW1 connects the 1, 10, 100 divider for the master clock (the switch has a center-off position that doesn't have a connection) for the LTC1799 oscillator. SW4 is the power switch; its center is off, and the other two positions are for either external supply or battery.

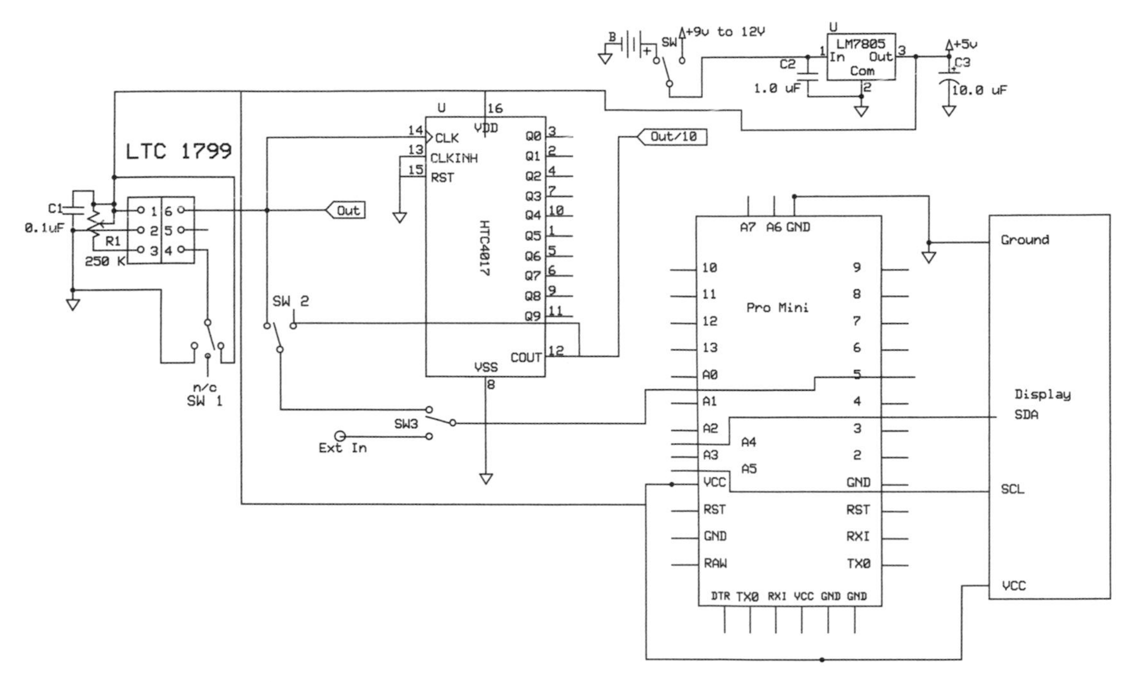

Figure 9-7: Schematic of the Square-Wave Generator

Important Notes on the Pro Mini

Before you build the breadboard, note that one major pinout difference among Pro Mini boards is the placement of pins A4, A5, A6, and A7. Some versions locate all four analog inputs on the short side of the board, while the Deek-Robot used in this project splits them (see Figure 9-8). It places A4 and A5 near the other analog pins, but not in line with them, and A6 and A7 are on the short side of the board. Pins A4 and A5 are used to drive the I^2C bus for the display.

There are some other minor differences between the Pro Mini and the Arduino Nano, but one of the most prominent is that the Pro Mini does not include a USB interface, so you have to program it using an external serial interface of some kind. There are several serial adapters on the market using FTDI technology. (*FTDI* is an abbreviation for *Future Technology Devices International*, a privately held Scottish semiconductor device company specializing in USB technology.)

An alternative to using FTDI-based, purpose-built serial adapters is to use another microprocessor board to program the Pro Mini. I use an Arduino Uno clone to program my Pro Mini because it's inexpensive, it allows me to remove the processor chip so I don't end up programming both boards, and it's easy to use. I use a simple breadboard setup to do the programming. Go to "Uploading Sketches to Your Arduino" on page 5 for connection details.

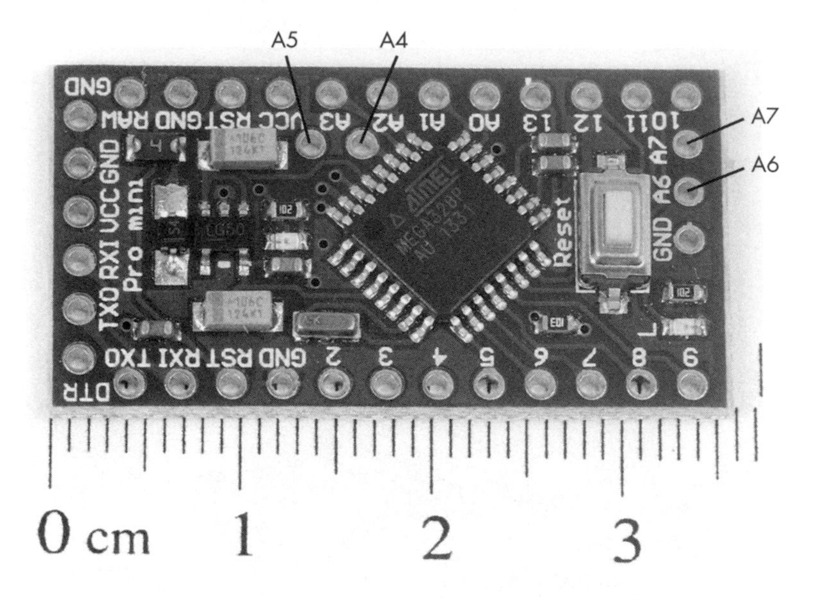

Figure 9-8: The Deek-Robot Pro Mini next to a centimeter ruler. The pinout is a little different from similar Arduino clones. For example, pin A4 is the unmarked pad between A2 and A3, while A5 is the one between A3 and VCC.

How the Square-Wave Generator Was Developed

This project was developed to solve a need that emerged when creating the Ballistic Chronograph in Chapter 8. While developing the chronograph, I needed some way to test it to assure it worked properly without using a weapon with live ammunition and shooting holes in my shop.

The Square-Wave Generator was my solution to the problem. With it and a small handful of other parts, I was able to simulate the signal the chronograph should receive as a projectile breaks a sequence of light beams. I decided to start with the time it took a projectile to travel an arbitrary distance of 3 inches, which turned out to be somewhere between 50 and 1,000 microseconds, depending on the speed of the projectile. I then used the Square-Wave Generator to generate a signal at frequencies between 20,000 Hz and 1,000 Hz, the reciprocals of those times. Once I figured out what I needed for the chronograph, the Square-Wave Generator project took on a life of its own, and the final version is what you see in this chapter.

Deciding How to Generate Signals

First, I looked for an easy way to satisfy my timing requirements. There were several DDS (direct digital synthesis) products and boards that would have easily solved the problem, but all of the solutions I found were a little pricier than I was hoping, and many had other shortcomings. Abandoning the DDS, I looked at several alternatives, some bringing me back to my old Radio Amateur days. One solution was to use a crystal oscillator and divide a fundamental frequency to achieve frequencies near the ones I needed. This presented several problems, not the least of which being that the circuit would likely need several divider chips.

Another solution was to create my own variable frequency oscillator (VFO) from scratch. While a possibility, that solution entailed more design work than I was prepared to do at the time, so I went back to the data sheets.

I found that Linear Technology's LTC1799 single-chip precision oscillator had just about what the doctor ordered—and more. According to the data sheet, this chip provides a square-wave signal from 1 KHz to 33 MHz with a single variable resistor and a switch to divide the fundamental oscillator frequency by 1, 10, or 100. It boasts good stability, too: nominally, it has less than 1 percent error. And it was a lot less pricey than the DDS solutions at just under $4.00.

The final part of the problem was to see what frequency the generator was creating. Without an external frequency counter or a calibrated oscilloscope, it would be extremely difficult to get even a close approximation of the frequency generated. So the project mushroomed to include a built-in frequency counter. Because the counter was there anyway, I included a switch to allow me to use the frequency counter as a stand-alone instrument.

Planning How to Display the Frequency

Now I could generate square waves, but there were still some other problems to be addressed. For example, how would I read the frequency from the outside? I could mark the potentiometer positions with calibrations—as many generators of yore have done—but that is at best a clumsy and inaccurate approach in today's digital age. A built-in frequency counter and bright display seemed most practical.

I went back to the drawing board—and to the Arduino Library. I found several approaches to Arduino frequency counters online, including at least two separate frequency counter libraries. The simplest and most convenient library for this application was *FreqCount.h*, developed by Paul Stoffregen. How I used this library is discussed more under "The Sketch" on page 271. For more information on the library itself or to get the latest updates, you can go to *https://github.com/PaulStoffregen/FreqCount/*.

A preliminary breadboard prototype indicated that the frequency counter worked well. I put the breadboard together using a 20×4 LCD display using the I^2C interconnect. After labeling the display with the word *Frequency* and displaying the frequency in Hz under that, I still had two lines of 20 characters left (see Figure 9-9).

Figure 9-9: The display of the Square-Wave Generator, showing the frequency and the impulse time

Waste not, I always say. Because at least two of the projects I planned to use the Square-Wave Generator with required evenly spaced pulses (more or less a pulse generator), I decided to use the second two lines of the display to indicate the time of the impulse. Calculating the time in the sketch would be relatively easy, as the time (in seconds) is a function of the frequency ($T = 1/f$).

Signal Integrity

Without getting involved with the higher math of signal composition, a square wave can be thought of as an infinite series of sine wave harmonics added together. As the frequency increases, so does the complexity and fragility of the waveform. If you connect your breadboard circuit to an oscilloscope, you can observe this yourself.

This project was initially developed as a square-wave generator/frequency counter that could operate in the area of 1,000 Hz to 1 MHz. This generator does its job with panache, but the fundamental oscillator chip has a range far in excess of that.

In developing the project, I had two options. The first was to intentionally limit the device's performance to the area that was initially proposed or to extend it closer to the limits of the oscillator and suffer some degradation at the higher end. I selected the latter. While the square wave starts rounding off at around 15 MHz or so, the performance at the lower—and intended—frequencies is not impacted whatsoever. Figure 9-10 shows four oscilloscope traces at different frequencies to demonstrate.

At 1 kHz, the wave pattern is close to perfect, as you can see from the display on my older analog oscilloscope (see Figure 9-10A). In Figure 9-10B, at 5 MHz, the edges of the square wave are compromised a very small amount, showing a slight overshoot on the rising edge. When the frequency is increased to 12 MHz, the signal begins to look a little ragged, with even more distortion (see Figure 9-10C). Some of the distortions of the wave are a result of tuning, or stray capacitive and inductive effects, by certain components used in the construction. This is an avoidable phenomenon, and I mention it primarily so you're aware of the shape of the waveform. I suspect that most of your applications for the generator will be in the lower-frequency area, at less than 1 MHz, where the wave pattern produced is as good as it gets. Further, I have used the generator in higher frequencies, and the slightly distorted waveform had virtually no effect on the result.

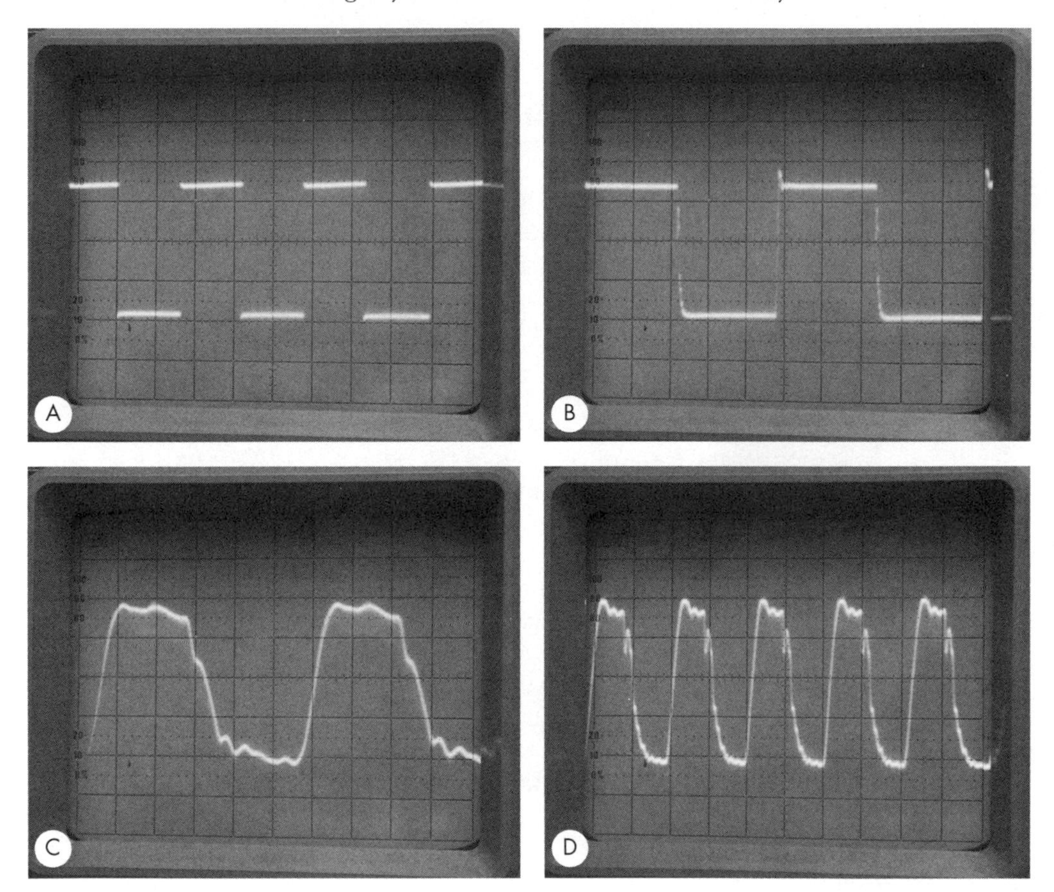

Figure 9-10: Four oscilloscope traces showing the output of the Square-Wave Generator at 1 kHz (A), 5 MHz (B), 12 MHz (C), and 20 MHz (D). It quickly becomes apparent that the square-wave signal starts to lose integrity at around 12 MHz.

While the waves continue to distort at frequencies above 20 MHz (see Figure 9-10D), they are fully recognizable as square waves and remain useful. At 30 MHz—the extent of the range of the Square-Wave Generator as built—the signal trace looks increasingly like a sine wave, but for most test purposes, it is still totally valid.

LISTENING TO SQUARE WAVES

An interesting experiment that will provide you with some idea of the harmonics present in a square wave is to set the Square-Wave Generator to the audible range of the frequency spectrum and plug its output into the input of an amplifier and loudspeaker. Listen to the quality of the sound. The "fuzz" you hear is a result of rich harmonics produced by the square wave, which essentially comprises a composite of all other sine waves within the frequency limit.

Fine-tuning with a Decade Counter

In designing the system, there were a couple final additions I settled on to add utility and improve performance. The frequency counter, as put together, had a frequency range of about 100 Hz to around 10 MHz tops. The LTC1799 oscillator offered a frequency range from 1 kHz to about 30 MHz, and for most applications, that would be far more than adequate. But there were some applications I had in mind that would need an AC source down to about 100 Hz.

Well, there turned out to be a way to kill both birds, so to speak, with one chip: a divide-by-10 counter—in this case, an HCT4017 or CD4017 decade counter and a couple of switches.

The decade counter accepts an AC signal, counts to 10, and then starts over. By looking at one of the counter outputs, it essentially divides by 10. It was possible to feed the output of the oscillator through the divide-by-10 counter and show output frequency on the LCD, while the actual output frequency would be 10 times the frequency shown. This workaround lets the Square-Wave Generator show frequencies well above 10 MHz on the LCD as long as you can mentally move the decimal point over one place. On the flip side, the switch (SW2) could be moved to take the output of the oscillator divided by 10 directly so that it could output a minimum output frequency as low as 100 Hz, or 1 kHz/10.

The Oscillator in Detail

The oscillator part of this project is pretty much self-contained in the LTC1799 and requires only an external variable resistor, a bypass capacitor, and a switch used to divide the fundamental oscillator frequency by 1, 10, or 100 times. The value that the frequency is divided by depends on what you're connecting to pin 4. When you connect pin 4, or the DIV pin, of the LTC1799 to GND, the frequency is divided by 1; when pin 4 is left floating or open, the frequency is divided by 10; and when pin 4 is connected to 5V, the frequency is divided by 100. This allows the unit to cover a range of frequencies from 1 kHz to 30 MHz.

Also, while I have chosen to use a 250-kilohm potentiometer between pins 1 and 3 of the LTC1799, any potentiometer between 3 kilohms and 1 megaohm is acceptable. The frequency decreases as this resistor value increases, and vice versa.

According to the manufacturer, the LTC1799 outputs a fairly crisp square wave throughout its frequency range. As frequencies increase, however, there are a variety of considerations that impact the integrity of the wave. These include stray capacitances and inductances due to the layout of the circuit, such as the output position, the hookup of the variable potentiometer, the switch, and other components. Because most applications I had in mind were in the lower end of the frequency spectrum offered by the LTC1799, I did not pay strict attention to the layout and thus probably have somewhat compromised integrity at the higher frequencies. See Figure 9-10 for actual signal traces.

The Breadboard

As in virtually all of my Arduino projects, somewhere during the design process, I end up making a breadboard layout. Figure 9-11 shows the prototype for the Square-Wave Generator.

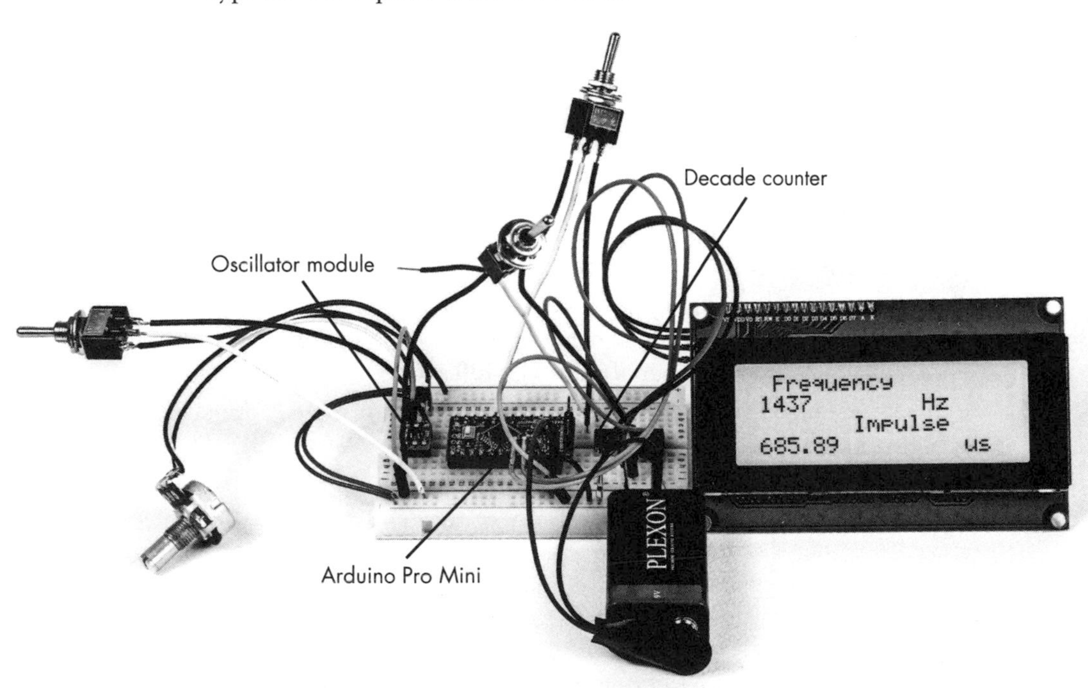

Figure 9-11: The breadboard for the Square-Wave Generator

Wiring the breadboards for testing posed no difficulty, with the exception that the oscillator became a little squirrely at the higher frequencies when using longer interconnect leads.

Here are the connections:

1. Connect all of the red positive rails together and blue negative rails together. Be careful not to connect the red and blue rails under any circumstances.
2. Insert the LTC1799 oscillator module as close as possible to one end of the breadboard. See the far left-hand side of Figure 9-11. The oscillator has to be mounted on an adapter board so it will fit in the 0.100 centers of the breadboard, as shown in Figure 9-12.
3. Insert the Arduino Pro Mini into the breadboard. Connect the 5V terminal of the Pro Mini to the red positive rails of the breadboard.
4. Connect the GND of the Pro Mini to the blue negative rails.
5. Insert the HCT4017 decade counter into the breadboard. (It's immediately to the left of the LCD in Figure 9-11.)
6. Insert the LTC1799 on its adapter board into the breadboard.
7. Connect one end of the three-position (center-off) switch SW1 to ground. Connect the other end to VCC, and connect the center pin of the switch to pin 4 of the LTC1799.
8. Connect one end of switch SW2 to pin 6 of the LTC1799 adapter board (or pin 5 of the LTC1799).
9. Connect one end of the potentiometer R1 to pin 3 of the LTC1799. Connect pin 1 of the LTC1799 to the red positive rail along with the other side of the potentiometer and the wiper, or the center pin of the potentiometer.
10. Pin 6 of the LTC1799 will be the output of the oscillator, which will go to pin 14 of the HCT4017 and to one leg of switch SW2.
11. Capacitor C1 should have been installed on the adapter board as described.
12. Connect pins 13 and 15 of the HCT4017 to ground.
13. Connect the other end of switch SW2 to pin 12 of the HCT4017 (CD4017).

Figure 9-12: The SOIC has been soldered to an adaptor board, which will fit into the 0.100-inch centers of the breadboard. The chip includes only 5 pins, but I used a 6-pin adapter.

14. Connect the center of switch SW2 to an empty row on the breadboard. (I used one between the Pro Mini and the HCT4017).
15. Connect one end of switch SW3 to the same empty row you used in step 8, which should be connected to the center of switch SW2.
16. Connect the center of switch SW3 to digital pin 5 (D5) on the Pro Mini.
17. Pin 3 of switch SW3 will serve as the input if you use the breadboard in the frequency-counter-only mode.

NOTE *SW4, the AC/battery on/off switch, does not need to be configured in the breadboard, as the circuit can receive power from the computer while programming the Arduino. The LM7805 is not used in the breadboard configuration for the same reason.*

18. Check your LCD. If it includes the I^2C subassembly board soldered to it, you're okay to continue. Otherwise, solder the I^2C board to the display as described in "Affixing the I^2C Board to the LCD" on page 3.
19. When your LCD is ready, you will need four male-to-female connector wires to hook it up. (In the finished version, you can make a small wire harness for it, including wires for VCC, GND, SCL, and SDA.) I usually color code these with red and black for positive and ground, green for SCL, and yellow for SDA. Plug the VCC wire into the red positive rail, the negative into the blue negative rail, SDA to A4 on the Pro Mini, and SCL to A5 on the Pro Mini.

Finally, build your programming circuit or plug in your FTDI adapter, load the sketch onto the Pro Mini, and you're all set to go.

The Sketch

The Square-Wave Generator sketch simplified somewhat as I iterated on the project. The result is a mercifully compact program, thanks to the integration of the LTC1799 and the *FreqCount.h* library (available from the Library Manager section of the Arduino IDE).

```
/* Square Wave Generator Sketch. Gives a proper reading with multiplier.
 * Parts of this sketch are derived from Paul Stoffregen's public domain
 * example code.
 */
#include <FreqCount.h>
#include <LiquidCrystal_I2C.h>
#include <Wire.h>

unsigned long freq = 0;
float impulse;

LiquidCrystal_I2C lcd(0x27,20,4);
```

```
void setup() {
  lcd.init();
  lcd.backlight();
  FreqCount.begin(1000);
}

void loop() {
  if(FreqCount.available()) {
    freq = FreqCount.read();
    lcd.clear();
    lcd.setCursor(1,0);
    lcd.print(" Frequency");
    lcd.setCursor(0, 1);
    lcd.print(freq);
    lcd.setCursor(10, 1);
    lcd.print("Hz");
    lcd.setCursor(0, 2);
    lcd.print("Impulse");
    lcd.setCursor(0,3);

    impulse = ((1/(float)freq)* 1000000);
    lcd.print(impulse);
    lcd.print("   uS        ");
  }
}
```

The sketch starts by including three libraries: *FreqCount.h* for the frequency counter and two others for working with the LCD. To add *FreqCount.h* to your Arduino IDE, go to **Sketch ▸ Include Library ▸ Manage Libraries...** and install the FreqCount library from the Library Manager. The `setup()` section prepares the LCD and starts the frequency counter. The `loop()` section fetches the frequency, calculates the impulse width, and displays both.

The Arduino doesn't actually have anything to do with generating the signal—that's all done at the oscillator and subsequently in the divider. The Arduino's function is to look at the signal and read out the frequency.

NOTE *Although pin 5 on the Arduino is connected to the LTC1799 oscillator, I do not set it as an analog input in this sketch. That is apparently taken care of in the FreqCount library. See "Frequency Input Pin" at* https://www.pjrc.com/teensy/td_libs_FreqCount.html *for various Arduino models. Incidentally, using this library renders analog pins 3, 9, 10, and 11 unusable as analog outputs (PWM).*

The Shield

For this project, I developed a small PCB shield to hold the various components. Although the shield could probably have been designed to use only a single layer, I elected to use a two-layer board. First, it greatly reduced layout time, and second, because I was producing another two-layer board at the same time, I could expose and etch them both at once with little additional effort. (It's more efficient to etch multiple boards simultaneously, when you can.) Figure 9-13 shows the top and bottom foil patterns of the shield.

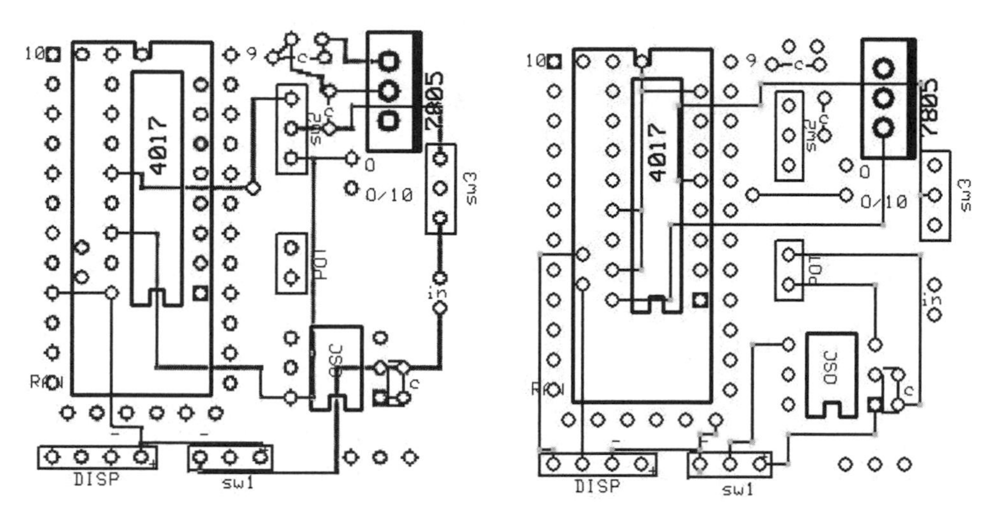

Figure 9-13: Top (left) and bottom (right) foil pattern for the Square-Wave Generator's PCB shield

Assembling the components on the shield requires special attention to the bypass capacitor. The shorter the leads of the 0.1 µF capacitor from pin 1 of the LTC1799 oscillator to ground, the better the oscillator works. I actually soldered the capacitor directly to the chip-mounting board.

Figure 9-13 shows the component placement on the shield. Notice that the HCT4017 (CD 4017)is located beneath the Pro Mini; the PCB was designed this way to conserve space and hold high-frequency traces to a minimum length.

As in other projects, it's necessary to populate the headers for the Pro Mini only where they actually connect to the board, in addition to a header at pin 1 to simplify aligning the Pro Mini on the shield. The LM7805 voltage regulator requires no heat sink.

Construction

Building the Square-Wave Generator was relatively straightforward, but note that I did not take particular care with wiring the leads for switch SW1, which provided the divider for the oscillator, or with the placing and wiring of the potentiometer. Shortening these wires—and perhaps adjusting the placement of the parts themselves—probably would have improved the integrity of the waveform somewhat at higher frequencies. Figure 9-14 shows the inside of my Square-Wave Generator; if you look carefully, you can see my hand-scribbled notations as to where things are located.

Figure 9-14: Everything fit easily in the slope-panel enclosure. The LCD was held in place with four screws, and the shield was mounted on top of the LCD with double-sided foam adhesive. The switches, I/O jacks, and potentiometer were soldered by hand.

Preparing the Enclosure

First, mark the front of the enclosure for the following holes:

- Two to aid in cutting out the space for the LCD
- Four for mounting holes for the LCD
- Four for the banana jacks
- Four for the switches
- One for the potentiometer

If you're using the Hammond 1595C sloped-front enclosure I recommend in "Parts List" on page 260 or an equivalent, you can follow the template in Figure 9-15. Just locate the PDF of the drawing in this book's resource files, print it out, lay it over the front of the enclosure, and carefully center punch for the holes. I also use a fine-tip Sharpie marker to indicate locations and to draw on the enclosure. Excess marker can be easily cleaned with isopropyl alcohol.

Figure 9-16 shows how to use the radius of the enclosure to determine the measurement for the display.

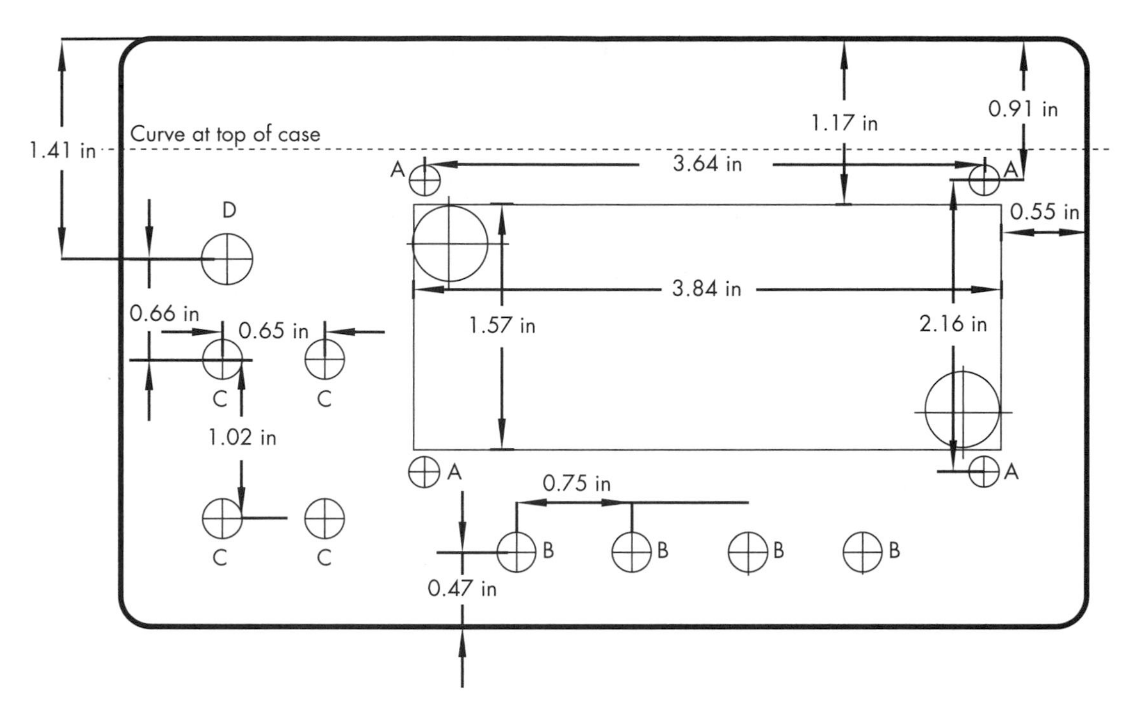

Figure 9-15: The drilling template for the Square-Wave Generator

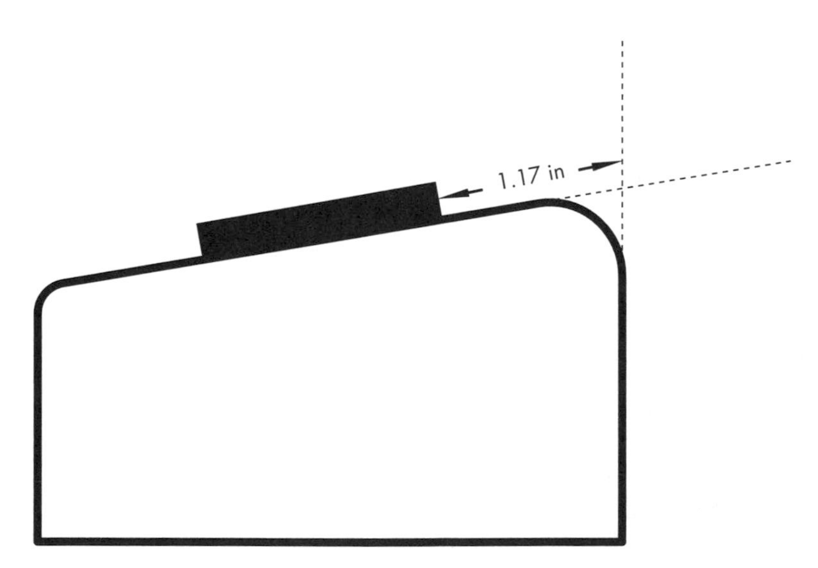

Figure 9-16: The LCD is placed using the radius of the enclosure.

I drilled the holes in the following order:

1. Carefully drill 1/2-inch holes at the corners of the cutout for the LCD. If you mark the centers of the holes correctly, the edge of the display will be tangential to the outer diameter of the hole. You can then draw lines connecting the edges of the holes to use as a guide to cut out the display.
2. Cut out the display using a keyhole saw or saber saw. The enclosure is made of a relatively soft ABS plastic, so you should have no difficulty making the cut.
3. Clean debris from the cutout with a file if necessary, and check that the display fits.
4. Drill the four mounting holes (labeled A in Figure 9-15) for the display.
5. Drill four 5/16-inch holes for the banana jacks (labeled B).
6. Drill four 1/4-inch holes for the switches (labeled C). It's a good idea to identify the switch locations with a permanent marker on the inside of the enclosure to simplify wiring them.
7. Drill a 9/32-inch hole for the potentiometer (labeled D).
8. Locate a position you like for the 3.5 mm power jack that is 3/4 inches from the edge of the enclosure and 1/2 inches from the bottom. Drill a 1/4-inch hole for a 3.5 mm jack there.
9. Mount the LCD with four 4-40 screws. Some LCDs have a protrusion on one side for the backlight. If yours does, you will have to raise the LCD off the surface of the enclosure to accommodate the backlight section on the right-hand edge (when looking at the LCD with the connections at the top). I simply put 4-40 nuts on the back of the screws to leave space. If a single nut is not enough (4-40 nuts can have different thicknesses), include a washer. Then, fasten the display to the enclosure. If your display does not include the protrusion, then just fasten the display to the case.
10. Mount the banana plug jacks and switches in the enclosure. It is sometimes easier to solder wires onto the switches and jacks to minimize damage to the enclosure from accidental contact with the soldering iron.

Wiring the Electronics

Before mounting the PCB shield in the enclosure, solder the components to it, and solder all the wires for the LCD, the potentiometer, the switches, the power jack, and the banana jacks. You may find it helpful to solder the wires to the switches and jacks first. When in doubt, leave some extra length on the wires, but abide by the axiom, "If it's too short, you can always splice it; if it's too long, you won't know what to do with it." I suggest using male and female headers for the LCD to make hookup easier.

I mounted the shield directly to the rear of the LCD using double-sided foam tape. This, however, could have contributed to the distortion of the waveform at higher frequencies. You might prefer to mount the module as far from the display as possible. The front of the enclosure needs no special treatment other than the placement of the labels at your discretion.

Design Notes and Mods

I toyed with several iterations of the Square-Wave Generator before arriving at the version described in this chapter. Along the way, I tweaked some aspects and considered other changes. While the ideas in this section didn't make it into this project, you may enjoy trying them yourself.

Displaying Frequency in Other Units

The sketch displays the square-wave frequency in Hz, as most of the applications I have planned are in the area of 100 Hz to 10 kHz. But if you find yourself using the device a lot in higher frequencies, looking at six or seven integers can be confusing. Never fear: it's easily possible to change the sketch to show the frequency in kHz or even MHz by simply truncating the display.

To truncate the display, simply add a comma and the number of digits you want it to show. For example, to change from Hz to kHz, change these lines in the sketch:

```
lcd.print(freq);
lcd.setCursor(10, 1);
lcd.print("Hz");
```

to these lines:

```
lcd.print(freq/1000);
lcd.setCursor(10, 1);
lcd.print("kHz");
```

If you want to reduce the number of digits appearing in the readout, change this line:

```
lcd.print(freq/1000);
```

to this:

```
lcd.print(freq/1000,3);
```

You can change `3` to the number of digits you want.

Reading External Input Frequencies

The schematic and the finished project include a switch to change from generating pulses to reading the frequency of an external input. The switch (SW3) brings the input jack directly to the input (pin 5) of the Arduino, where the oscillator would normally connect. I have used this very successfully for a variety of applications, particularly when I wanted a quick frequency reading somewhere outside my shop.

There is no circuitry to protect the processor, so just be careful. The unit is meant to use inputs that are standard TTL levels—0 to 5V. The Arduino is relatively sensitive and can detect signals at somewhat lower levels. If you plan, however, to use it with very low-level inputs—that is, less than 0.5V—then you should build some kind of prescaler or preamp circuit.

If you plan to use the Square-Wave Generator as an independent frequency counter, you might want to consider using a preamp to provide the amplification and prevent damage to the processor. A simple one appears in Figure 9-17, using one-sixth of a 74HC14 Hex Schmitt-trigger inverter.

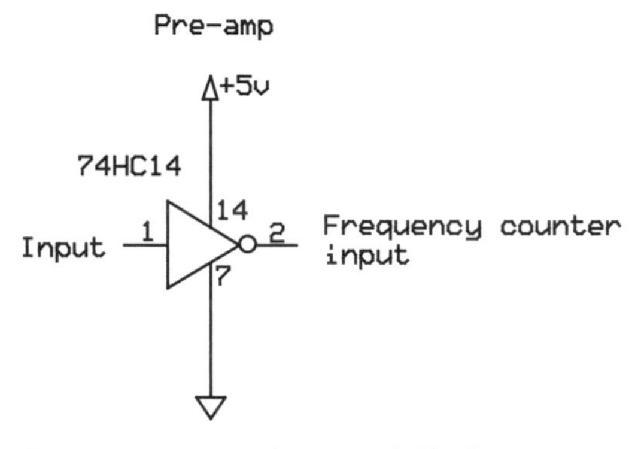

Figure 9-17: Optional preamp/buffer for input to frequency counter

Using a preamp will help protect the input of the Arduino because the output of the preamp will be limited to the supply voltage.

Battery Power

While this project was initially designed for use with an external power supply, it can be easily converted to battery power. The total current drain with the LCD backlight lit is just under 100 mA. The capacity of a zinc manganese battery is approximately 500 mA hours. Thus, you can expect a life of about 5 hours. Alkaline batteries will tend to do better.

To accommodate the battery power, I simply replaced the power switch with one that had a center-off position. I wired one outer terminal to the AC-based power jack, the other outer terminal to the positive terminal of the battery, and the center terminal to the positive rail on the PCB shield. The negative terminal of the battery goes to ground. That configuration is shown in the current schematic. If you use the enclosure I suggest, a battery should fit conveniently. You can use double-sided adhesive to attach a battery holder.

10

THE CHROMATIC THERMOMETER

This project was initially created to provide a quick visual indication of local temperature. At its simplest, it is a thermometer that displays the temperature by turning on a sequence of LEDs of different colors. During development, however, the project gained more features, including an LCD readout to supplement the basic color readout. And while experimenting, I came across an IC that provides extremely accurate measurement without special calibration, which improved the device greatly.

The Chromatic Thermometer includes 10 different colored LEDs, each of which lights up when the sensor detects a particular temperature. The original version was designed to measure from 68 to 78°F, and each LED represented a 1-degree Fahrenheit change in temperature. I subsequently varied that for different applications. The finished project shown in Figure 10-1 can measure a wide range of temperatures. It also includes a waterproof probe for measuring the temperature in liquids, which is useful for fish tanks, swimming pools, and so forth, and can be constructed in a variety of physical configurations. In the version that appears in the sketch, the temperature ranges from 76 to 86°F in 1-degree increments.

Figure 10-1: The finished Chromatic Thermometer

You could also program an alarm by flashing a lamp at a specific temperature, or with a minor hardware addition, you could add an audible alarm. I am sure you can think of even more hardware or software modifications to make this thermometer a very practical device.

Choosing a Temperature Sensor

The key ingredient in any electronic thermometer is the temperature sensor. There are many kinds of temperature sensors available to choose from.

Thermistors change resistance with temperature and range from inexpensive to very pricey, depending on how they are made and tested. *Resistance temperature detectors (RTDs)* employ a coil of pure wire, such as silver, platinum, or copper, wrapped around a glass core. Combined in a resistance bridge, RTDs can be extremely accurate, but they're somewhat expensive. *Thermocouples* are still an industry-standard sensing technology, particularly at higher temperatures (that is, greater than 500°C). At lower temperatures, thermocouples are being replaced by RTDs because of accuracy, precision, consistency, and linearity.

Semiconductor temperature sensors—that is, dedicated integrated silicon sensor circuits—continue to gain popularity because of their accuracy, precision, ruggedness, and convenience. In preparing for this project, I looked at virtually all the approaches and elected to use a semiconductor sensor, as it provided sufficient accuracy with a relatively simple hookup and a modest price.

With any thermometer, *accuracy* and *precision* are issues. Consider accuracy the ability to measure temperature as close to some standard value established by the NIST, with some deviation. For this casual definition, precision can be referred to as *repeatability*—that is, the ability to read the same temperature consistently in the same environment.

The Custom pH Meter in Chapter 7 used the temperature of boiling water and ice in solution to set boundaries of 100°C and 0°C, respectively, to calibrate a thermometer. I checked the Chromatic Thermometer with the same approach, but I used the high-accuracy MCP9808 module described here as a standard because it was extremely close.

Accuracy and precision are ultimately a system—not necessarily a sensor—issue. This project discusses two different sensors with essentially the same accuracy and precision. The simplest sensor, an LM35 analog temperature sensor, depends on other parts of the system for its accuracy and precision. The second sensor, an MCP9808 IC, provides accurate results with or without the associated breakout board because it includes the other variable components as an on-chip subsystem.

Both the LM35 and the MCP9808 boast a maximum accuracy of 0.25°C and a precision of 0.0625°C. To achieve this kind of accuracy, they use a *silicon band-gap temperature sensor*, which takes advantage of the forward voltage of a silicon diode. However, in addition to the sensor, the MCP9808 includes its own on-chip ADC, voltage reference, and other internal circuitry to assure accuracy.

NOTE *If you want to dig deeper on the band-gap sensor technology, there is a wealth of information on the web, including background on Bob Widler, who is largely credited with discovering the phenomenon.*

In this project, the LM35 depends on the ADC and voltage reference of the Arduino, which, while pretty good, does not match that of the MCP9808's monolithic system and thus can require calibration, as illustrated in "Sketch for the LM35 System" on page 289. When I built a version with the MCP9808 instead, I used an Adafruit breakout board for the chip because it significantly simplified assembly—no need to fuss with the MicroSMT package's tiny leads.

While the MCP9808 does cost more than the LM35D, it's worked out well. I included traces in the project's PCB shield for the chip itself, if you elect to solder it using one of the techniques suggested in "Using SOICs" on page 20.

Required Tools

Soldering iron and solder

(Optional) Electric drill with bits for 1/4 inches (used to drill a hole for the jack in order to connect a remote temperature sensor or to make a hole for a power adapter)

Parts List

First, decide which temperature sensor you want to use: the LM35 or the MCP9808. If you just want to use an LM35, here's what you'll need to make a basic Chromatic Thermometer:

One Arduino Nano or clone

One LM35 temperature sensor

Ten different colored LEDs (see "Mod: Try Different LEDs" on page 300)

Ten ZTX649 transistors

Ten 470-ohm resistors

One 7.5 to 9V wall adapter, or equivalent (or 9V battery)

One plastic enclosure (see "Construction" on page 298)

28- or 30-gauge hookup wire

One printed circuit board (Use the provided shield template, design your own shield, or use any other prototyping board you feel comfortable working with.)

I also describe several variations on the Chromatic Thermometer in this chapter. Give the project a skim before you go shopping, and if you want to make one of the variations, also buy the following:

If you plan to use the temperature sensor as described in "Design Decision: Remote the Temperature Sensor," get one 3.5 mm stereo jack.

For a Chromatic Thermometer with a digital readout, also buy one 16×2 I^2C LCD.

For a high-accuracy Chromatic Thermometer, replace the LM35 with either one MCP9808 Adafruit Breakout Board or one Microchip MCP9808 IC with a 100 nF capacitor and two 10-kilohm resistors.

For the breadboard prototype, make sure you have one large breadboard (as opposed to the smaller ones used in much of this book) and at least 30 jumper wires.

DESIGN DECISION: REMOTE THE TEMPERATURE SENSOR

If your application takes you in another direction, you can modify the shield to *remote the chip*. That is, you can connect long wires directly to the chip (you'd need only four wires) and place the chip in a location separate from the readout. If you remote the chip, just include a small capacitor (around 100 nF) between pins 4 and 8, very close to the chip, as shown in Figure 10-2.

Figure 10-2: The MCP9808 with wires soldered directly

However, remember that I^2C stands for *inter-integrated circuit* and is intended for chip-to-chip communications. Therefore, the MCP9808 can be moved only a limited distance from the Arduino. While some hobbyists online claim success with wires as long as 100 cm, the longest that I have been able to do reliably is about 50 cm. The LM35, on the other hand, can be remoted and made waterproof for longer distances with only three wires and without needing miniature hands and the dexterity of a watchmaker. (If you can tolerate only two wires, there's a solution to that; check the data sheet for the LM35.)

The MCP9808 could likely be encapsulated as I did with the LM35 in the Custom pH Meter from Chapter 7, though I have not tried that. Trying to insulate the connections from each other and keep the delicate pins of the chip from breaking off can be a problem when making a remote sensor with the MCP9808.

This is a design decision to make before you build the final Chromatic Thermometer, so I suggest reading through the full chapter to decide before you put the device together.

Downloads

Sketch for the LM35 version *LM35Thermo.ino*

Sketch for the MCP9808 version *9808Thermo.ino*

***Adafruit_MCP9808* library** *https://www.adafruit.com/product/1782* (for the MCP9808 temperature sensor only)

PCB pattern for the shield *Thermo.pcb*

How the Chromatic Thermometer Works

In operation, the Chromatic Thermometer is quite straightforward. For starters, let's look at the basic configuration using the LM35 sensor.

The sensor generates a voltage of 10mV/°C. For example, at 28°C (around 82°F), the chip outputs 0.280V. You can easily check that with your multimeter. To make a usable Arduino thermometer, all you have to do is change that voltage to something the Arduino can understand and then have the Arduino translate it to something you want to see on the LEDs or LCD.

The first step is to convert the analog voltage to a digital value so the Arduino can work with it. To do that, connect the output of the sensor to one of the Nano's analog inputs. (I tend to use A0, but any analog input can be used. Just don't use analog pins A4 or A5, which are used for the I^2C portion of this project.) The output voltage of the LM35 is pretty low compared with the 5V the Nano is working with, and the ADC divides the 5V of the supply into 1,024 parts (range = 0 to 1,023) to determine the analog value of an incoming voltage. If you use the LM35 output as is, each degree Celsius change in temperature will change the output voltage by 0.010V and therefore change the result of the ADC by 2.046 parts (units) out of the 1,024 total.

That works, but small increments of the ADC at the very low end of the reference voltage are subject to random amounts of error. There is also significant error from rounding, as the Nano's ADC outputs only whole digits.

To reduce the impact of error, you'll change the reference voltage of the ADC from 5V, where each increment in the 1,024 represents 0.004882V, to 1.1V, where each increment represents only 0.00107V. A single degree Celsius change will then represent only 9.345 of the 1,024 units. Thus, the 0.280V the LM35 outputs at 28°C will correspond to about 261 of the 1,024 units, rather than only 52.

The Schematic

Figure 10-3 shows the schematic for this project.

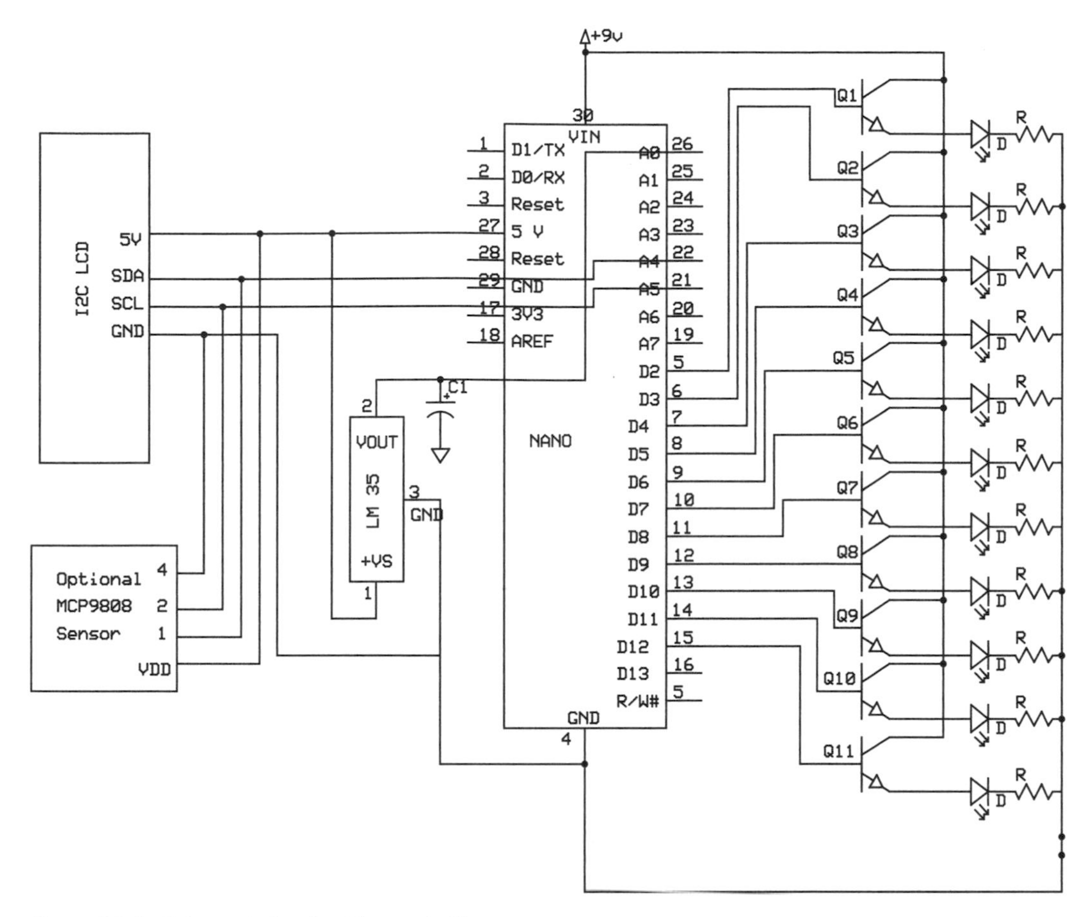

Figure 10-3: Schematic for the Chromatic Thermometer

Note that both the LM35 and the MCP9808 sensors are wired up in this sketch. That is not a problem, as you can select which one to use in software by changing the code that you upload to the Arduino. You can wire up either one or both. This schematic also shows the Chromatic Thermometer with an LCD, though that is optional if you just want to read the temperature based on the LEDs.

The Arduino processor could probably drive the LEDs unaided, but I elected to use transistor drivers for each LED. This assures that if you elect to use higher-output LEDs—or even incandescent lamps—there will be no problem driving them. The transistors used are capable of sinking as much as 1 A.

An 11th LED-transistor-resistor group (Q11) is shown connected to pin D12 on the Nano, though the final Chromatic Thermometer uses only 10 LEDs. I show this extra pair and even include it in the shield PCB file to give you a built-in customization option. You can add another temperature digit, a buzzer alarm, or any other output you like.

The Breadboard

As in the other projects in this book, I suggest starting with a breadboard to sound out the design and exercise the sketch before committing to the final assembly. Because the project uses 10 LEDs and 10 driver transistors, I used a large-format breadboard to comfortably fit all the components (see Figure 10-4).

Figure 10-4: The Chromatic Thermometer's completed breadboard

The LEDs are along the middle of the breadboard, and one is shown lit. I assembled a wire harness for the LCD. The LM35 temperature sensor (left) is on a tether and held in heat shrink tubing so it is waterproof.

These are the steps I used to assemble the breadboard:

1. Place the Arduino Nano toward the top left of the breadboard.
2. Connect VIN (pin 30) of the Nano to where the 9V input from the battery or other power source will go.
3. Connect the 5V pin of the Nano to the red positive rail.

4. Connect GND (pin 4) of the Nano to the blue negative rail on the breadboard.
5. Identify locations for the 10 driver transistors (Q1 to Q10), and place them on the breadboard. I placed them so the beveled edge of the transistor faced right when looking from the bottom of the board.
6. Connect the collector of all the transistors to where the 9V input will go (VIN of the Nano). See Figure 10-5 for the pinout of the transistor.
7. Connect the base of Q1 to D2 (pin 5) on the Nano.
8. Connect the base of Q2 to D3 (pin 6) on the Nano.
9. Connect the base of Q3 to D4 (pin 7) on the Nano.
10. Connect the base of Q4 to D5 (pin 8) on the Nano.
11. Connect the base of Q5 to D6 (pin 9) on the Nano.
12. Connect the base of Q6 to D7 (pin 10) on the Nano.
13. Connect the base of Q7 to D8 (pin 11) on the Nano.
14. Connect the base of Q8 to D9 (pin 12) on the Nano.
15. Connect the base of Q9 to D10 (pin 13) on the Nano.
16. Connect the base of Q10 to D11 (pin 14) on the Nano.
17. Connect each of the emitters for all the transistors Q1 through Q10 to the positive lead of each colored LED.
18. Connect the negative lead of the LEDs to an empty location on the breadboard.
19. Connect one side of the 470-ohm resistors to the negative leads of each LED.
20. Connect the other side of the 470-ohm resistors to ground (the blue negative rail).
21. Connect pin 1 (4–20V) of the LM35 to the red positive rail. (See Figure 10-6 for the pinout of the LM35.)
22. Connect pin 3 (GND) of the LM35 to the blue negative rail.
23. Connect the output pin (center) of the LM35 to pin A0 (26) on the Nano.

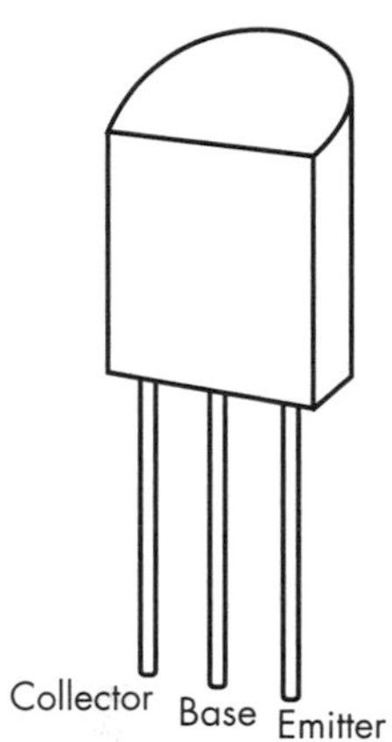

Figure 10-5: Pinout of the ZTX649 transistor

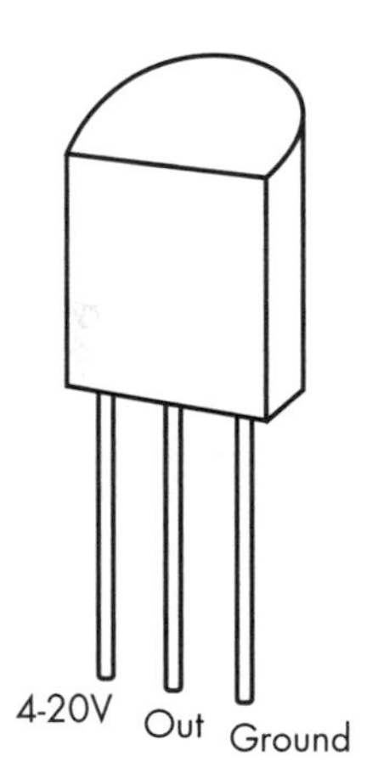

Figure 10-6: Pinout of the LM35 temperature sensor

24. Connect power and ground to the respective red positive and blue negative rails.

Now, you're all set to go with the most basic configuration. To add the digital display:

1. Connect the power and ground of the LCD's I^2C adapter to the respective red positive and blue negative rails.
2. Connect the SDA pin of the LCD's I^2C adapter to A4 (22) on the Nano.
3. Connect the SCL pin of the LCD's I^2C adapter to A5 (21) of the Nano.
4. Upload the appropriate software code to the Nano.

If you want to play with the high-accuracy temperature sensor:

1. Remove the LM35 from the circuit (or you can leave it in—it will not affect anything).
2. Place the high-accuracy sensor in a location where the pins will not be affected by anything.
3. Connect the power pins of the I^2C connection on the MCP9808 sensor to the blue negative rail and 5V—pin 27, or the red positive rail—on the Nano, respectively.
4. Connect the SDA pin of the I^2C connection on the MCP9808 sensor to A4 (22) on the Nano.
5. Connect the SCL pin of the I^2C connection on the MCP9808 sensor to A5 (21) on the Nano.
6. Upload the software to the Nano with the high-accuracy sensor version of code.

Now you're all set to go.

NOTE *This particular breadboard's positive and negative power rails are not continuous. The first 15 are connected, the next 20 are connected, and the last 15 are connected—but not to each other. I used jumpers to connect the rails as needed.*

The Sketches

There are two versions of the sketch: one for the LM35 version of this project and one for the MCP9808 version. You can pick a sketch to use based on which version of the Chromatic Thermometer you'd like to build, but both sketches comprise three basic sections.

After the boilerplate section of loading libraries and preliminary setup, the first component of each sketch deals with the temperature sensor itself. The second section deals with setting up the temperature and LCD readout (if used). The final element of each sketch details the conditions to turn the LEDs on and off to indicate the temperature. I have annotated the code with comments where appropriate.

Sketch for the LM35 System

Here is the sketch for the Chromatic Thermometer with the LM35 sensor, including the digital LCD readout.

```
//Chromatic Thermometer sketch for the LM35, with an alarm
//around 78 to 79 degrees Fahrenheit (26 degrees Celsius)

#include <Wire.h>                      //Set up Comm wire library
#include <LiquidCrystal_I2C.h>         //Set up LCD library with I2C

LiquidCrystal_I2C lcd(0x27, 16, 2); //16x2 display I2C address 0x27

//Establish the number of readings to average at 10
❶ const int numReadings = 10;
float readings[numReadings];           //The readings from the analog input
int index = 0;                         //The index of the current reading

float total = 0;                       //The running total
float average = 0;                     //The average
int tempPin = A0;
float TempC;
float tempF;

void setup() {
  lcd.init();                          //Initialize LCD and turn on backlight
  lcd.backlight();

  //Initialize serial communication with computer:
  Serial.begin(9600);

  //Initialize all the readings to 0:
  for(int thisReading = 0; thisReading < numReadings; thisReading++)
    readings[thisReading] = 0;

  analogReference(INTERNAL);           //Set voltageReference = 1.1V
}

void loop() {
  total = total - readings[index];          //Subtract the last reading
  readings[index] = analogRead(tempPin);    //Read from the sensor
  total = total + readings[index];          //Add the reading to the total

  //Advance to the next position in the array:
  index = index + 1;

  if(index >= numReadings)         //If we're at the end of the array...
    index = 0;                      //...wrap around to the beginning

  average = total / numReadings;  //Calculate the average temperature
  TempC = average / 9.81;         //Adjust/calibrate average
                                  //based on empirical measurement
  tempF = (TempC * 9 / 5) + 32;   //Convert to Fahrenheit
```

```
  //Set up serial readout if desired
  Serial.println();
  Serial.print("TempC        ");
  Serial.println(TempC);
  Serial.print("TempF       ");
  Serial.println(tempF);
  Serial.print("Average       ");
  Serial.println(average);
  delay(10);                //Delay in between reads for stability

  //Set up for LCD
  lcd.setCursor(0, 0);
  lcd.print("Temp            ");
  lcd.setCursor(7, 0);
  lcd.print(TempC, 1);
  lcd.print((char)223);   //Degree symbol see below
  lcd.print(" C");
  lcd.setCursor(0, 1);
  lcd.print("Temp    ");
  lcd.setCursor(7, 1);
  lcd.print(tempF, 1);   //Truncate second decimal place
  lcd.print((char)223);
/*May need to use (char) 178 if LCD displays the greek alpha character.
Different LCDs display different special characters*/

  lcd.print(" F");
  delay(50);

  //Beginning of conditional statements for display
  if((tempF > 85.00 ) && (tempF < 200)) {
    //85 degrees and 200 degrees are arbitrary.
    digitalWrite(2, HIGH);
  }
  else {
    digitalWrite(2, LOW);
  }

  if((tempF > 84) && (tempF < 85)) {
    digitalWrite(3, HIGH);
  }
  else {
    digitalWrite(3, LOW);
  }

  if((tempF > 83.00 ) && (tempF < 84)) {
    digitalWrite(4, HIGH);
  }
  else {
    digitalWrite(4, LOW);
  }

  if((tempF > 82) && (tempF < 83)) {
    digitalWrite(5, HIGH);
  }
```

```
  else {
    digitalWrite(5, LOW);
  }

  if((tempF > 81.00 ) && (tempF < 82)) {
    digitalWrite(6, HIGH);
  }
  else {
    digitalWrite(6, LOW);
  }

  if((tempF > 80) && (tempF < 81)) {
    digitalWrite(7, HIGH);
  }
  else {
    digitalWrite(7, LOW);
  }

  if((tempF > 79.00 ) && (tempF < 80)) {
    digitalWrite(8, HIGH);
  }
  else {
    digitalWrite(8, LOW);
  }

  if((tempF > 78) && (tempF < 79)) {
    //Code for blinking LED as an alarm
    digitalWrite(9, HIGH);
    delay(50);
    digitalWrite(9, LOW);
    delay(50);
  }
  else {
    digitalWrite(9, LOW);
  }

  if((tempF > 77.00 ) && (tempF < 78)) {
    digitalWrite(10, HIGH);
  }
  else {
    digitalWrite(10, LOW);
  }

  if(tempF < 76) {
    digitalWrite(11, HIGH);
  }
  else {
    digitalWrite(11, LOW);
  }
}
```

After including libraries, the LM35 sketch defines the number of samples to keep track of at ❶. The higher the number, the more the readings will be smoothed, but the longer it will take to settle. Using a constant rather than a normal variable for the number of samples allows this value to determine the size of the `readings` array.

The `setup()` code initializes the LCD, turns on serial communication for debugging, initializes the readings array with all zeros (because the sketch hasn't read anything yet), and sets the ADC reference voltage to 1.1V. The `loop()` code stores sensor data in the `readings` array and averages the readings to calculate a temperature to display. If you have an LCD, the sketch shows the temperature on it, and then it checks various temperature ranges with `if` statements to see which LEDs to turn on.

The big difference between this sketch and the next is that the LM35 version sets up the system to accept the analog voltage from the sensor and direct it to an analog input. It also establishes a reference voltage of 1.1V with the `analogReference(INTERNAL)` command.

Sketch for the MCP9808 System

The sketch for the version with the high-accuracy MCP9808 chip (or the board from Adafruit) uses much of the same code; the conditional statements that turn on the LEDs are identical. The only part that differs is how the Arduino gets the temperature information from the sensor. Here is the full sketch:

```
/*Sketch for MCP9808 version of Chromatic Thermometer

 This version of the Chromatic Thermometer uses the Adafruit MCP9808 breakout
 Board (or Microchip MCP9808 chip) and the Adafruit_MCP9808 library
 available from Adafruit ----> https://www.adafruit.com/product/1782

 The averaging may not be necessary. It tends to slow the
 response a little, but smoothes out the display.

 The result is truncated to a single decimal for both Celsius and Fahrenheit.
 This version includes the code to turn on and off 10 LEDs.
*/

#include <Wire.h>                     //Set up wire/serial library
#include "Adafruit_MCP9808.h"         //Set up MCP9808 library
#include <LiquidCrystal_I2C.h>        //Set up library for LCD

LiquidCrystal_I2C lcd(0x27, 16, 2); //16x2 display; define LCD

// Create the MCP9808 temperature sensor object
Adafruit_MCP9808 tempsensor = Adafruit_MCP9808();

const int numReadings = 10;         //Once again averaging 10 readings
float readings[numReadings];        //The readings from the analog input
int index = 0;                      //The index of the current reading
float TempC = 0;
float tempF = 0;
float total = 0;                    //The running total
float average = 0;                  //The average
```

```
void setup() {
  Serial.begin(9600);

  lcd.init();  //Initiate LCD and backlight
  lcd.backlight();


/*Make sure the sensor is found. You can also pass a different I2C address
from the default, as in tempsensor.begin(0x19).*/
  if(!tempsensor.begin()) {
    Serial.println("Couldn't find MCP9808!");
    while(1);
  }
}
void loop() {
  //Read and print out the temperature; then convert to Fahrenheit
  float c = tempsensor.readTempC();
  total = total - readings[index];
  //Read from the sensor
  readings[index] = c;
  //Add the reading to the total
  total = total + readings[index];
  //Advance to the next position in the array
  index = index + 1;
  if(index >= numReadings)       //If we're at the end of the array...
    index = 0;                    //...wrap around to the beginning:
  //Calculate the average
  average = total / numReadings;

  TempC = average;
  tempF = (TempC * 9 / 5) + 32;
  delay(100);

  //Set up LCD to print out temperatures
  lcd.setCursor(0, 0);
  lcd.print("Temp           ");
  lcd.setCursor(6, 0);
  lcd.print(TempC, 1);          //Truncate to one decimal place
  lcd.print((char)223);
  lcd.print(" C");
  lcd.setCursor(0, 1);
  lcd.print("Temp   ");
  lcd.setCursor(6, 1);
  lcd.print(tempF, 1);          //Truncate to one decimal place
  lcd.print((char)223);
/*May need to use (char) 178 if LCD displays the greek alpha character.
Different LCDs display different special characters*/

lcd.print(" F");
  delay(100);

  //Beginning of conditional statements for display
  if((tempF > 85.00 ) && (tempF < 100)) {
    digitalWrite(2, HIGH);
  }
```

```
else {
  digitalWrite(2, LOW);
}

if((tempF > 84) && (tempF < 85)) {
  digitalWrite(3, HIGH);
}
else {
  digitalWrite(3, LOW);
}

if((tempF > 83.00 ) && (tempF < 84)) {
  digitalWrite(4, HIGH);
}
else {
  digitalWrite(4, LOW);
}

if((tempF > 82) && (tempF < 83)) {
  digitalWrite(5, HIGH);
}
else {
  digitalWrite(5, LOW);
}

if((tempF > 81.00 ) && (tempF < 82)) {
  digitalWrite(6, HIGH);
}
else {
  digitalWrite(6, LOW);
}

if((tempF > 80) && (tempF < 81)) {
  digitalWrite(7, HIGH);
}
else {
  digitalWrite(7, LOW);
}

if((tempF > 79.00 ) && (tempF < 80)) {
  digitalWrite(8, HIGH);
}
else {
  digitalWrite(8, LOW);
}

if((tempF > 78) && (tempF < 79)) {       //Code for blinking LED
  digitalWrite(9, HIGH);
  delay(50);
  digitalWrite(9, LOW);
  delay(50);
}
else {
  digitalWrite(9, LOW);
}
```

```
  if((tempF > 77.00 ) && (tempF < 78)) {
    digitalWrite(10, HIGH);
  }
  else {
    digitalWrite(10, LOW);
  }

  if(tempF < 76) {
    digitalWrite(11, HIGH);
  }
  else {
    digitalWrite(11, LOW);
  }
}
```

The MCP9808 sketch's version of the `setup()` code checks to make sure you have the MCP9808 temperature sensor plugged in. In the `loop()` section, it calls the `readTempC()` function from the *Adafruit_MCP9808* library to fetch the current temperature, instead of calling `analogRead()` directly, as the LM35 code does. Unlike the LM35 code, this sketch doesn't need to set up an external voltage reference: one advantage of using the MCP9808 is that the chip contains its own internal reference. Otherwise, apart from a few differences in variable names, the rest of the sketch is the same as the LM35 code.

How the Temperature Readouts Work

In both sketches, the basic temperature readout comprises a series of 10 different colored LEDs with transistors driven in an emitter-follower configuration from outputs D2 through D11 on the Nano.

The outputs of the Nano are activated by the sketch, and each output corresponds to a conditional statement of the form, "If temperature is between *X* degrees and *Y* degrees, turn on an LED. If not, turn off the LED." The `if` statements for these commands are the same in both the LM35 and the MCP9808 high-accuracy version.

When I first completed the project, I used it in a saltwater fish tank, where I wanted to accurately view the temperature at a glance and at a distance. I set up the LEDs to blink at unacceptable temperature extremes to get my attention so I could take corrective action. The blinking effect required only turning the LED off and on with a delay in between. I modified the sketch to read as follows for the warning condition:

```
  if((tempF > 78) && (tempF < 79)) {
    digitalWrite(9, HIGH);
    delay(50);
    digitalWrite(9, LOW);
    delay(50);
  }
  else {
    digitalWrite(9, LOW);
  }
```

This example blinks the display when, and only when, the temperature is between 78 and 79°F, as indicated in the full sketch. The length of the delay, combined with any further delays you might add, determines the blinking rate.

The general readout system and this simple, silent alarm worked extremely well together. I could set alarms to make sure that the temperature was above or below the threshold temperature.

I also thought it might be valuable to have a digital readout, however, for two reasons:

1. I wanted the option to see exactly what the temperature was.
2. Should the temperature go out of range and the "blink" alarm execute, I wanted to see how far from the limit the temperature was.

Adding the digital display was relatively easy, as I used a standard 16×2 LCD with I^2C interface requiring only four wires. I bought an LCD with a built-in I^2C adapter this time. It was larger than I had hoped for, but I was unable to find a smaller display easily. If you can find one, go for it.

The Shield

The PCB shield for this project was designed with two copper layers rather than a single layer. The extra layer makes fabricating the board a little more difficult, but it saves a lot of effort in identifying and wiring jumpers. I initially etched the board in-house, but I had the finished board produced by Express PCB. The plated through-holes on the professionally finished board made assembly a lot easier. The layers of the PCB are shown in Figure 10-7.

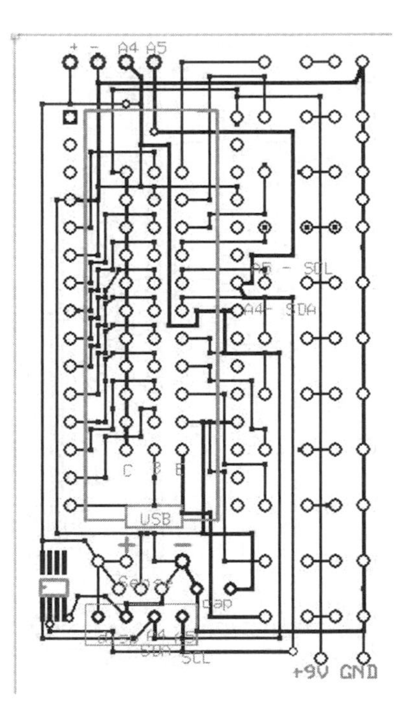

Figure 10-7: The top layer of the PCB shield with silkscreen image. You can see the pattern for the MCP9808 in the bottom left.

Notice that the LED driver transistors are located below the Nano board to save on surface area. The only penalty to this is that the Nano sits a little higher on the board. You can see the transistors in Figure 10-8, which shows the populated PCB with the Nano board next to it.

The shield also includes provisions for two I^2C devices—one at either end of the board. Because I thought this board might find its way into many different projects, I made it as flexible as possible. The I^2C

connections make it easy to hook up both the high-accuracy Adafruit breakout board as well as a digital display. You can either wire them in directly or solder a straight or right-angle female header into the board for use with a crimped-connector housing with cable (see Figure 10-9).

Figure 10-8: Populated shield next to an Arduino Nano. Notice that I used full-length female headers for the Nano. Also note the holes and traces for an 11th LED in the top right.

Figure 10-9: The assembled thermometer using the MCP9808 chip soldered directly to the shield (lower right). One I^2C connector is above and to the right of the MCP9808 chip; connections for another I^2C connector are at the lower left.

The shield includes pads for the MCP9808 chip (installed in Figure 10-9) if you elect to solder the chip directly to the board. If you use the MCP9808 chip rather than the breakout board without another connection to the I^2C interface, you will have to provide 10-kilohm pull-up resistors to the SDA and SCL lines. These are not included in the shield layout but can easily be added to the I^2C port connections since they would be unused. In the implementation in Figure 10-9, the 10-kilohm resistors are on the LCD adapter because there will be an LCD adapter plugged into the I^2C port, eliminating the need for the pull-up resistors.

While the shield files included for this project provide for direct attachment of the LEDs to the PCB, there are many instances where you may want to separate the PCB from the LEDs. For example, I made one version where I spaced out high-intensity, 10 mm LEDs an inch apart on a decorative piece of wood to create a more dramatic look. To do something like that, you'd want to solder long wires to the LED leads and then solder those to the PCB. The cathodes of all the LEDs can be wired together, and only the anodes need be connected to the board individually.

Construction

If you haven't soldered your components to the shield PCB or a prototyping board, do so now, and remember to assemble the LEDs with the correct polarity. The final configuration of the Chromatic Thermometer depends very much on your final application, so I will not go into a lot of specific detail on constructing this project.

For example, if you want to use a remote temperature sensor, you will want a jack or another appropriate connector. You may also want to cut a slit in the enclosure to accommodate a wire for the sensor, as well as another for a long power connection. And if you are going to build the Chromatic Thermometer with only the LEDs and no LCD, you would probably mount the electronics inside the enclosure differently. Similarly, if you build the high-accuracy version, you may need to adjust how you fit your parts into the enclosure. The Chromatic Thermometer shown in Figure 10-10 uses the LM35 sensor with the 16×2 LCD.

I used a small box I found on Amazon (originally sold to hold baseball cards) to enclose the whole thing and mounted the sensor directly to the shield in the three holes to accommodate it in the PCB. If you don't want the LED on the Nano to show, you can unsolder it, or simply cover it with a small piece of black electrical tape.

Whether you use the 3 mm LEDs or the heftier, high-output 5 mm LEDs, using some kind of spacer works well for making sure the LEDs are lined up at the height you like. When I mounted the readout LEDs to the shield, I placed a sliver of unused PCB material to hold all of the LEDs at a uniform height (see Figure 10-11). You could use cardboard just as easily, and if you want a different height, just use a taller or shorter spacer.

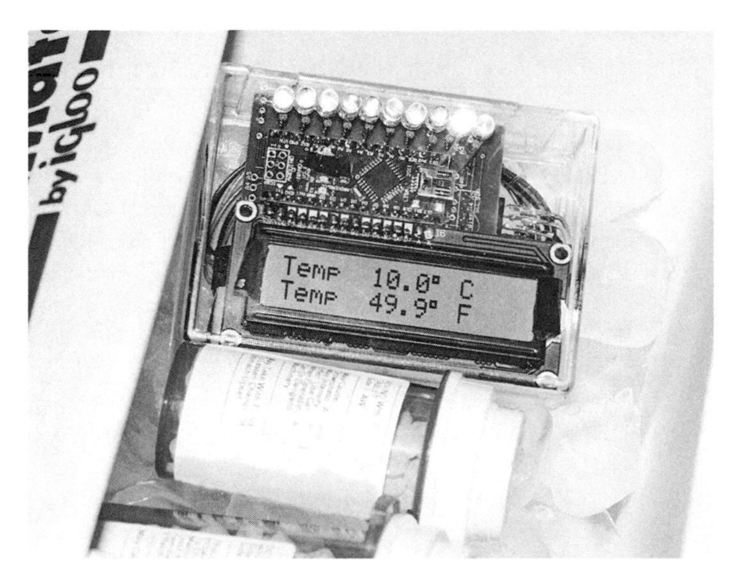

Figure 10-10: The completed Chromatic Thermometer in an acrylic enclosure, monitoring the temperature of medicine in a cooler when traveling. This version uses a 9V battery, which is inside the enclosure. I used a display with a backlight, but the battery will last a lot longer without the backlight. Turn on/off thresholds for LEDs were modified for cold temperature use.

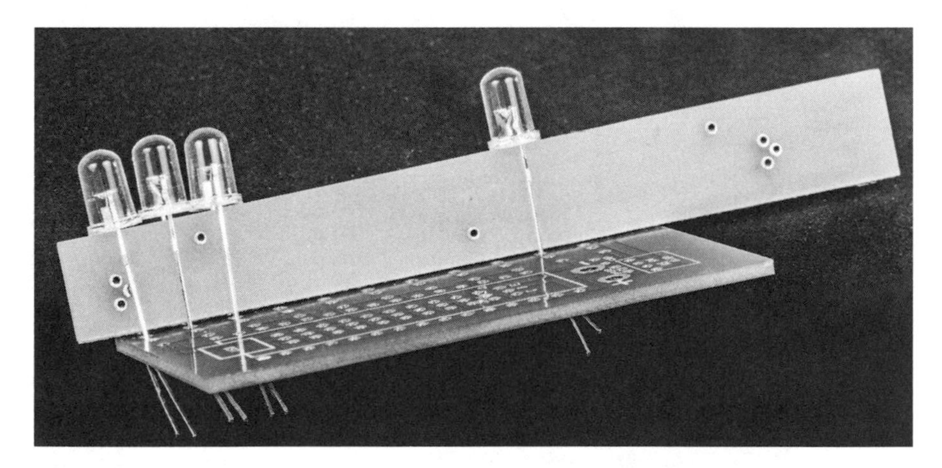

Figure 10-11: I used a discarded strip of PCB material to hold LEDs at uniform height

If you use 5 mm LEDs, you may have difficulty fitting them through the holes in a standard PCB, as some 5 mm LEDs have wider leads. I expanded the holes with a drill to fit the larger LEDs; because there were no connections on the other side of the PCB, the plate-through was not needed.

MOD: TRY DIFFERENT LEDS

The Chromatic Thermometer shown in Figure 10-8 uses 3 mm LEDs. I ordered 3 mm LEDs on eBay in 10 different colors, but the leads were a little short. Subsequently, I ordered 200 5 mm LEDs in a selection of 10 different colors. These also had short leads, but they were brighter and worked well. I did have to file the edges of some of the LEDs where there was a little extra flashing from the mold so they would fit within the 0.200-inch spacing (a little more than 5 mm) on the PCB. However, I could not find enough different-colored high-output 5 mm LEDs.

For one experimental version, I used six different-colored high-output 5 mm LEDs, repeated the colors at the extremes, and went back to the sketch to create blinking patterns to differentiate the similar colors. That version of the Chromatic Thermometer worked well, and the high-output LEDs made it quite noticeable.

To attract even more attention, you can use high-output 10 mm LEDs. They will not fit on the shield PCB shown here, however, as I only spaced the LEDs out by 0.200 inches. That spacing allows for most 3 mm and 5 mm LEDs, but not 10 mm ones. See Figure 10-12 for a size comparison.

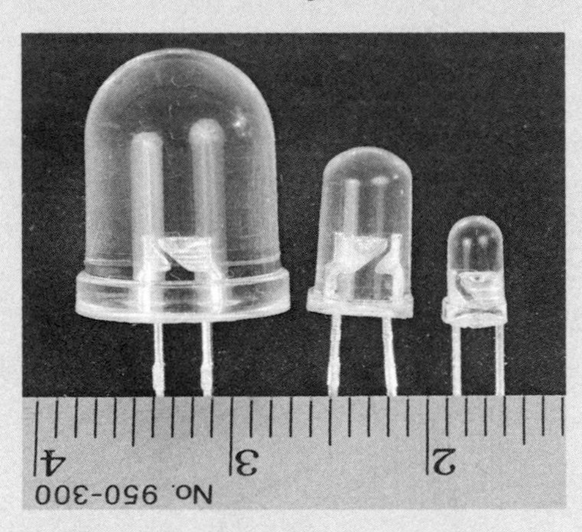

Figure 10-12: From left to right, 10 mm, 5 mm, and 3 mm LEDs shown next to a metric ruler

To use 10 mm LEDs, you will have to mount them elsewhere and connect them to the PCB with wires. For inspiration, look at the lightbar for the Watch Winder in Chapter 4.

Using the Chromatic Thermometer

How you place the thermometer in the environment you want to monitor is limited only by your imagination. The version I use on my fish tank had no LCD at first, so I simply attached a couple of wire hooks to the bare board and hung it from the edge of the tank with the LEDs facing out. (I could have mounted the entire thing in a small acrylic box, but I had difficulty finding a box with the right dimensions.) This configuration worked well.

I eventually replaced that Chromatic Thermometer with a version including an LCD that I enclosed in an acrylic box. I'm thinking of modifying it again to use high-intensity LEDs that shine through the tank. That should create an interesting effect!

INDEX

A

D

E

F

G

H

I

J

K

L

M

N

O

P

R

S

T

U

V

W

Z

Arduino Playground is set in New Baskerville, Futura, Dogma, and TheSans Mono Condensed. The book was printed and bound by Sheridan Books, Inc. in Chelsea, Michigan. The paper is 60# Finch Offset, which is certified by the Forest Stewardship Council (FSC).

The book uses a layflat binding, in which the pages are bound together with a cold-set, flexible glue and the first and last pages of the resulting book block are attached to the cover. The cover is not actually glued to the book's spine, and when open, the book lies flat and the spine doesn't crack.

RESOURCES

Visit *https://www.nostarch.com/arduinoplayground/* for resources, errata, and more information.